Oral Delivery of Therapeutic Peptides and Proteins

ORAL DELIVERY OF THERAPEUTIC PEPTIDES AND PROTEINS

Edited by

PUNEET TYAGI

Dosage Form Design and Development, BioPharmaceuticals R&D, AstraZeneca, Gaithersburg, MD, United States

J. ANAND SUBRAMONY

Biologics Engineering & Targeted Delivery, Oncology R&D, AstraZeneca, Gaithersburg, MD, United States

ELSEVIER

ACADEMIC PRESS

An imprint of Elsevier

Academic Press is an imprint of Elsevier
125 London Wall, London EC2Y 5AS, United Kingdom
525 B Street, Suite 1650, San Diego, CA 92101, United States
50 Hampshire Street, 5th Floor, Cambridge, MA 02139, United States
The Boulevard, Langford Lane, Kidlington, Oxford OX5 1GB, United Kingdom

Notices

Knowledge and best practice in this field are constantly changing. As new research and experience broaden our understanding, changes in research methods, professional practices, or medical treatment may become necessary.

Practitioners and researchers must always rely on their own experience and knowledge in evaluating and using any information, methods, compounds, or experiments described herein. In using such information or methods they should be mindful of their own safety and the safety of others, including parties for whom they have a professional responsibility.

To the fullest extent of the law, neither the Publisher nor the authors, contributors, or editors, assume any liability for any injury and/or damage to persons or property as a matter of products liability, negligence or otherwise, or from any use or operation of any methods, products, instructions, or ideas contained in the material herein.

ISBN: 978-0-12-821061-1

For Information on all Academic Press publications
visit our website at https://www.elsevier.com/books-and-journals

Publisher: Stacy Masucci
Acquisitions Editor: Andre G. Wolff
Editorial Project Manager: Sara Pianavilla
Production Project Manager: Kiruthika Govindaraju
Cover Designer: Christian J. Bilbow

Typeset by MPS Limited, Chennai, India

Contents

List of contributors

Meenakshi Arora
College of Community Health Sciences, The University of Alabama, Tuscaloosa, AL, United States; The Center for Convergent Bioscience and Medicine (CCBM), The University of Alabama, Tuscaloosa, AL, United States; Department of Biological Sciences, The University of Alabama, Tuscaloosa, AL, United States; Alabama Life Research Institute, The University of Alabama, Tuscaloosa, AL, United States

Annette Bak
Advanced Drug Delivery, Pharmaceutical Sciences, R&D, AstraZeneca, Waltham, MA, United States

Nigel M. Davies
Advanced Drug Delivery, Pharmaceutical Sciences, R&D, AstraZeneca, Gothenburg, Sweden

Ahil N. Ganesh
Novartis Institutes for BioMedical Research, Cambridge, MA, United States

Sudhakar Garad
Novartis Institutes for BioMedical Research, Cambridge, MA, United States

Eric Gruff
E4 Consulting, Poway, CA, United States

Tariq Islam
BDD Pharma, Glasgow, United Kingdom

Mark J. Jackman
Advanced Drug Delivery, Pharmaceutical Sciences, R&D, AstraZeneca, Cambridge, United Kingdom

Swapnil Khadke
BDD Pharma, Glasgow, United Kingdom

Fiona MacDougall
BDD Pharma, Glasgow, United Kingdom

Sanyogitta Puri
Advanced Drug Delivery, Pharmaceutical Sciences, R&D, AstraZeneca, Cambridge, United Kingdom

M.N.V. Ravi Kumar
College of Community Health Sciences, The University of Alabama, Tuscaloosa, AL, United States; The Center for Convergent Bioscience and Medicine (CCBM), The University of Alabama, Tuscaloosa, AL, United States; Department of Biological Sciences, The University of Alabama, Tuscaloosa, AL, United States; Alabama Life Research Institute, The University of Alabama, Tuscaloosa, AL, United States; Chemical and Biological Engineering, University of Alabama, Tuscaloosa, AL, United States

Jefferson D. Revell
AstraZeneca R&D, Discovery Sciences, Cambridge, United Kingdom

Manuel V. Sanchez-Felix
Novartis Institutes for BioMedical Research, Cambridge, MA, United States

J. Anand Subramony
Biologics Engineering & Targeted Delivery, Oncology R&D, AstraZeneca, Gaithersburg, MD, United States

Ruchit Trivedi
Loxo Oncology at Lilly, Louisville, CO, United States

Puneet Tyagi
Dosage Form Design and Development, BioPharmaceuticals R&D, AstraZeneca, Gaithersburg, MD, United States

Dianxiong Zou
College of Community Health Sciences, The University of Alabama, Tuscaloosa, AL, United States

About the editors

Puneet Tyagi, Ph.D.
MedImmune, AstraZeneca

Puneet Tyagi is an associate director with the Dosage Form Design and Development group at AstraZeneca (AZ), based in Maryland, United States. He leads a team of scientists in the area of biologics formulation development, with a focus on long-acting formulations, transdermal delivery, and oral delivery of biologics. He received his bachelor's and master's degrees in pharmaceutical sciences from India and his Ph.D. from the University of Colorado. He is a coinventor of several US and international patents and has published extensively in the field of drug delivery.

J. Anand Subramony
R&D, AstraZeneca

Anand Subramony is the vice president within the Biologics Engineering group, R&D, at AZ and leads various cross-functional programs and teams in drug delivery and nanomedicine (for targeted delivery, enabling new modalities for genomic medicine, and creating technology platforms). In addition, he provides leadership to AZ's external innovation initiatives and CoSolve, an AZ open innovation platform that brings compelling technologies to the portfolio. Before serving in his current role, he headed the Drug Delivery & Device Development group at MedImmune and built an internal team that focused on combination product development for biologics and oral peptide delivery and other novel controlled release technologies for Biologics. One of the early pioneers in digital health technology solutions in pharma, Dr. Subramony led several novel connected health initiatives in the pharma value chain, including remote patient monitoring, adaptive therapy, and continuous monitoring with diagnostic sensors, wearables, and use of smartphone apps. In prior roles, he was the principal fellow and head of Novel Delivery Technologies at the Novartis Institutes for BioMedical Research (NIBR, Cambridge, MA), head of the Materials Science group (E-TRANS) at J&J (Alza, Mountain View, CA), and director of Materials Science & Drug Delivery at Dr. Reddy's Laboratories (Bridgewater, NJ). With his broad experience in pharma biotech, he brings innovation, scientific rigor, and execution excellence, with a sense of urgency toward drug development to create patient-centric products. He holds an M.S. degree in materials science engineering (IIT, Bombay) and a Ph.D. degree from Purdue University and was a research scientist at the University of Washington, Seattle. With several refereed publications and patents, he is a key opinion leader for bringing innovations through drug delivery, devices, and digital health.

Foreword

This book on *Oral Delivery of Therapeutic Peptides and Proteins* edited by Drs. Puneet Tyagi and J. Anand Subramony is a very timely addition to the literature in the exciting field of oral drug delivery. Oral delivery is a highly preferred and most convenient mode of drug administration. Historically, oral delivery has been associated with high patient compliance due to patient familiarity, ease of use, and painless administration. Unfortunately, oral delivery is restricted to small molecules due to the poor stability of proteins in the gastrointestinal tract and their poor intestinal permeability. This is a particular current and future limitation since biologics represent a dominant component of the therapeutic landscape. New peptides and antibodies continued to be approved for the treatment of various diseases, including diabetes and oncology, among others. Nucleic acids also represent a new emerging therapeutic modality. Currently, these biologics are delivered by injections that suffer from pain, inconvenience, and poor patient acceptance.

Over the years, extensive research has been done to overcome the challenges associated with oral delivery of biologics. These efforts have come in a variety of forms, including studies to understand the mechanisms of oral absorption and means to improve oral absorption. Several strategies including the use of physical methods and chemical approaches have been developed and studied for improving oral drug delivery. While the road to develop oral biologics has been long and challenging, this journey has reached an exciting point in recent times with the approval of two oral peptide products, semaglutide and octreotide. The success of these products provides validation of the approach and motivates further research and development. Future research should aim at expanding the scope of biologics that can be delivered orally, including those that are substantially larger than the currently approved peptides and those that are needed in larger doses.

Innovation is the key to the future success of this field. For oral delivery to succeed for a broad class of therapeutics, the field needs additional means of addressing the biological barriers of the gastrointestinal tract. New formulations and devices are needed to meet this goal. Innovative strategies must be accompanied by careful and critical analysis of their clinical and commercial translatability. Safety of the administered formulation must be studied and addressed. These new strategies not only need to address the fundamental issues of improved bioavailability but also address practical challenges associated with variability and food uptake.

Ultimately, to truly leverage the benefits of oral delivery, the administration needs to be robust and easy to use. Design and development of modified drugs that are designed for oral delivery will also play an important role in the development of orally administrable biologics products.

This book covers basic and translational aspects of oral peptide delivery that covers the entire spectrum of details for the development of oral drug products for peptide-based therapeutics. This book will be a valuable resource for students and scientists working toward oral delivery strategies in industry and academia. It will provide practical information about the design strategies of oral delivery products and their translation to the clinic. To those researchers that are already working in the field, this book will provide a valuable guide. At the same time, it will also inspire researchers with creative solutions that are currently not working in the field to enter the field.

Samir Mitragotri
Harvard University

Preface

It has been a privilege and honor to be an editor of this book on oral peptide/biologics delivery and to reflect upon the many years of work in this field. We have tried to cover the decades of delivery approaches and the various strategies that researchers have used to realize the vision of an oral dosage form for peptides, as opposed to parenteral injections. Clinical translation and key learnings, as well as formulation design, disruptive dosage form, chemistry, manufacturing, and controls, and regulatory perspectives, have been summarized in their own chapters. We have also tried to capture some of the emerging ingestible devices for oral delivery in addition to the use of traditional permeating enhancers. It is humbling to see that oral peptide delivery remains a challenge after decades of work, particularly as researchers consider new modalities, such as RNA-based therapeutics. With renewed interest in the treatment of chronic diseases with peptides and the emergence of several new approaches, we sincerely hope that this collection of chapters will serve as a complete handbook for the development of oral peptides/biologics.

Finally, sincere thanks to all the authors and contributors for the various chapters.

Puneet Tyagi
J. Anand Subramony
April 2022

Acknowledgments

It has been a privilege and honor to be an editor of this book on oral peptide/biologics delivery and to reflect upon years of work that have happened in this field. We have tried to bring in decades of delivery approach that researchers have taken and the various strategies that have been looked at different times to realize the dream of an oral dosage form for peptides as opposed to parenteral injections. Summary of clinical translation and key learnings have been brought in as a chapter in addition to formulation design; disruptive dosage form; chemistry, manufacturing, and controls; and regulatory perspectives. It is humbling to see that oral peptide delivery remains as a delivery challenge after decades of work and given the fact that researchers are looking at new modalities such as RNA-based therapeutics. We have also tried to capture some of the evolving ingestible devices for oral delivery in addition to traditional permeating enhancers. With renewed interest to treat chronic diseases with peptides and the emergence of several new approaches, we really hope that this collection of chapters will serve as a complete handbook for the development of oral peptides/biologics.

Finally, sincere thanks to all the authors and contributors for the various chapters.

J. Anand Subramony
February 2022

CHAPTER ONE

Landscape for oral delivery of peptides

Mark J. Jackman[1], Nigel M. Davies[2], Annette Bak[3] and Sanyogitta Puri[1]
[1]Advanced Drug Delivery, Pharmaceutical Sciences, R&D, AstraZeneca, Cambridge, United Kingdom
[2]Advanced Drug Delivery, Pharmaceutical Sciences, R&D, AstraZeneca, Gothenburg, Sweden
[3]Advanced Drug Delivery, Pharmaceutical Sciences, R&D, AstraZeneca, Waltham, MA, United States

1.1 Introduction

Peptides represent a class of molecules that have enabled the expansion of the "druggable" space beyond that possible using traditional small molecules and further provide increased target selectivity due to their larger surface areas and conformational flexibility (Craik et al., 2013). However, these beneficial features of peptides are also linked to their known limitations, being metabolic instability, poor membrane permeability, and negligible oral bioavailability (Matsson et al., 2016). As a result, of their negligible oral bioavailability, peptides are typically administered by injection which is often associated with patient discomfort. To offer patients an alternative to injections with the aim of improving convenience, compliance [c.20% of the population reportedly have fear of needles (Wright et al., 2009)], and ultimately therapeutic outcomes, significant efforts have been made into investigating strategies that can enable oral delivery of peptides. This chapter will cover the current landscape of oral peptide delivery and present trends, developability considerations for the various strategies/technologies, and emerging innovations in the area, as well as an industrial perspective on the requirements to develop a commercial oral peptide product.

1.2 Barriers to oral delivery of peptides

Nature has designed the gastrointestinal (GI) tract to limit the oral absorption of peptides (>3 amino acids; MW >c.500 Da). Gastric and

Oral Delivery of Therapeutic Peptides and Proteins.
DOI: https://doi.org/10.1016/B978-0-12-821061-1.00001-0

intestinal enzymes cleave peptides at specific sites into constituent amino acids or di/tripeptides for active carrier-mediated systemic absorption (see Section 1.3.5 for more details). The viscous mucus layer can present a diffusional barrier, particularly when using delivery approaches such as nanoparticles to enable oral peptide delivery (see Section 1.3.3 for more details). Furthermore, the intestinal epithelium presents a formidable barrier to the absorption of peptides. Peptides are typically too hydrophilic to partition and diffuse across the lipid cell bilayer (transcellular absorption) and too large to pass via the paracellular route, at least when the tight junctions are unperturbed (see Section 1.3.1 for more details). The intestinal epithelium is also rich in efflux transporters (see Section 1.3.5 for more details) and together with intracellular lysosome trafficking ensures that peptides are effectively excluded from crossing the epithelial barrier intact. Fig. 1.1 summarizes the main barriers present in the intestine that limit the systemic absorption of peptides following oral administration.

As a result of the formidable barriers presented within the GI tract to the absorption of peptides, only two peptides (at least with MW >1000 Da) have been widely approved up until 2019 for oral administration, being cyclosporin and desmopressin (other recent approvals are discussed later). Cyclosporin is rather unique as it can demonstrate what has been referred to as "chameleonic" behavior (Rossi Sebastiano et al., 2018). It is an 11-residue, macrocyclic peptide having a molecular weight of 1203 Da. It contains *N*-methylated, nonstandard, and D-amino acids and is a lipophilic molecule (log P of 3.6). It is able to adopt its molecular

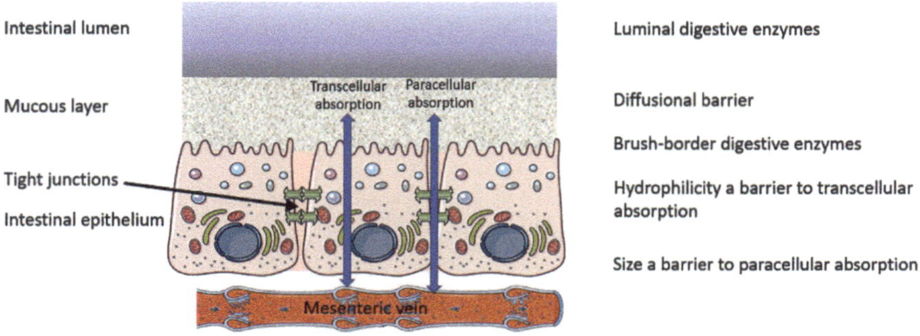

Figure 1.1 Schematic illustration of the main barriers to the oral delivery of peptides.

conformation and surface polarity to fit the surrounding physiological environment, exposing a more hydrophilic surface in extracellular fluids and a more lipophilic surface when entering the cell membrane (Augustijns et al., 2000). When combined with an appropriate formulation strategy, such as a self-micro-emulsifying drug delivery system, oral bioavailabilities of 20%–50% have been achieved (Neoral® Soft Gelatin Capsules, SmPC, 2020), despite it being a substrate for efflux transporters and being metabolized in the gut wall (Tjia et al., 1991). Desmopressin (DDAVP; 1-deamino-8-D-arginine vasopressin) also has some unique features, in that it is a synthetic analog of the antidiuretic hormone vasopressin, again having a cyclic structure, a deamidated cysteine residue, and including an unnatural D-amino acid. It has a reported oral bioavailability of only 0.08%–0.16% (DDAVP® tablets 0.2 mg, Desmopressin patient Leaflet 0.2 mg, 2012), but oral administration is made possible by the high potency of the peptide such that tablet strengths of only 0.1 or 0.2 mg are needed, thereby making it economically feasible to commercialize despite the low bioavailability of the peptide.

1.3 Strategies to improve the oral bioavailability of peptides

1.3.1 Transient permeation enhancers

The most investigated, and to date probably the most successful strategy (based on the number and progressions of clinical trials) for improving the oral bioavailability of peptides having a molecular weight of >1000 Da is to formulate the peptide together with transient permeation enhancers (TPEs) (with or without protease inhibitors), typically in a single-unit dosage form suitable for oral administration. TPEs temporarily and reversibly increase the permeability of the intestinal epithelium in such a way to allow the transient absorption of peptides that would otherwise be excluded. First-generation TPEs, as coined by Maher et al. (2016), increase membrane permeability either via opening of the tight junctions (directly or indirectly) allowing paracellular absorption or (since many are surfactants) by transiently perturbing the epithelial cell membrane enabling transcellular absorption, or possibly a combination of both effects (Maher et al., 2016). Table 1.1 lists examples of classes of such TPEs, together with examples of specific compounds and proprietary permeation-enhancing technologies using such enhancers (list is not exhaustive, but examples selected as these proprietary technologies have progressed into clinical studies—see Table 1.2).

As mentioned earlier, the use of TPEs has resulted in some degree of success, with a number of peptide/TPE combinations (with and without enzyme inhibitors) entering clinical studies, some advancing to late state clinical testing and regulatory "new drug" applications together with a couple of recent approvals (Rybelsus, in 2019; Mycapssa in 2020). Table 1.2 summarizes some of the more advanced clinical studies that have been reported in the literature using TPEs. Despite the progress made using TPEs to enable oral delivery of peptides and the success (albeit still limited) in advancing these to clinical studies, oral bioavailabilities are still low (single digit at best) and variability is high; both of which limit the commercial "developability" of peptides for oral administration.

Another issue associated with using TPEs for promoting the oral bioavailability of peptides is the known effect of food on the performance of the enhancer (Karsdal et al., 2009; Tuvia et al., 2014). Tuvia et al. (2014) reported an approximate 90% reduced bioavailability of octreotide in subjects receiving a high-fat, high-calorie meal immediately before dosing compared to when dosing after a 10-h fast. In this study, octreotide was formulated in an oily suspension containing enhancers and administered as an enteric-coated capsule. Similar findings were reported by Karsdal et al. (2009) for calcitonin formulated together with 8-(*N*-2-hydroxy-5-chlorobenzoyl)-amino-caprylate as the enhancer, where an approximate 70% reduction in oral bioavailability was observed when food was administered within 4 h of dosing. The effect of food on the performance of TPEs is believed to be a result of the TPE being sequestered in lecithin/bile mixed micelles present in the intestinal lumen, particularly under fed conditions, and consequently no longer available to execute their effects on the intestinal membrane (Gradauer et al., 2019). It would not be unreasonable to assume that the presence of food itself can also reduce the luminal concentration of the TPE, thereby diminishing its effect on membrane permeability, and that food could also negatively affect peptide absorption in a similar manner to other biopharmaceutical classification system (BCS) class III, low permeability compounds (Custodio et al., 2008). Such food interactions, which can be particularly expected for

Table 1.1 Examples of transient permeation enhancers (TPEs); classes and examples of specific compounds and proprietary permeation-enhancing technologies using such enhancers

TPE class	Examples	Proprietary technologies
Medium-chain fatty acids and salts	Sodium caprylate, sodium carate, sodium laurate	GIPET I[a] from Merrion Pharmaceuticals—sodium caprate; TPE[b] from Chiasma—sodium caprylate and glycerides
Medium-chain mono-, di-, and triglycerides and derivatives thereof	Caprylic, capric, propyleneglycol monocaprate, caprylocaproyl polyoxyglycerides	GIPET II[a] from Merrion Pharmaceuticals—mono/di/triglycerides of caprylic and capric acid; TPE[b] from Chiasma—mono/triglycerides of caprylic acid and sodium caprylate
Acylated amino acids	*N*-[8-(2-hydroxybenzoyl) amino] caprylate (SNAC), 8-(*N*-2-hydroxy-5-chloro-benzoyl)-amino-caprylate (5-CNAC), (4-[(4-chloro-2-hydroxy-benzoyl)-amino]butanoic acid (4-CNAB)]	Eligen enhancers from Emisphere Technologies—SNAC, CNAC, 4-CNAB
Acyl carnitines	Lauroyl carnitine, myritoyl carnitine	Petelligence from Enteris Biopharma—Alkyl carnitines with organic acid pH modifier
Bile salts	Chenodeoxycholic acid, ursodeoxycholic acid, taurocholate, and salts therof	POD[c] from Oromed Pharmaceuticals—bile salt with protease inhibitor, for example, soybean trypsin inhibitor (early prototype; later information indicates switch to use of longer chain fatty acids as TPE)
Aromatic alcohols	Propyl gallate, phenoxyethanol, benzyl alcohol	Axcess from Diabetology—aromatic alcohols and protease inhibitor

[a]*GIPET*, gastrointestinal permeation-enhancing technology.
[b]TPE, transient permeation enhancer
[c]POD, protein oral delivery.

Table 1.2 Examples of peptide/TPE formulations that have advanced to clinical testing/registration.

Peptide	TPE/delivery technology	Oral bioavailability in humans (%)	Variability (CV%)	Number of subjects	Clinical stage	References
Basal Insulin 338	Sodium caprate (GIPET I, Merrion)	Not evaluated	82–132	44	Phase II	Halberg et al. (2019)
Desmopressin	Mono-, diglycerides of C8 and C10 (GIPET II, Merrion)	2.4	124	18	Phase I	Leonard et al. (2006)
Salmon calcitonin	Acylated amino acid (Eligen, Emisphere)	0.41–1.44 (dose dependent)	59–99 (dose dependent)	8	NDA accepted for review 2015	Buclin et al. (2002)
Parathyroid hormone (PTH1–34)	Acyl carnitine/citric acid (Peptelligence, Enteris)	0.58 (week 0)	99 (week 0)	32	Phase II	Sturmer et al. (2013)
Mycappsa Octreotide	Oily suspension containing sodium caprylate, glyceryl mono- and tricaprylate (TPE, Chiasma)	0.8	62	71	Approved, 2020	Tuvia et al. (2012)
Rybelsus Semaglutide	Acylated amino acid, SNAC (Eligen, Emisphere)	0.4–1	65–84	11–32 (for variability)	Approved, 2019	Granhall et al. (2019), Rybelsus (2021)

single-unit enteric-coated dosage forms that may not immediately empty from the stomach following administration (Davis et al., 1986), may lead to inconvenient dosage regimens with regard to dosing in relation to food (see, e.g., Mycapssa Prescribing Information, n.d.). Similar food interactions could also be expected for some of the other "enabling" technologies highlighted in this chapter such as nanoparticles, ingestible injectable devices as well as possibly when aiming to exploit transporter systems. However, as few of these strategies have reached clinical development, nothing is yet known around the effect of food on their performance for enhancing oral delivery of peptides.

To overcome some of the issues associated with the use of TPEs, an emerging trend is to chemically engineer peptides to render them more suitable for oral administration. This includes modifications to increase peptide stability toward enzymatic degradation in the GI tract as well as increasing the systemic half-life of the peptide to better accommodate variability. Having a long half-life (e.g., >100 h), allows the high variability associated with oral delivery of peptides to be somewhat accommodated because the resulting drug accumulation will "keep drug onboard" even over days when no or minimal drug is absorbed. This strategy has been adopted for the successful development of the oral glucagon-like peptide-1 (GLP-1) receptor agonist semaglutide product, Rybelsus, approved by the FDA in September 2019 and represents the first successful commercial oral peptide product delivered using a TPE.

As the oral bioavailabilities of peptides using traditional TPEs remain low, considerable efforts are ongoing to both identify more promising TPEs. Recent, notable examples of such TPEs include the ionic liquid based on choline and geranate (CAGE) and the strawberry extract, pelargonidin. Dosing CAGE with insulin via intrajejunal administration yielded an oral bioavailability of insulin of approximately 50% compared to subcutaneous administration (Banerjee et al., 2018). Pelargonidin also yielded impressive effects on insulin oral bioavailability yielding levels similar or better to those achieved following subcutaneous administration (Lamson et al., 2020b).

Efforts are also being made to identify "next-generation" TPEs that more specifically work on modulating the integrity of paracellular tight junctions, the so-called "second-generation" TPEs as coined by Maher et al. (2016). The development of these second-generation enhancers is largely based on the understanding of how first-generation enhancers modulate tight junctions and also elucidation of proteins that regulate tight junction structure and function (e.g., actomycin, occludin, claudin, and tricellulin). Second-generation TPEs, many based on the discovery of protein-based toxins, directly disrupt cell adhesion recognition sequences between adjacent intestinal cells to increase paracellular permeability to enable absorption of macromolecules (Maher et al., 2016). Examples of such second-generation TPEs include AT1002, a short peptide sequence from the zonula occludens protein (Gopalakrishnan et al., 2009) and viral protein 8 derived from the outer capsid protein of the rotavirus (Nava et al., 2004). Second-generation TPEs are considered to be more specific in their mechanism of action and hence more selective in their ability to promote absorption of co-delivered peptides than first-generation enhancers that have a multimodal mechanism of action.

Another recent trend is the investigation of lipid-based formulations (LBFs) for enabling the oral bioavailability of peptides. These systems have typically been investigated for the oral administration of poorly soluble, lipophilic, BCS Class II/IV compounds, and hence this approach may seem counter-intuitive for peptides that are typically hydrophilic. However, it is known, and can be suspected, that many of the first-generation medium-chain surfactant-like enhancers are rapidly absorbed from the intestine within minutes of release from the oral dosage form (Raoof et al., 2008). As such, it can be expected that the permeability enhancing effects of these types of enhancers are short lived as the intestinal membrane remodels and recovers. The use of LBFs may allow for the inclusion of "TPE precursors" within the formulation; for example, medium-chain glycerides and derivatives thereof that are digested to medium-chain fatty acid TPEs. These could function as "sustained release" depots of the TPE, thereby prolonging the permeability enhancing effect. This has been speculated as a potential mechanism for the permeability enhancing effects of Labrasol, a mixture of caprylocaprroyl polyoxyyl-8 glcerides (Fernandez et al., 2013) and also possibly the Mycapssa formulation, which includes a combination of a traditional TPE (sodium caprylate) together with excipients that could be considered TPE precursors in the form of glyceryl monocaprylate and glyceryl tricaprylate (Mycapssa Prescribing Information, n.d.).

As mentioned, the use of LBFs for enabling oral delivery of hydrophilic peptides seems rather counter-intuitive. To incorporate peptides into LBFs, they must be rendered lipophilic. This can be accomplished by covalently linking lipids to the peptide or by cyclization (Rossi Sebastiano et al., 2018); or by the formation of hydrophobic ion pairs with suitable counterions (Griesser et al., 2017) or hydrophobic H-bond pairs with nonionic surfactants (Nazir et al., 2020). Such modifications result in the masking of H-bond donors or H-bond acceptors, altering the characteristics of the peptide to better comply with Lipinski's "Rule of 5" that predicts a molecule's suitability for oral absorption (Lipinski et al., 1997). It can therefore be anticipated that such modifications will improve membrane permeability of the peptide and consequently oral bioavailability (Bonengel et al., 2018; Menzel et al., 2018). The approach of incorporating hydrophobically modified peptide within LBFs such as self-emulsifying drug delivery systems (SMEDDS or SNEDDS) reportedly also protects the peptide against luminal peptidases (Hetényi et al., 2017) can improve transport across the mucous layer (Griesser et al., 2018) as well as possibly improving the distribution of the TPE over the intestinal membrane. The interest in using SMEDDS/SNEDDS for improving oral delivery of peptides has naturally made headway in the investigation of "solid" lipid nanocarriers/suspensions for oral peptide delivery (Dumont et al., 2018), but these systems seem to yet deliver on their promise (see Section 1.3.3 for more details).

A key and ongoing controversial consideration to the use of TPEs to enable absorption of peptides from the gastrointestinal tract is safety. As discussed, first-generation TPEs enable oral absorption of peptides by nonspecifically increasing intestinal permeability either via opening of the tight junctions enabling paracellular absorption or by transiently perturbing the epithelial cell membrane enabling transcellular absorption, or possibly a combination of both effects (Maher et al., 2016). It can be speculated that for many indications requiring chronic therapy that modulating the permeability of the intestinal epithelium to enable absorption of peptides could also allow for the entry of "bystander" molecules such as lipopolysaccharides, toxins, and even pathogens that otherwise would also be excluded from being absorbed. However, there is some evidence to suggest that there is a relationship between permeation-enhancing effects of TPEs and the molecular weight of the moiety that can be absorbed. For the TPE, sodium caprylate, permeation enhancement of fluorescein isothiocyanate (FITC)-dextran across the rat intestinal epithelium was found to be inversely correlated with molecular weight, and no significant permeation-enhancing effects were observed at a molecular weight of greater than 70 kDa (Tuvia et al., 2014). As the molecular weight of many of these xenobiotics is greater than this (e.g., LPS has a molecular weight of ~100 kDa and exists in an oligomeric state having a molecular weight of >1000 kDa), it can be assumed that TPEs are unlikely to "open-up" tight junctions or modulate epithelial permeability to such an extent to enable coabsorption of this kind of "bystander" molecules (Cox et al., 2008). Further, it can be speculated that the repeated "insult" of the intestinal epithelium upon exposure to TPEs, as would be the situation with daily administration of solid oral dosage forms containing TPEs, could lead to local intestinal inflammation or injury. However, it is highly unlikely that a solid dosage form containing a TPE will dissolve and release the TPE at the same place in the intestine over consecutive days of dosing (the small intestine having a length of at least 3 m). Release of the TPE at different sites of the intestine over consecutive days of dosing would allow the intestinal epithelium time to repair and recover before being exposed to the next dose of TPE. The safety of repeat, daily dosing of TPEs is best supported by the recently approved products, Rybelsus and Mycapssa. To support approval of Rybelsus, the safety of daily oral dosing of semaglutide in combination with the TPE, *N*-[8-(2-hydroxy-benzoyl) amino] caprylate (SNAC) was investigated in studies with a mean duration of c.60 weeks (Rybelsus Prescribing Information, n.d.) and for Mycapssa (Octreotide with TPE based on sodium caprylate) with a mean duration of c.29 weeks (Mycapssa Prescribing Information, n.d.), both exhibiting a safety profile allowing for FDA approval.

1.3.1.1 Industrial development perspective of transient permeation enhancers

As discussed, even with the use of TPEs (with or without protease inhibitors), oral bioavailabilities of peptides are still low (single digit at best) and variability is high (at least from reported clinical studies; see Table 1.2). As oral bioavailabilities can be expected to be low, then from an industrial development perspective, the peptide itself needs to be potent and have a low dose to make development economically feasible. Further, the cost of manufacture of the peptide must be considered in relation to the indicated pharmacological target/disease area, with the cost increasing with the number of amino acid residues in the peptide sequence and particularly upon inclusion of peptide modifications or unnatural amino acids. To accommodate the expected high variability (CV typically of 50%–100%; see Table 1.2), the peptide either needs to have a wide therapeutic index or have a sufficiently long half-life to enable drug accumulation to gradually attain a relatively constant "steady-state" plasma/tissue concentration and even "keep drug on board" over days when no or minimal drug can be expected to be absorbed.

In terms of the dosage form, and in addition to the usual considerations for developability and manufacturing of single-unit dosage forms of small molecules for oral administration (e.g., excipient compatibility, material characteristics, etc.), it must be borne in mind that TPEs are most likely to be considered as novel excipients by regulatory authorities and further, not have a Pharmacopoeial monograph. A "novel excipient" is defined as one which is being used for the first time in a drug product, or being used by a new route of administration, or has not yet been used for human administration and/or particular human administration pathway (EMEA, 2007). Administration of doses in excess of those previously approved may also confer "novel" status. The likely categorization of a TPE as a "novel excipient" brings with it additional regulatory requirements over "standard Pharmacopoeial" excipients in that full details of manufacture, characterization, and controls (CMC), together with safety data (nonclinical and/or clinical) are required in regulatory dossiers in a similar format to the active substance (EMA guideline 2018), that is, the enhancer almost needs to be regarded as an "Active Pharmaceutical Ingredient." Further, sources of such novel excipients/TPEs manufactured according to "Good Manufacturing Practice" may be hard to come by and even if sources/suppliers of appropriate quality material can be identified, there could still be issues in accessing CMC proprietary information confounding support to regulatory submissions. These considerations regarding novel excipients also apply to other oral peptide delivery systems such as nanoparticles.

Regarding the use of more traditional "first-generation" TPEs, it must also be borne in mind that the dose of TPE to be included, and necessary for permeation-enhancing effects, may constitute a significant portion of the dosage form. For example, GIPET I from Merrion (sodium caprate as TPE) has been formulated in some instances with 550 mg of sodium caprate in a tablet (Halberg et al., 2019), and even up to 700 mg per tablet (Tillman et al., 2008). To keep a single-unit dosage form of an acceptable size for patient acceptability then there seems little room for the inclusion of other traditional tablet such as binders, disintegrants, lubricants, etc., that may facilitate tablet manufacture and performance. Hence, unless the TPE has suitable material properties to enable processing and manufacturing, the high proportion of TPE required could make scale-up and manufacturing of commercial quantities challenging.

Finally, when developing dosage forms containing TPEs to enable oral peptide delivery, it is recommended to transition early, or as soon as possible, to "first time in human" trials to assess the translatability of preclinical findings, particularly regarding bioavailability.

1.3.2 Oral nanoparticles for peptide delivery

1.3.2.1 Background and introduction

Nanoparticles (NPs) for oral delivery of peptides continue to be of interest due to their unique properties such as controlled particle size, drug loading, and tunable surface properties that can be used to increase peptide GI stability and intestinal uptake into the systemic circulation. Transport of NPs via the transcellular pathway has been studied in relation to NP size, surface charge, and mucoadhesivity. Different mechanisms may coexist to allow NPs to be taken up by the intestinal epithelium, depending on the materials used to form the NPs (Griffin et al., 2016; Mahato et al., 2003). A number of reviews (Florence, 1997; Lundquist & Artursson, 2016) have suggested that nanoparticles are mainly absorbed across the intestinal epithelia via M cells on the domes of the Peyer's patches in small animals. Peyer's patches are the intestinal surfaces least protected by mucus and most exposed to chyme, which supports the proposed pathway of entry across the intestinal barrier. It is widely recognized that polymeric NPs cannot diffuse through the healthy/intact intestinal barrier via the paracellular route. To improve the paracellular transport of NPs, tight junctions must be opened (reversibly) by using either cationic (e.g., chitosan and its derivatives) or anionic TPEs (polyacrylic acid and its derivatives), polymers, or calcium chelators (Plapied et al., 2011). Designing NP carriers to enhance the absorption of protein drugs via the GI tract has involved adapting the NP physicochemical properties either to optimize M cell uptake or attaching ligands to the particle surface to specifically target cells and enhance endocytosis. Here, we discuss a number of different NP systems such as polymeric, mucoadhesive, lipid, and inorganic carrier systems that have been extensively researched and those that have progressed into clinical. We also highlight key, learnings, and the developability assessment of an NP formulation for clinical translation and improved probability of success.

1.3.2.2 Polymeric nanoparticles

Polymeric nanoparticles are composed of both synthetic and natural origin materials, such as gelatin, hyaluronic acid, cellulose, chitosan,

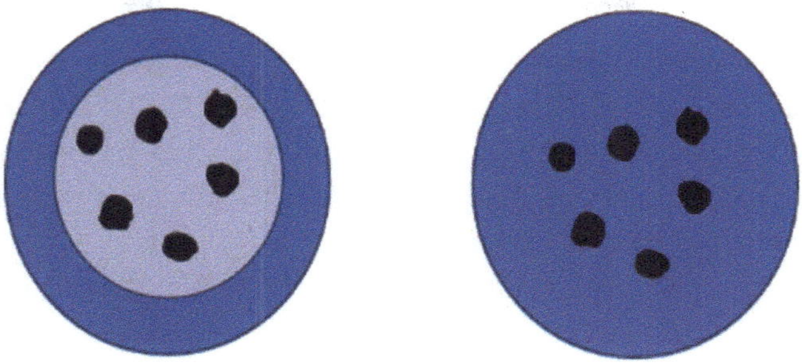

Figure 1.2 Nanocapsules and nanospheres. In nanocapsules (left), the drug/peptide is encapsulated in a central aqueous core surrounded by a polymer shell, and in nanospheres (right), the drug/peptide is embedded in a solid polymeric matrix, which differs mainly in drug release behavior and also loading potential for the same dug/peptide.

poly(lactic-co-glycolic) acid (PLGA), polycaprolactones, polyanhydrides, and β-cyclodextrin-based polymers have been reported in the literature (Lai et al., 2009; Xiong et al., 2007). Fig. 1.2 illustrates the concepts of nanocapsules or nanospheres widely used for formulating polymer-based peptide-loaded NPs. Early work in the 80s by Couvreur et al. reported oral delivery of insulin in diabetic rats using poly(alkylcyanoacrylate) nanocapsules (mean size 220 nm and 55% w/w insulin loading) at 12.5, 25, and 50 U/kg insulin doses, decreased fasted glycemia in a sustained manner, with bioavailability of roughly 36% (Damgé et al., 1988). The authors attributed the enhanced efficacy to the peptide protection against enzymatic degradation and also to the internalization of the nanocapsules into the intestinal epithelium. Subsequent studies led the authors to conclude that the nanocapsules were taken up by M cells (Peyer's patches) and that only a small portion could cross the intestinal layer via the paracellular pathway that also showed a highly variable PK profile in rats (Michel et al., 1991). They did not report any further progress in the formulations.

Polymeric nanoparticle research soon shifted focus to ligand targeted systems designed to enhance active transport across the gut epithelial barrier. Some early examples included the neonatal Fc receptor FcRn, Vitamin B12, and lectin, amongst others. Pridgen et al. loaded insulin in nanoparticles made of poly(lactic acid)–poly(ethylene glycol) (PLA-PEG) block copolymers and targeted FcRn in the intestinal epithelium. FcRn is well known for its ability to transport Immunoglobulin G (IgG) in breast milk from mother to infant. The authors utilized ^{14}C-labeled nanoparticles to assess uptake and bioavailability and reported a 11.5-fold higher absorption efficiency (13% pharmacological activity (PA)) in rats compared to the nontargeted particles along with a prolonged hypoglycemic response (at 1.1 IU/kg insulin) (Pridgen et al., 2013). These particles also had a much smaller size (53 nm) compared to the PLGA prototypes investigated early on, which could also highlight the impact of particle size on receptor-mediated internalization.

Oral insulin delivery of 150 nm PLGA-PEG NPs decorated with an engineered human albumin variant and improved FcRn binding was reported by Azevedo et al. At insulin doses of 50 IU/kg, functionalized NPs were effective in glucose lowering with a PA of around 3% and a hypoglycemic decrease similar to a subcutaneous injection of insulin. The low PA was attributed to the incomplete release of insulin, which could have increased at later timepoints, but this was not studied extensively (Azevedo et al., 2020). FcRn is expressed in many tissues in addition to the intestine. Using IgG Fc to target the FcRn could potentially also result in uptake in tissues other than the targeted tissue postsystemic administration. IgG immunogenicity and bioavailability studies in higher species for further trafficking insights need to be conducted.

Some reports of using vitamin B12 conjugated to insulin or octapeptides reported increased oral uptake across the intestinal barrier in vitro and in vivo (Alsenz et al., 2000). Vitamin B12 (VB_{12})can be bound by an intrinsic factor (IF) in the small intestine to form VB_{12}−IF complex; this complex can be recognized by an IF-specific receptor located on the luminal surface of intestinal cells and facilitate transport across the intestinal enterocytes (Chen et al., 2011). After oral administration of $VitB_{12}$-conjugated dextran nanospheres and encapsulated insulin (20 IU/kg), rapid reduction in plasma glucose and a prolonged second phase were reported. The % PA of NPs containing 2%−4% w/w insulin was only 1−2.6-fold higher, compared to NPs without VB_{12} thus confirming uptake was only modestly boosted by the vitamin B_{12} transport (Chalasani et al., 2007). It must be noted that VB_{12} uptake capacity in humans is limited to approximately 2.5 μg/day for an adult human and is a slow and delayed process, over 3−4 h. Hence, the amount of peptides that can be introduced through the VB_{12} pathway is limited (Clardy et al., 2011). Recently, Xu et al. developed targeted nanoparticles toward increased intestinal L cell stimulation to improve oral exenatide delivery. Enteroendocrine L cells express various G protein-coupled receptors that are activated by different dietary nutrients such as short-chain free fatty acids present in the lumen to modulate the secretion of hormones (e.g.,

GLP-1). However, they concluded that there was no additional PD advantage of propionate fatty chains coatings compared to the pegylated versions of both lipid nanocapsules and PLGA nanoparticles in vivo (Xu, Keersmaecker, et al., 2020). The limited success of such examples highlights gaps in the understanding of relevant endosomal-mediated cell uptake pathways involved, with regard to uptake capacity, disease translation, and stability, particularly if the peptides end up in lysosomal compartments prone to acidic degradation. Optimization of ligand density and steric stability in relation to the absorption kinetics and bioavailability of the system is another important design and characterization aspect that should be conducted and is missing in a number of reported prototypes.

In general, PLGA-based prototypes also have a complex manufacturing procedure and reproducibility on scale-up is challenging. Particle sizes between 100 and 200 nm are generally reported but with high polydispersity indexes and poor loading (Mohammadi-Samani & Taghipour, 2015). The addition of targeting ligands would add further complexity requiring in-depth surface characterization and stability studies ahead of prototype selection. Besides, it has been well documented that PLGA-based systems produce an acidic microenvironment postdegradation in vivo and this could be highly detrimental for peptide sequences liable to degradation at low pH. Besides, they commonly show an unusually high burst release followed by a long lag phase and finally diffusion and erosion-controlled release (Cui et al., 2007; Mohammadi-Samani & Taghipour, 2015). This profile and limitations need to be factored into the early technology selection stages, particularly for peptides requiring a high plasma Cmax for efficacy.

Chitosan-based NPs form the majority of systems researched for mucoadhesive oral peptide delivery, primarily due to their charge-based mucoadhesive properties (see Fig. 1.3) and ability to reversibly alter the tight junction barrier. However, the extent to which this is reproducibly achieved in chitosan solutions versus chitosan-based NPs is debatable and also mainly performed in studies in cell culture or rat intestinal sac models (Prego, Torres, et al., 2005). Other mucoadhesive systems include thiolated polymers, methacrylates, and alginates (Chuang et al., 2013; Prego, García, et al., 2005). Similar effects confirming higher mucoadhesion of chitosan-coated PLGA nanospheres to the intestinal tract were reported by Kawashima et al. (2000). Thiolated polymers or thiomers are generated by the immobilization of thiol-bearing ligands to mucoadhesive polymeric excipients. They form disulfide bonds with mucus glycoproteins and can

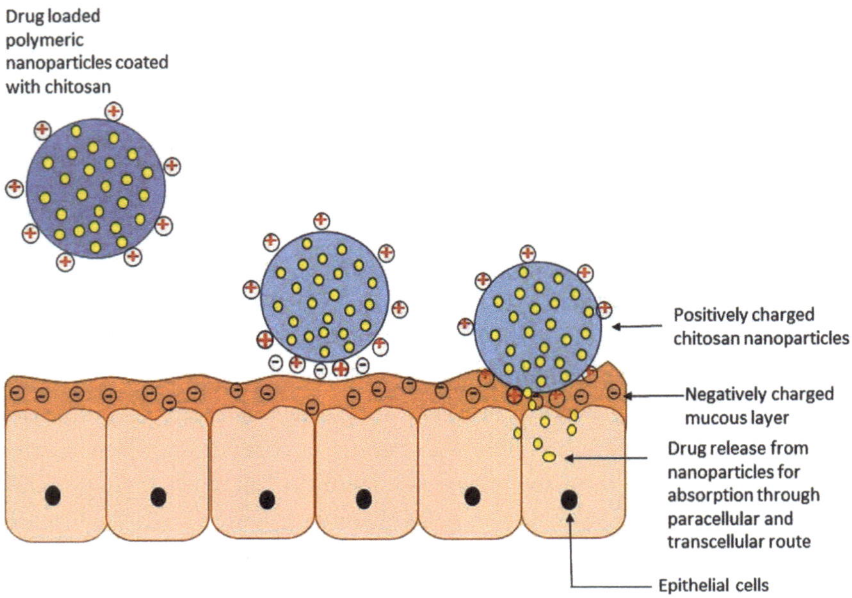

Figure 1.3 Chitosan-coated nanoparticle interaction with gut epithelia. Illustration of chitosan-coated nanoparticle interaction with gut epithelia (Mohammed et al., 2017). *Figure taken from Mohammed, M. A. S., Syeda, J. T. M., Wasan, K. M., & Wasan, E. K. (2017) An overview of chitosan nanoparticles and its application in non-parenteral drug delivery.* Pharmaceutics, *9(4), 53.*

improve mucoadhesion by 130-fold, however only reported via traditional Caco-2 cell models. Tablets comprising thiomer and pegylated insulin, for instance, resulted in a pharmacological efficacy of 7% in diabetic mice. Thiolated polymer chitosan-4-thio-butylamidine conjugate (chitosan–thio-butylamidine (TBA) conjugate) and glutathione compressed to tablets were dosed orally to rats. The absolute and relative bioavailability was determined to be 1.1% and 3.2%, respectively (Bernkop-Schnürch et al., 2004, 2005). Although the data previously discussed is on tablets and not NPs, it can be concluded that overall results with this class of polymers have been variable and only with modest improvements in bioavailability.

Mucoadhesive delivery systems still have limitations for oral drug delivery. From rat in situ loop experiments, intestinal mucin turnover time is around 50–270 min (Lehr et al., 1991) thus limiting the transit and residential time of mucoadhesive particles in the GI tract. Furthermore, as mucoadhesive systems efficiently adhere to mucus, they are largely incapable of penetrating across the mucus layer and entering the underlying epithelia. An important design aspect for chitosan-based systems to overcome this limitation is to have a formulation charge promoting interactions with mucus and the apical membrane of GI epithelia. To facilitate uptake and cell penetration, achieving the balance of positive charges on the nanoparticle surface with a strong electrostatic affinity to anionic mucus glycoproteins, while not restricting mucus permeation is critical. Also, ensuring electrostatic interaction with anionic glycoproteins (e.g., glycocalyx) and lipid rafts in the cell membrane is key (Bernkop-Schnürch, 2018). An elegant example of altering the zeta potential of nanoparticles was reported by Shan et al. They produced a nanocomplex core composed of insulin and cell-penetrating peptide (CPP), and a dissociable hydrophilic coating of *N*-(2-hydroxypropyl) methacrylamide copolymer (pHPMA) derivatives. They confirmed 20-fold higher absorption across mucus due to the "mucus-inert" pHPMA coating, as well as high epithelial absorption mediated by the CPP. The investigation of NP behavior (in Caco-2 models with low mucin turnover) showed that the pHPMA molecules gradually dissociated from the NP surface as it permeated through mucus, revealing the CPP/insulin complex rich core in time for subsequent transepithelial transport through the secretory endoplasmic reticulum/Golgi pathway and endocytic recycling pathway. Although a prominent hypoglycemic response and an increase in the serum insulin concentration were reported, no bioavailability measurements were performed (Shan et al., 2015). Unfortunately, not many of the research prototypes belonging to the above class of materials have advanced to the clinic.

1.3.2.3 Inorganic nanoparticles

Inorganic drug carriers, such as gold, iron oxide, metal-organic frameworks, silicon, and silica, have emerged as attractive alternatives. Among these, silica NPs are particularly interesting because of their highly tunable surface chemistry (via silanol groups), proven biocompatibility, cost-effective synthesis, and chemical and enzymatic stability. Mesoporous silica nanoparticles offer further advantages of tunable porosity, pore diameters, pore morphology, and particle size and morphology (Abeer et al., 2020). Lamson et al. dosed negatively charged 50 nm silica particles with insulin, in rats and reported 20–30% bioactivity versus subcutaneous injection of 1/Ukg insulin. The permeation-enhancing effect of the nanoparticles was reversible, nontoxic, and attributable to the binding to integrins on the surface of epithelial cells in an unusual way: opening of tight junctions and specifically by the 50 nm negatively charged particles only (Lamson et al., 2020a). Hristov et al. recently produced insulin and exenatide formulations with a silica nanoparticle shell with L-Arginine/$ZnCl_2$ as permeation enhancers and complexing agent, respectively. They hypothesized that the NP will release payload in high concentration close to the epithelium relatively quickly, ideally triggered by the higher pH of the small intestine followed by intestinal epithelial permeability assisted by co-formulated excipients to enable delivery into plasma. Both formulations were bioactive albeit exenatide was less active than free form and bioavailabilities have not been reported (Hristov et al., 2020). Another example of a hybrid multicomponent NP is from Bao et al. where exenatide-loaded zein nanoparticles (with exenatide embedded in Zein) were dosed orally to T2D mice and showed a significant and sustained glucose-lowering effect. The formulation also contained cholic acid as an absorption enhancer (via bile acid transporters), casein as a stabilizer, and hypromellose phthalate as a gastric fluid protectant. Despite having several components and manufacturing complexity, the oral pharmacological activity was 18% while oral bioavailability was only around 5% (Bao et al., 2020), illustrating that it may not always be sufficient to measure the pharmacological activity of an oral drug delivery system and a systematic investigation of the absorption contribution of each component needs to be undertaken.

1.3.2.4 Lipid-based nanoparticles

Delivery of peptides in liposomal formulations has not been extremely successful by the oral route. This is mainly because of poor lipid stability in gastrointestinal pH, peptide proteolysis bile salts and enzymatic lipid degradation, and poor permeation across the intestinal epithelium (Chiang & Weiner, 1987). Also, the loading capacity of peptides into liposomes is limited by the capacity of the aqueous core and the lipid bilayer to hold the peptide molecules. Molero et al. encapsulated insulin within dipalmitoyl phosphatidylcholine, cholesterol, phosphatidic acid, and liposomes. At 5 IU/kg, significantly reduced glucose and raised insulin in rats and rabbit were reported. However, some batches failed to reduce glucose despite apparently adequate incorporation, and some animals failed to respond with the same batch of liposomes that had previously produced a response in other animals. Thus their stability (storage shelf life and during in use handling/dosing/temperature excursions), manufacturing reproducibility, and effectiveness are unpredictable (Arrieta-Molero et al., 1982) and highlight destabilization of the nanostructure in the acidic environment as a major contributor to this (Choudhari et al., 1994).

Recent advances have focused on modulating the compositions of the lipid bilayers or modifying the liposomal surfaces with mucoadhesive polymers or ligands to modulate the in vivo fate and stability of liposomes after oral administration (Agrawal et al., 2014; He et al., 2019). One such modification is the addition of lectin-targeting insulin-loaded liposomes first reported by Zhang et al., but still only achieved modest bioavailabilities of around 8% (Zhang et al., 2005).

Addressing the stability limitation, recently, liposomal formulations containing dipalmitoylphosphatidylcholine and tetraethyl lipids (TELs) have been reported. Due to the ether bonds and the unsaturated alkyl chain, TELs are less susceptible to hydrolysis and oxidation than normal ester phospholipids. A mixture of TELs with conventional phospholipids results in a combination of mono- and bilayers, whereby the membrane-spanning character and the rigidity of TELs lead to a membrane stabilization. Further, their low glass transition temperature ($<0°C$) allows for easy production and handling at room temperature. Parmentier et al. showed that octreotide-loaded liposomes with 25% TEL could improve the oral bioavailability of octreotide 4.1-fold, and a cationic TEL derivative could improve bioavailability 4.6-fold, indicating that vesicle charge had only a minor influence on the absorption enhancement. However, the main contribution to this was peptide protection in the GI tract as radiolabelled uptake studies showed very low permeation (Parmentier et al., 2011). Further improvement in permeability across mucus would be key to advance these forwards.

Solid lipid nanoparticles (SLN) made of solid lipids (lipids being solid at room and body temperatures) are colloidal carriers (50−1000 nm) dispersed either in water or in an aqueous surfactant solution (Rawat et al., 2006) and could overcome some stability limitations of conventional liposomes. SLN formulation approaches have mainly centered on avoiding high-temperature melts due to the thermal sensitivity of peptides that has also led to issues with insufficient loading. Hence for peptides, they have involved the use of counterions (see Section 1.3.1) to increase the hydrophobicity of the peptide before encapsulation or in the process of encapsulation, the formation of peptide-loaded reverse-micelles, and the use of the double emulsion approach to facilitate the physical entrapment of the peptide molecules. Peptide release rates have been tuned by virtue of the lipids' inherent degradation profiles and also through the control of the porosity of the lipid matrices. Sarmento et al. produced a cetyl-palmitate SLN loaded with insulin based on a w/o/w double emulsion using poloxomer 407 as a stabilizer. They reported a considerable hypoglycemic effect (at a dose of 50 IU/kg) in vivo. Release occurred in a biphasic manner with an initial burst of surface-bound drug and prolonged release from the lipid matrix of the SLN. The oral bioavailability of the SLNs was around 5% (Sarmento et al., 2007). However oral administration of insulin (25 IU/kg) loaded Chitosan-coated Witepsol 85E solid lipid nanoparticles (SLN) to diabetic rats, resulted in a considerable hypoglycemic effect resulting in relative pharmacological bioavailabilities of 8% and 17% for uncoated and chitosan-coated SLN, respectively and at half the dose versus Sarmento et al. pointing to the benefit of the combination approach (Fonte et al., 2011). Zhang et al. also demonstrated ca. 7% bioavailability with lectin-modified insulin SLNs thus highlighting the value of the targeted SLN approach (Zhang, Ping et al., 2006). The specific drug release profile of SLNs needs to be factored in for other peptide applications, especially if needing a more immediate Cmax effect. To date, the majority of the research on SLNs is focused on their particle characteristics, and their contribution to the gastrointestinal stability of embedded biopeptides is not well elucidated. Hence, when it comes to efficacy, the lack of an in-depth understanding of their interaction with biological barriers might justify the fact that they have not reached the clinical trials with regards to utility for oral peptide and protein delivery (Niu, 2016). Besides, they are still immature with respect to successful scale-up and critical temperature and raw material control strategies.

Brayden et al. presented key learnings from the EU FP7 oral peptide nanoparticle project, TRANS-INT consisting of 19 partners researching a variety of constructs (Brayden & Alonso, 2016). They cited insufficient stability and need for lyophilization, reproducibility issues for phys-chem characterization, poor drug loading, and lack of consistency in bioassays across labs as key areas that needed addressing. There were also a number of reflections posted by Brayden et al. around the data published across a number of papers on oral insulin delivery (Brayden et al., 2020). Notably, most of the PD data was established in induced diabetic animal models, none of which led to human translation. The doses required to reduce blood glucose concentration by 50%, using some of these promising NP systems, ranged from 30 to 100 IU/kg, whereas the standard of care for injectable insulin doses is only 1 IU/kg, thus making most reported NPs competitively disadvantageous in the diabetes market.

1.3.2.5 Nanoparticle formulations to have reached clinical trials

Despite the early optimism in the potential of NPs for increasing oral bioavailability of peptides, only a couple of polymeric NPs have made it to clinical trials but no longer being pursued actively. The first polymeric NP to reach the clinic was OR14−1, oral insulin from Oshadi Drug administration Ltd. (Israel). The company completed two phase I trials in patients with type 1 diabetes in 2013 followed by phase 2 clinical trials in 2017 (Rachmiel et al., 2019). The formulation consists of noncovalently associated insulin and proinsulin encapsulated within silica nanoparticles or Aerosil, coated in a hydrophobic polysaccharide, suspended in oil. Very little info is available in the literature on the formulations; inclusion of mucoadhesive polymer classes (pectin/chitin) in the patents may suggest their inclusion for enhancing oral absorption in addition to the suspension matrix of oily cosolvents or lipidic excipients (Alexander & Orna, 2011). A substantial glucose-lowering effect and tolerability of Oshadi-Icp (insulin, proinsulin, and C-peptide) formulation in patients were demonstrated. The second polymeric NP technology to reach the clinic was developed by NOD Pharmaceuticals (Shanghai, China) and went to Phase I in 2011. The technology consisted of enteric-coated capsules with calcium phosphate and insulin (NODLin) or exenatide nanoparticles (5−200 nm in size) in the presence of PEG salts of fatty acids (e.g., caprylate, caprate) or bile salts as precipitating agents that may also be contributing to further permeation enhancement (Li et al., 2012). One phase I clinical trial in 12 healthy volunteers tested three different doses of oral insulin (50, 100, 200 IU) against a 6 IU dose of subcutaneously injected insulin as the active comparator. Overall, the patterns of the pharmacodynamic response to oral and subcutaneous insulin were rather similar. In conclusion, further trials were warranted to demonstrate a clinically relevant metabolic effect of this oral insulin formulation (Zijlstra et al., 2014). With very little information available on the early preclinical or clinical performance of these products with no recent published progress into late clinical development, it may be fair to assume these are struggling to meet the target product profile (TPP).

From the lipid-based NPs, Diasome Pharmaceuticals Inc. (USA) has progressed a liposomal insulin formulation termed hepatic directed vesicle insulin (HDV-I) to Phase 2b trials. The phospholipid bilayer of HDV-I vesicles ($<$150 nm) consists of insulin bound to biotin (hepatic targeting molecule). HDV-I has been tested in a number of small-scale clinical trials, where preprandial oral delivery of HDV-I to type I diabetics was comparable to postprandial parenteral delivery (Geho & Lau, 2009). Limited details on the preclinical oral HDV-1 formulations are publicly available. However, the most recent third-generation products have been manufactured by microfluidization (as opposed to sonication). Improved stability in aqueous and acidic media has been reported.

Endorex Corp (USA) had developed a stable liposome delivery system that was found to improve the oral bioavailability of a number of peptides including insulin and hGH. The liposomes that constitute the Orasome technology were stabilized by interfacial polymerization to prevent deterioration in presence of bile salts and at acidic pH. Monomeric liposomes composed of 1,2-di (2,4-octadecadienoyl) phosphatidylcholine were cross-linked in the presence of radicals and demonstrated enhanced GI fluid stability (Okada et al., 1998). Oral delivery of hGH in low polymerized Orasome liposomes in mice boosted the plasma concentration of hGH to levels similar to the subcutaneous format (2001). No further reports for these systems have been published post 1999.

1.3.2.6 Industrial development perspective of oral nanoparticles for peptide delivery

To date, NP products have not offered any benefits over the simpler permeation enhancer strategies when it comes to improving bioavailability or in terms of numbers progressed into clinical trials. With the more complicated design features and raw materials of NPs versus the simple lipid-based excipients/TPEs, NP-based systems have a lot more critical quality attributes and controls for reproducible scale-up and manufacturing. The size, shape, and surface chemistry of nanoparticles can greatly impact their cellular uptake and overall efficacy. A number of studies have reported the criticality of particle sizes for NP permeation/uptake along with the requirement for specific surface charges and shapes (Banerjee et al., 2016; Lamson et al., 2020a; Wang et al., 2018). Thus a more defined acceptable target range needs to be identified for each parameter with accompanying analytically sensitive methods to accurately measure and highlight deviations should be developed early in development. Some of the newer trends involve looking into combining permeation enhancers with local mucus adhesion/penetration and designing in a flexible surface charge/targeting ligand amenable system. These approaches could prove to be more successful in the future and should be extensively studied for a range of peptide chemistries and disease conditions. The SLNs and inorganic carriers may offer advantages over the more conventional polymer and lipid carriers. For example, sourcing chitosans at the right degree of deacetylation, molecular weight, and purity has proved to be a challenge. Similarly, polydispersity index control and reproducibility with a number of PLGA-based systems or controls on aldehyde impurities in PEG, etc., are critical control parameters for starting materials. Early considerations should be given to starting material purity and supplier sourcing strategy. This is also applicable for a number of polymer and lipid systems reviewed earlier. The challenges of poor and nonreproducible drug loading need to be revisited by employing more advanced and scalable manufacturing methods such as microfluidic-based parallel scale-up, or inverse flash precipitation that could improve batch-to-batch consistency, and reduce wastage of API. In addition, all starting materials should be deemed sufficiently nontoxic, nonimmunogenic, noninflammatory, and biodegradable, with consideration toward green/sustainable materials. As a target, a significant improvement in drug loading (<25% w/w) should be set. Exploring formulations with hydrophobic ion pairing/lipidization/counterions will be key to achieving this. Another critical liability to be considered in selecting lipid-based systems (liposomes/LBFs, SLNs) is their potential to degrade oxidation labile peptide sequences that could be accelerated by oxidative impurities in the lipids. In an early screening of prototypes, efforts should be made to generate bioavailability data in vivo in addition to in vitro insights across a range of cell/tissue models. Oral bioavailability of at least 10% needs to be targeted, with studies carried out in a range of species. This will offset the potential increase in the cost of manufacturing and developability of nanotechnology-based products and yet be economically feasible for broad patient access.

1.3.3 Transporter-mediated oral peptide drug delivery

It has been known for decades that the intestinal brush borders contain a peptide carrier system that can transport di- and tripeptides with various structural modifications. Several marketed small peptidomimetic drugs including β-lactam antibiotics and angiotensin-converting enzyme (ACE) inhibitors are claimed to be transported by this system (Bai & Hu, Subramanian, et al., 1992; Bai & Amidon, 1992; Bai et al., 1991). Fig. 1.4 provides a recent overview of the transporters in the human small intestine that can generally be grouped into two classes: efflux and uptake transporters.

As shown in Fig. 1.4, uptake transporters in enterocytes include, on the luminal/apical side, proton-dependent peptide transporter 1 (PepT1),

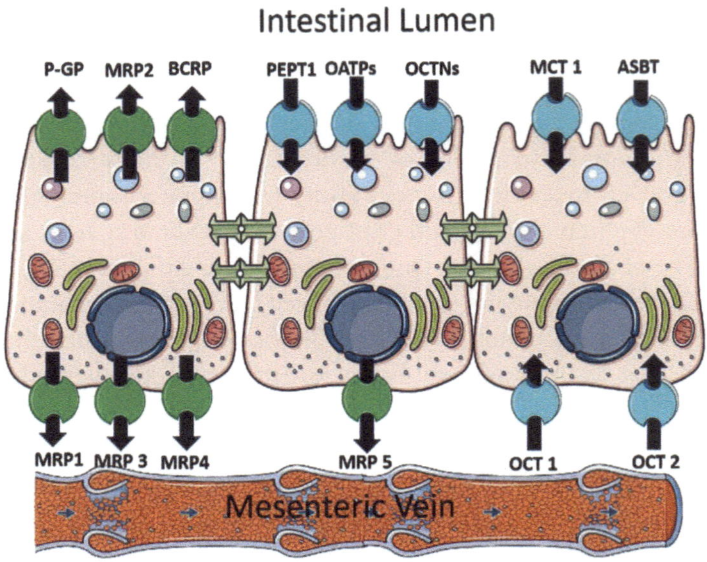

Figure 1.4 Major drug transporter proteins. Major drug transporter proteins expressed by the intestinal epithelia, including intestinal uptake (*blue*) and efflux (*green*) transporters (Estudante et al., 2013; Murakami & Takano, 2008; Tamai, 2012; Müller, 2017, Wang et al., 2020). *Figure prepared from information in Estudante, M., et al. (2013). Intestinal drug transporters: An overview.* Advanced Drug Delivery Reviews, *65(10), 1340–1356; Murakami, T., & Takano, M. (2008). Intestinal efflux transporters and drug absorption.* Expert Opinion Drug Metabolism Toxicology, *4(7), 923–939; Tamai, I. (2012). Oral drug delivery utilizing intestinal OATP transporters.* Advanced Drug Delivery Review, 64*(6), 508–514. https://doi.org/10.1016/j.addr.2011.07.007; Wang, G., Zhao, L., Jiang, Q., Sun, Y., Zhao, D., Sun, M., He, Z., Sun, J., & Wang, Y. (2020). Intestinal OCTN2- and MCT1-targeted drug delivery to improve oral bioavailability.* Asian Journal of Pharmaceutical Science, 15*(2), 158–173. https://doi.org/10.1016/j.ajps.2020.02.002.*

organic-anion-transporting polypeptides (OATPs), carnitine/organic cation transporters (OCTNs), monocarboxylate transporter protein (MCT 1), and the sodium-dependent bile acid transporter (ASBT). The serosal/basolateral side of the enterocyte has the organic cation transporters 1 and 2 (OCT1 and OCT2). The physiological function of uptake transporters is to deliver nutrients and other desirable substances into the cell.

Fig. 1.4 also shows major efflux pumps such as, on the luminal/apical side, P-glycoprotein also known as multidrug resistance protein 1 (MDR1), multidrug resistance-associated protein 2 (MRP2), and breast cancer resistance protein (BCRP). These pumps can also be found on the serosal/basolateral side of enterocytes, specifically MRP 1, 3, 4, and 5.

The function of efflux pumps is to keep xenobiotics out of the cell, and can also, along with passive diffusion, be important contributors to pumping drugs out of the cell and into the bloodstream, thereby completing the absorption process. The details around expression, regulation, and function of intestinal drug transporters are outside the scope of this book chapter and the reader is encouraged to consult recent reviews for more details (Müller et al., 2017; Tamai, 2012; Wang et al., 2020). This section will mainly concentrate on the PepT1 transporter, the key transporter involved in the absorption of smaller peptides, and provide a brief overview of its structure, function, regulation, and application for drug delivery (in terms of molecular- and formulation-related targeting). Other transporters such as ASBT will be included in case studies where relevant for formulation-related targeting.

1.3.3.1 Molecular targeting of PepT1 to increase intestinal absorption

PepT1 and PepT2 are members of the proton-dependent oligopeptide transporter (POT) family. PepT2 is mainly expressed in the kidney, and while it is highly relevant for peptide excretion, PepT1, widely expressed in the intestine, is an attractive target for oral peptide delivery. The POT family uses an inwardly directed electrochemical proton gradient to drive the uptake of essential nutrients into the cell (Newstead, 2017), as illustrated in Fig. 1.5.

The PepT1 gene is up- and downregulated by the prevalence of its nutritional substrates and is a high-capacity and low-affinity peptide transporter (Spanier, 2014). A PepT1 crystal structure revealed that the protein contains 12 membrane-spanning domains and a large extracellular domain (ECD) (consisting of two immunoglobulin-like domains connected in series) (Wang et al., 2017). The transporter has the ability to transport a wide range of substrates, recently explained by multiple binding pockets and modes in the ECDs, thereby giving multiple possibilities for di-/tripeptides or peptidomimetics to bind to the transporter (Newstead, 2011, 2017). The substrate promiscuity of PepT1 allows transport of closely related drugs with passive poor absorption. This can result in high oral bioavailability, making PepT1 targeting an attractive approach for pharmaceutical development (Spanier, 2014).

When designing PepT1 substrates, knowledge of critical substrate features for binding is paramount (summarized in Table 1.3). Thorn et al. (2007) used a subset of these structural features to design 29 tripeptides as potential PepT1 promoieties and evaluated them in PepT1 affinity and transport studies using Caco-2 cells leading to the selection of a lead

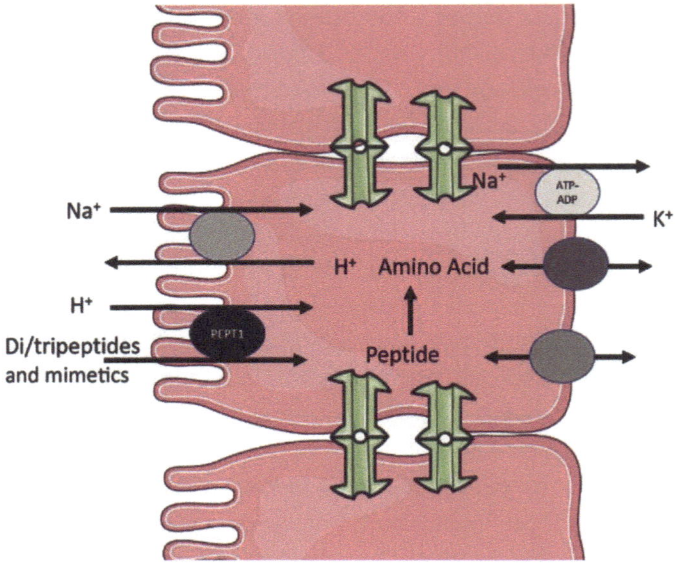

Figure 1.5 Di- and tripeptide transport across the gastrointestinal epithelium. Model for di- and tripeptide transport across the gastrointestinal epithelium (Terada & Inui, 2004; Yang et al., 1999). *Figure prepared with info from Terada, T., & Inui, K. (2004). Peptide transporters: Structure, function, regulation and application for drug delivery.* Current Drug Metabolism, *5(1), 85–94; Yang, C. Y., Dantzig, A. H., & Pidgeon, C. (1999). Intestinal peptide transport systems and oral drug availability.* Pharmaceutical Research, 16*(9), 1331–1343.*

Table 1.3 Critical substrate features for binding to PepT1 (Foley et al., 2010; Thorn et al., 2007).

Design feature	PepT1 affinity requirement/impact
A peptide bond	Not required
Peptide size	Di- and tripeptides
Free amino terminus	Required
Free carboxy terminus	Desirable
Cis/Trans peptide bond conformation	Trans conformation has higher affinity
L/D- amino acids	Preference for L-configuration
Lipophilicity	Hydrophobic peptides have higher affinity

Source: Adapted from Thorn, K., Andersen, R., Christensen, J., Jakobsen, P., Nielsen, C. U., Steffansen, B., & Begtrup, M. (2007). Design, synthesis, and evaluation of tripeptidic promoieties targeting the intestinal peptide transporter hPEPT1. *Chem Med Chem, 2*(4), 479–487. https://doi.org/10.1002/cmdc.200600200.

promoiety, H-Phe-Ser-Ala-OH. These, and other studies to determine ligands for PepT1, are important, as the information can be used in modern drug design and discovery to attain high oral bioavailability of smaller highly polar/BCSIII drugs, peptides, and otherwise.

A recent example is peramivir, a potent neuraminidase inhibitor, that has been approved for the treatment of influenza as an IV drug (Alame et al., 2016). It exhibits a very low oral bioavailability due to the high polarity and the low permeability across the intestinal mucosa. The structure has peptide resemblance and is shown in Fig. 1.6.

As illustrated in Fig. 1.6, Sun et al. (2018) utilized a PepT1-mediated prodrug strategy to improve GI absorption, testing permeability of 14 prodrugs, 7 amino acid esters, and 7 amino acid amides in Caco-2 cells. Peramivir-$(CH_2)_2$-l -Val and peramivir-l-Ile were the highest permeability ester and amide prodrugs, respectively. The bioactivation and oral absorption of peramivir-$(CH_2)_2$-l -Val was rapid and extensive with a bioavailability of 65% in rats (vs only 4% of native Peramivir). Peramivir-l-Ile was not completely converted to the parent drug in vivo due to the more stable amide bond. This prodrug has a 37% bioavailability in rats, comparatively.

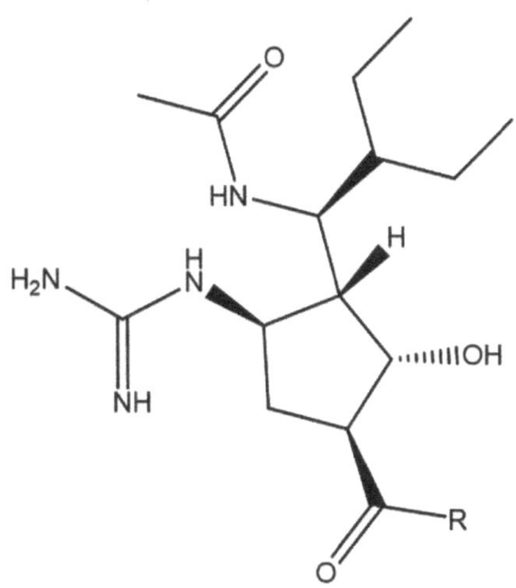

Figure 1.6 Structure of peramivir and prodrugs. Structure of peramivir and prodrugs (Alame et al., 2016; Sun et al., 2018). Ester prodrugs: R = -O-(CH2)2-L-amino acid; amide prodrugs: R = -N-L-amino acid; free form: R = H. *Source:* Adapted from Alame, M. M., Massaad, E., & Zaraket, H. (2016). Peramivir: A novel intravenous neuraminidase inhibitor for treatment of acute influenza infections. Frontiers in Microbiology, 7, 450; Sun, Y., Gan, W., Lei, M., Jiang, W., Cheng, M., He, J., Sun, Q., Liu, W., Hu, L., & Jin, Y. (2018). PEPT1-mediated prodrug strategy for oral delivery of peramivir. Asian Journal Pharmaceutical Sciences, 13(6), 555-565. https://doi.org/10.1016/j.ajps.2018.05.008.

1.3.3.2 Industrial development perspective of transporters

The use of PepT1 to facilitate the absorption of peptide drugs is not new. Enalapril, the ester prodrug of the ACE-inhibitor enalaprilat, a dipeptide, is claimed to be a substrate for PepT1, and so does many β-lactam antibiotics. Enalapril was marketed in 1981. Oral bioavailability is increased from 3% to 12% for enalaprilat to 60%–70% when dosing the Enalapril prodrug (Zhang, Phelps et al., 2006). In the interest of full transparency, there are some disagreements in the literature whether this increase in bioavailability is due to PepT1 targeting. For example, two studies both using Xenopus laevis oocytes point in different directions; Zhu et al. (2000) find that enalapril has affinity and is transported by PepT1, whereas Knütter et al. (2008) draws the opposite conclusion. In addition, there are recent case studies in the literature showing the benefit for other small polar drugs. For example, Foley et al. have shown that thiopeptides can be used to target a variety of drugs to the transporter including ibuprofen, gabapentin, propofol, nabumetone, atenolol, zanamivir, baclofen, and mycophenolate (Foley et al., 2009, 2018). However, a search of clinical trials (Medicine, 2020) mainly returns studies to probe drug–drug interactions with the transporter, and hence it is the underutilized route to oral absorption in clinical development. This could be due to several developability concerns. Many of the molecular-targeted PepT1 designs are prodrugs. Peptide promoieties have the advantage of high solubility, but, subject to design, can suffer from classical peptide chemical and enzymatic instability. In addition, prodrugs have increased development complexity in terms of tracking both the active and the promoiety. Furthermore, to the authors' knowledge, there are no examples of larger peptide molecular designs (beyond catripeptides) aimed at absorption via PepT1 targeting, likely due to the transporter specificities outlined in Table 1.3. Therefore the use of molecular design to target PepT1 in terms of oral peptide drug delivery is going to be limited to smaller peptides or for the right drug, and assuming the right control strategies in place molecular targeting PepT1 should still have a place in modern drug design, discovery, and development.

1.3.3.3 Transporter-targeted nanoparticle delivery

An advantage of delivery system targeting, rather than molecular targeting is that the interaction of the therapeutic molecule with the transporter in question is not necessary. In addition, a broad range of targeting ligands, not just transporters, can be explored. However, for the purpose of this chapter, we will focus on transporters (i.e., the PepT1 and ASBT outlined in Fig. 1.4), nanoparticles (NPs), and cargos with a peptide nature. This is, by no means, a comprehensive review, and the reader should refer to reviews by Xu, Shrestha, et al. (2020) and Su et al. (2019) for additional details.

A recent example of transporter-mediated GI-targeting was carried out by Wu et al. (2019). They developed a colon-targeted NP with poly(lactic-co-glycolic acid) (PLGA), MMT (montmorillonite), and chitosan (CS) for local treatment of ulcerative colitis. They used the KPV tripeptide (Lys-Pro-Val) to target PepT1, expressed in colonic epithelial cells and macrophages, to deliver cyclosporine A (CyA), an immunosuppressive. A biodistribution experiment confirmed that the NPs adhered to colonic tissues as intended and was retained for up to 36 h. The MMT/CS coating was introduced to minimize leakage of CyA in the upper intestinal tract to minimize systemic absorption. Therefore, while the study was certainly intended for local use, it illustrates an important PepT1 NP design features: immediate release in the upper intestinal tract is needed for systemic absorption and shows how NPs can be tailored for various purposes throughout the intestinal tract with creative delivery system design.

Gourdon et al. (2015) researched the systemic delivery of peptides via PepT1 targeting. They formulated polylactic acid, polyethylene glycol, and valine (PLA-PEG-Val) NPs. These NPs were shown in competition experiments in vitro to inhibit the transport of known PepT1 substrates. Oxytocin was encapsulated in the Val-NPs, and it was shown in cell culture experiments to transport twice the amount of peptide compared to the naked drug. In vivo, after oral administration, oxytocin plasma levels were significantly higher when encapsulated in Val-NPs compared to free drugs. This study shows that PepT1-targeted NPs can moderately increase the transport and systemic availability of smaller peptides (i.e., oxcytocin is a nonapeptide with an MW of 1007 Da; it has a cyclic nature imposed by a disulfide bond).

Jain and Jain (2015) prepared L-Val-PLGA insulin-loaded NPs. Ex vivo intestinal studies showed that the L-Val conjugated NPs gave rise to a greater insulin uptake compared to their nonconjugated counterparts, and in vivo rabbit pharmacodynamic studies showed that the conjugated NPs gave rise to faster and greater glucose lowering as compared to the unconjugated particles. Glucose concentrations were measured and showed a pharmacokinetic–pharmacodynamic correlation. This study illustrates that the PepT1 transporter-targeted NP approach can be used for larger peptides (insulin has 51 amino acids organized in two chains held together by disulfide bonds and an MW of 5808 Da).

As mentioned earlier, the advantage of using nanoparticles for peptide drug delivery is that other transporters besides PepT1 can be utilized for targeting. Bile acid transporters such as ASBT (see Fig. 1.4 for details) are widely expressed throughout the intestinal tract and have been investigated for oral delivery of insulin (Wong et al., 2017) among other peptides. Niu, Tan, et al. (2016) investigated the mechanism of bile salt-containing liposomes for oral delivery of insulin. These were found to be more stable and have a longer GI residence time than conventional liposomes and were more effective for cellular uptake and transport of the insulin cargo. Furthermore, a research study by Fan et al. (2013) formulated insulin into deoxycholic acid conjugated chitosan nanoparticles. These NPs were shown to be internalized by the ASBT transporter in Caco-2 cells, facilitated endosomal escape in cells and mediated release across the basolateral membrane. Finally, rat PK studies showed an oral bioavailability of 16%, significantly higher than when insulin was not formulated in these NPs (Chen et al., 2011; Petrus et al., 2007). This is just one example out of many showing that other transporter than PepT1 can be used to target NPs to transporters.

1.3.3.4 Industrial development perspective of transporter-targeted nanoparticles

The transporter-targeted NPs are still in the early research phase, and many drug delivery, distribution, and efficacy/safety aspects still need to be addressed. Briefly, the transport mechanism across and within the GI mucosa still needs to be investigated, and the benefits of a targeted NP need to be weighed against the added development complexity. Often, such systems require additional toxicology, DMPK and CMC studies to be conducted to outline the effect of the targeting ligand, the distribution of the NP, the release of the drug, and many other aspects covered in Section 1.3.3.

1.3.4 Ingestible injection devices

As we have discussed in this book chapter, the traditional methods for improving bioavailability have been well tried and tested, and the success limited at best. The oral delivery of macromolecules with double-digit bioavailability and low variability would offer a truly disruptive potential for the oral delivery of peptides and other biomolecules. An emerging area is based on physically disrupting the absorption barrier to deliver the peptide to the underlying lamina propria/submucosa with a subsequent microinjection and transfer of the macromolecule. These techniques may offer the ability to deliver higher quantities across the GI mucosa with more acceptable bioavailability and variability. There are two classes of active devices that are currently in development: microneedles and jet capsules. Below, we will discuss the different technologies and the challenges that need to be overcome to produce a marketable product.

1.3.4.1 Microneedles

Microneedles were first used for transdermal drug delivery (Henry et al., 1998). The major goal for all transdermal microneedle designs is to penetrate the 10–15 μm stratum corneum, the outermost layer of skin (Jeong et al., 2017). Various needle designs and systems have been developed for use with transdermal patches, including solid, coated, hollow, and dissolvable needles. Advances in the technology and successes in the clinic (Bhatnagar et al., 2017) led to broader applications, including deploying microneedles to deliver large molecules to the intestinal wall. This also has the advantage of moving the invasive administration site from the subcutis, which has a high density of sensory neurons, to the gastrointestinal tract, which is devoid of such receptors. Traverso et al. demonstrated the potential of microneedle-based oral delivery in pigs, showing a faster drop in blood glucose levels after injections to the GI tract versus subcutaneous injection (Traverso et al., 2015). As shown in Fig. 1.7, hollow and solid microneedles could be utilized for oral delivery applications. In most cases a pH-responsive coating has been used to target the capsule to the relevant GI site. Once the coating has dissolved, the microneedles need to be exposed and interacted with the luminal wall of the small intestine. In the case of hollow microneedles, once they have penetrated the mucus and reached the intestine wall, the drug reservoir is compressed through peristalsis of the small intestine, and the drug is released through the needles. In the case of solid biodegradable microneedles, the drug is formulated into the microneedles, which penetrate the tissue and break off from the delivery device, leaving the microneedle to release the drug in a controlled manner.

Below, we discuss a number of microneedle technologies, all of which use the biodegradable solid needle approach. To the best of our knowledge, there have been no hollow microneedle delivery devices that have progressed for in vivo studies.

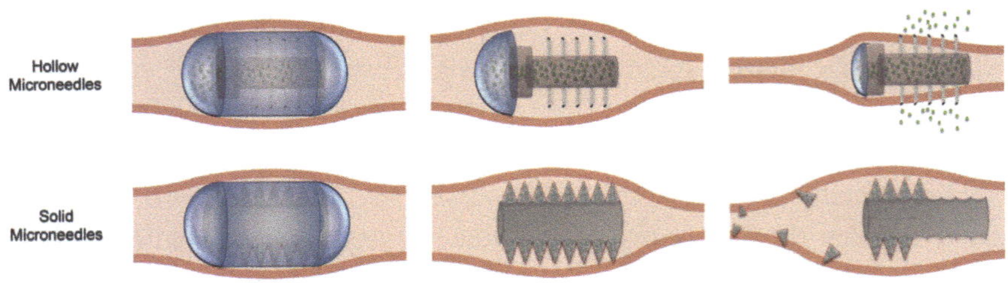

Figure 1.7 Microneedle GI drug delivery, solid versus hollow microneedles. Microneedle GI drug delivery, solid versus hollow microneedles. In both cases, the pill's needles are initially coated with an enteric coat to aid ingestion (left). When the pill reaches the desired location in the GI tract, the coating dissolves to reveal the microneedles (middle). For hollow microneedles (top right), peristalsis compresses the drug reservoir, and the drug is released through the needles. For solid microneedles (bottom right), the drug is formulated into the microneedles. The microneedles break off after penetrating the tissue, and the needle releases the drug. *Reprinted from Traverso, G., Schoellhammer, C. M., Schroeder, A., Maa, R., Lauwers, G. Y., Polat, B. E., Anderson, D. G., Blankschtein, D., & Langer, R. (2015). Microneedles for drug delivery via the gastrointestinal tract.* Journal of Pharmaceutical Sciences, 104*(2), 362–367. https://doi.org/10.1002/JPS.24182, with permission from Elsevier.*

In January 2020, Rani Therapeutics announced the results of their phase 1 study, demonstrating the safety and tolerability of octreotide loaded into sugar needles, which the RaniPill deploys into the intestinal wall. Fig. 1.8 outlines the concept behind the RaniPill and its mode of action. Rani Therapeutics achieved bioavailability of greater than 70% (100 μg dose), although no details of variability were shared (Newswire, 2020). Over a hundred animal studies have been carried out using the RaniPill. Rani Therapeutics has investigated loading basal insulin, GLP-1, parathyroid hormone, human growth hormone, TNFα inhibitors, interleukin antibodies, PCSK9, Factor VIII, IX, and X (Therapeutics, n.d.). The RaniPill is the most advanced ingestible injectable in development.

The self-orientating millimeter-scale applicator (SOMA) has been developed at MIT by Abramson et al. (2019). The design is based on a tortoise shell, which is designed to correct its orientation in the stomach where it injects 0.3 mg of drug contained in a microneedle (called a millipost). The mode of action is outlined in Fig. 1.9. A pig study was carried out where the devices were administered to the animals via an orogastric tube and dropped in the insufflated stomach at a height of 5–7 cm. The study was carried out on anesthetized pigs, but the authors used a "walking simulation" (see supplementary information of Abramson et al. (2019) for details) on the anesthetized pigs to show the SOMA device still found

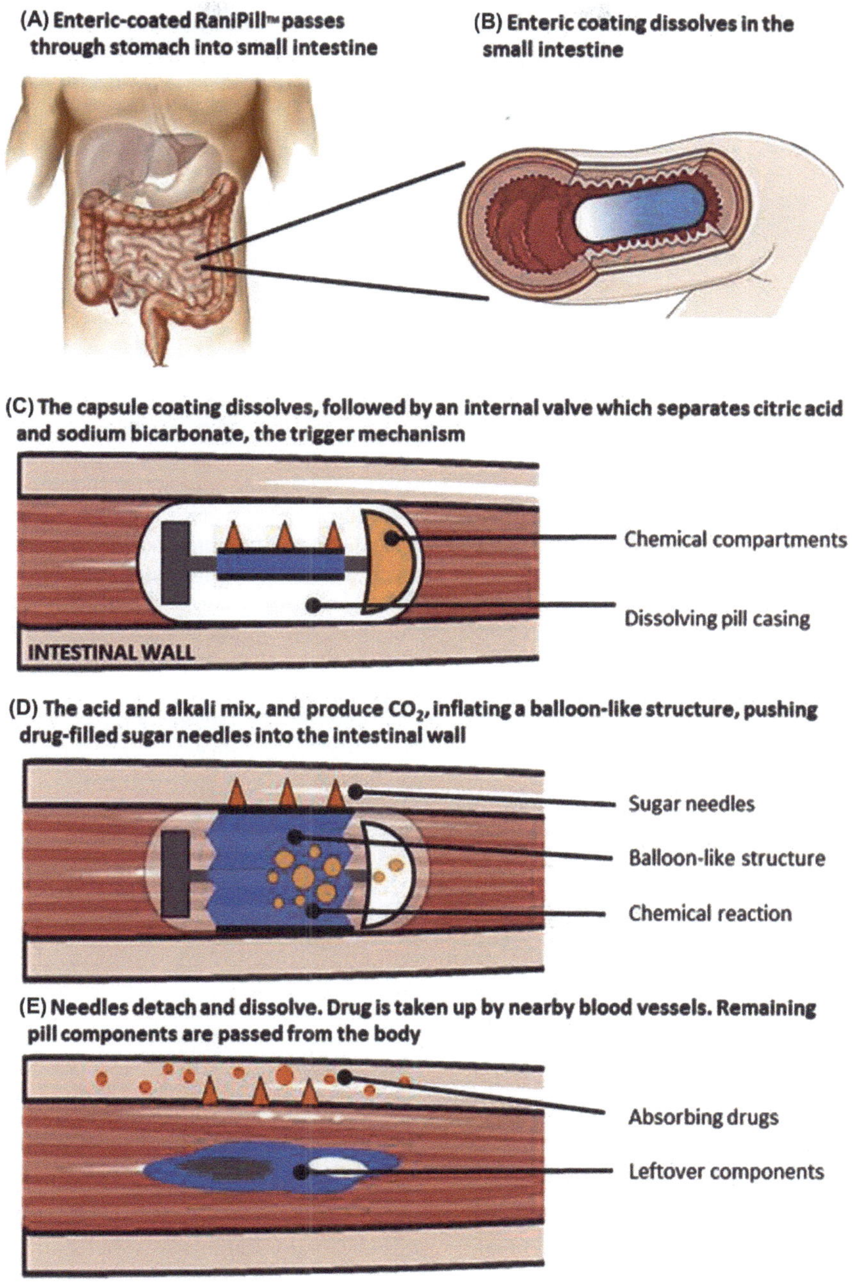

Figure 1.8 The RaniPill. Intended mode of action of the RaniPill.

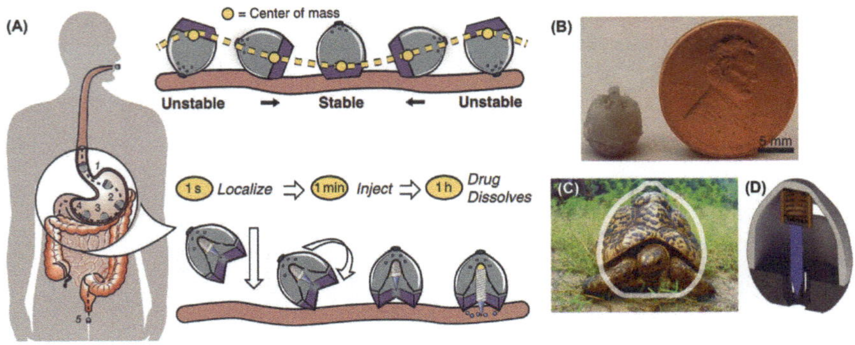

Figure 1.9 The self-orientating millimeter-scale applicator (SOMA). (A) The SOMA localizes to the stomach lining and orients its injection mechanism toward the tissue wall. It injects a drug payload through the mucosa. The drug dissolves and the device passes out of the body. (B) A fabricated SOMA. (C) Comparison between the shape of the leopard tortoise and the SOMA. The SOMA orients and remains stable in the stomach after reaching its preferred orientation (photo: M. M. Karim/ Wikimedia Commons, CC-BY-SA 2.5). (D) The SOMA uses a compressed spring that is fixed in caramelized sucrose (*brown*) to provide a force for a drug-loaded millipost (*blue*) insertion. The spring remains encapsulated within the device after actuation. *Figure taken from Abramson, A., Caffarel-Salvador, E., Khang, M., Dellal, D., Silverstein, D., Gao, Y., Frederiksen, M. R., Vegge, A., Hubálek, F., Water, J. J., Friderichsen, A. V., Fels, J., Kirk, R. K., Cleveland, C., Collins, J., Tamang, S., Hayward, A., Landh, T., Buckley, S. T., & Traverso, G. (2019). An ingestible self-orienting system for oral delivery of macromolecules.* Science, *363(6427), 611–615. https://doi.org/10.1126/science.aau2277. Reprinted with permission from AAAS.*

a position within 30degrees of normal in the stomach, a suitable angle for injection. Delivering insulin with the SOMA yielded similar insulin plasma levels to those achieved from delivering the SOMA milliposts subcutaneously. However, the SOMA only worked on animals in the fasted state.

Abramson et al. also developed the luminal unfolding microneedle injector (LUMI) device (Abramson et al., 2019). After the enteric coating dissolves and the trigger mechanism is fired, a microneedle patch is propelled from the device. 0.3 mg of drug can be loaded into a 0.5 cm^2 microneedle (dissolvable polymer) patch. A study was carried out on anesthetized pigs, comparing insulin delivery subcutaneously versus the LUMI. The LUMI devices were administered to the animals via an orogastric tube and placed in the small intestine. The device provided a faster pharmacokinetic uptake profile and yielded >10% of the bioavailability of the subcutaneous injection over a 4-h sampling period.

Biograil is a preclinical stage pharmaceutical company who have developed a spiked device that is contained in an enteric-coated standard capsule. When the coating and then the capsule dissolves, peristalsis of the GI tract pushes the spike into the gut wall. The drug-loaded spike dissolves inside the gut wall (Biograil Website, 2020). The device is being developed using injection molding techniques, with line of sight to low cost and high-volume commercial manufacturing.

1.3.4.2 Jet capsules

Devices are in development to produce jet streams of suitable pressure to penetrate the mucosal layer of the GI tract. Jet capsules are at much earlier phases of development than microneedle systems. The first jet system was not designed for the intestine, but the oral cavity. The MucoJet is a needleless capsule designed and built by Aran et al. to deliver vaccines in the oral cavity (Aran et al., 2017). As detailed in Fig. 1.10, the MucoJet is a microelectromechanical drug delivery technology. It uses a self-contained gas-generating

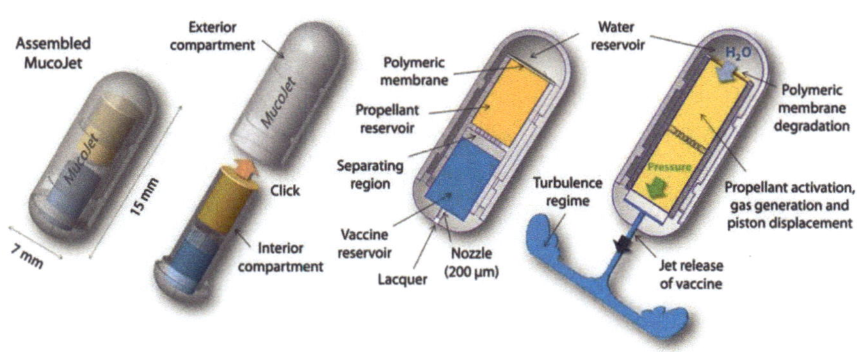

Figure 1.10 The MucoJet. The MucoJet 3D-printed plastic device consists of interior and exterior compartments, with the interior compartment holding the propellant and vaccine reservoirs, separated by a porous membrane and a moving piston. Before administration, in this case in the buccal cavity, the exterior compartment is filled with water, and the interior compartment is inserted into the exterior compartment. The polymeric valve membrane that seals the propellant reservoir dissolves, allowing water within the exterior compartment. Water contacts the chemical propellant and a chemical reaction generates CO_2, forcing the piston toward the vaccine reservoir, breaking the nozzle membrane, and ejecting the vaccine as a high-pressure liquid. *Figure taken from Aran, K., Chooljian, M., Paredes, J., Rafi, M., Lee, K., Kim, A. Y., An, J., Yau, J. F., Chum, H., Conboy, I., Murthy, N., & Liepmann, D. (2017). An oral microjet vaccination system elicits antibody production in rabbits.* Science Translational Medicine, *9(380), eaaf6413–eaaf6413. https://doi.org/10.1126/scitranslmed.aaf6413. Reprinted with permission from AAAS.*

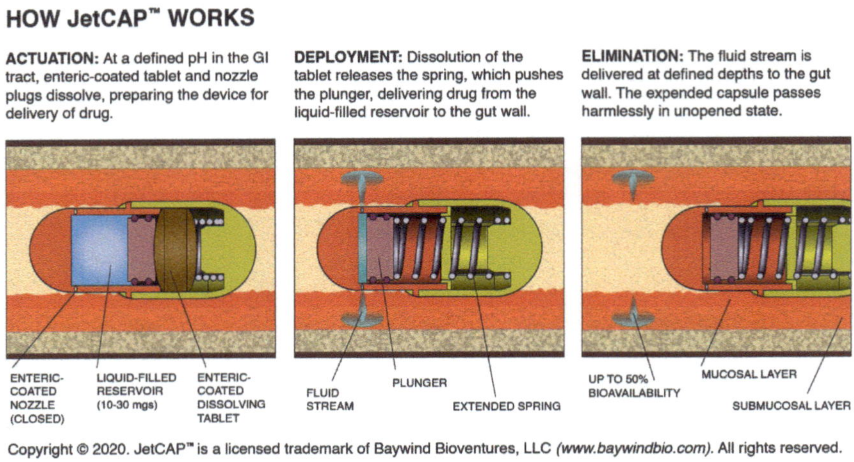

Figure 1.11 The Baywind Bioadventures JetCap. Intended method of action of JetCAP drug delivery system. *Figure reprinted with permission from Baywind Bioventures, LLC.*

chemical reaction within its two-compartment plastic housing to produce a high-pressure liquid jet of vaccine which fires into the oral cavity.

Jet capsule technology was soon modified for delivery to the GI tract. BaywindBio developed the JetCap device, an enterically coated capsule as shown in Fig. 1.11. Once the coating dissolves, the needle-free liquid jet is triggered by the dissolution of an enteric-coated tablet, which then releases a spring and pushes the formulation solution into the intestinal epithelium. No in vivo results have been reported. The dose of the device is determined by the size of the liquid reservoir (300 μL) and the solubility of the drug, but offers the capability of a larger dose than the microneedle devices.

1.3.4.3 Industrial development perspective of ingestible injectable devices

The challenges of the oral delivery of peptides and large molecules are well documented (Brown & Whitehead, 2020; Durán-Lobato et al., 2019; Morales et al., 2017), and while the jet capsules and microneedle devices have the potential to enable double-digit bioavailability, to become a pharmaceutical product, these systems have unique challenges to overcome, as well as the more conventional challenges of enteric coatings, mucus variabilities, and food effects.

Both microneedles and jet capsules require a consistent release mechanism to be effective. For the jet capsule, this means the release of payload with sufficient pressure, oriented to align the jet stream into the intestinal wall. The microneedles must be deployed so that they can penetrate the mucus lining and the GI tract and remain in place for a sufficient time for drug release. Transit of food and mucus variability may have a considerable effect on both kinds of devices. While injections into the GI tract should be relatively painless (Brown & Whitehead, 2020), the safety and tolerability of the microneedles must be established for long-term use.

With the exception of the SOMA delivering to the stomach, all systems rely on an enteric coating to reach the GI tract. Jet capsules can deliver much larger doses than microneedles (300 μL of formulation compared to 3–5 mg of biologic for the dissolvable microneedles). The larger dose may be advantageous, but this depends on the bioavailabilities of both systems. Dosing a liquid formulation enables the codosing of TPEs, NPs, etc., although the purpose of the device is to deliver across the epithelial barrier, and therefore should not require TPE or NP assistance. Using solid biodegradable microneedles could eliminate cold-chain logistics in some cases, and each device will have to demonstrate suitable peptide compatibility as well as performance stability. Both systems have their own inherent advantages/disadvantages and could offer differentiated product profiles beyond the improvements of bioavailability, which is the primary focus. For both microneedles and jet capsules, ensuring the device releases in the correct location is key, and the variability of both techniques will be greatly affected by the conditions of the patient's GI tract, mucus thickness, undigested food, etc. Another thing to consider is that these complex delivery systems would likely be considered as a device rather than a medicine and would yield an extra set of regulatory challenges. Both technologies have the potential to change the landscape of biologics formulations if they can introduce double-digit bioavailability with low variability.

1.4 Progress versus an oral peptide target product profile

Delivery strategies that would enable the switch between current injectable therapies into oral dosage forms remain an unmet need despite several advances across formulation, delivery, and device technologies. Peptide new chemical entities are being designed and developed, especially in diabetes, obesity and metabolism disease areas, and a step change in the design and delivery of peptide modalities is warranted. Aligning the peptide features with the formulation, the excipients, nanoparticles, or device to ensure compatibility and performance is key for designing a successful product. However, broader population diseases, patient accessibility, and delivery system complexity are also key considerations. The manufacturing cost of the delivery system is, therefore, an unavoidable consideration and needs to be carefully assessed in relation to the target population early in product development. Throughout this chapter, we have discussed technologies with increasing complexity, culminating in the development of ingestible injectable devices, and there will be a tradeoff in the performance of the product in terms of improved bioavailability and reduced variability versus economic sustainability.

Industrial drug discovery programs typically start with a clear definition of the critical attributes of the intended product, also called the target product profile (TPP), which is developed with consideration of the needs of the end users (patients and payers) (Lambert, 2010). The TPP, as described by the FDA, can encompass a fairly broad range of topics that may lead to inclusion in the product labeling. For the present perspective, the focus will be on oral peptide product properties, also called critical quality attributes (CQAs) that will directly influence the safety and efficacy of the product. Table 1.4 serves as an example of a high-level TPP for an oral peptide product. In an industrial development program, each attribute would have a defined acceptance limit and success criteria to enable decision making across the development cascade. Table 1.4 serves as an example of a high-level TPP for an oral peptide product.

Of the various approaches reviewed in the preceding sections, TPEs, transporters, nanoparticles, and ingestible injectables, in Table 1.5, we

Table 1.4 Example target product profile (TPP) of an oral peptide product.

Target product profile
Amenable for solid oral dosage forms (e.g., tablet)
Tolerable at the therapeutic doses, no adverse effects due to excipients
Patient-friendly frequency of dosing/pill burden
Efficacious at intended doses, sufficient oral bioavailability, acceptable PK variability, established PK/PD relationship
Line of sight to scalable and reproducible manufacturing process and analytical controls
Acceptable cost of development and competitive pricing in the intended market
IP/patentable

Table 1.5 Target product profile (TPP) assessment of the delivery technologies described in this chapter.

Delivery technology	Gaps
TPEs	With the low and highly variable bioavailability that can be expected when using TPEs to enable oral delivery of peptides, careful consideration needs to be paid to the candidate drug target profile (or what is commonly referred to as a "CDTP") to ensure its suitability for development of such a drug product. As discussed in Section 1.3.2, the peptide itself needs to be potent (thereby low dose) and either has a sufficiently wide therapeutic index or long half-life allowing for drug accumulation. To limit the effects of digestible enzymes on oral bioavailability, the peptide needs to have some degree of stability against luminal peptidases but modifications to achieve such stability need to be considered with respect to possible additional costs for peptide manufacture. First-generation TPEs are not specific in their mechanism of action and generally require significant doses (e.g., several hundred milligrams) to achieve their permeation-enhancing effects. A TPP must therefore take into consideration the "target patient population" to ensure that a "large" single-unit dosage form is acceptable. As also discussed, TPEs can be expected to suffer significantly from food interactions, and hence a likely dosing regimen of at least 30 min before food must also be acceptable to the target patient population. Safety is always a concern when perturbating the barrier properties of a membrane that separates the body from the extracorporeal environment, despite long-term safety being demonstrated for two TPEs that are used in recently approved oral peptide products. Hence, any development program should pay particular attention to safety and use the developed dosage form in such studies, rather than investigating individual components. Developability, ability to commercially manufacture the dosage form, as well as securing supply chain must also be considered, bearing in mind the high amount of TPE likely needed to be included in the dosage form as well as challenges to source novel excipients with appropriate quality.

(*Continued*)

Table 1.5 (Continued)

Delivery technology	Gaps
Oral nanoparticles (passive and receptor/transporter targeted)	NP delivery systems are complex with significant drug development hurdles with regard to timelines and costs involved. Aligned with the CDTP considerations as mentioned in TPE gaps earlier, potency, stability, and disease indication alignment are critical in choosing the correct nanoparticle technology (polymeric, lipidic, and targeting ligand format). NP CQAs, bioavailability, and characterization are largely underdeveloped and their regulatory validation remains a gap. Establishing the exact mechanistic insights and species translational aspects along with the safety aspects of the targeting moiety adds to cost of the delivery system, dose to man considerations, and regulatory complexity across early discovery and clinical development. On the upside, selective receptor targeting can result in a significant increase in oral bioavailability, relatively reduced variability, and therefore increased patient convenience and compliance. Both molecular and transporter-targeted NPs are largely underutilized, and hence further focused work to overcome the key limitations associated with these could result in novel IP in a very crowded competitive landscape.
Ingestible injectables	A technology in its infancy with many challenges to be faced. High bioavailability, low variability, and long-term safety data must be demonstrated for these complex delivery systems to be sustainable. Peptide-device compatibility and device performance over a suitable timeframe must be demonstrated.

present progress against the desirable TPP (Table 1.4) and highlight gaps, specific to the technology, that need to be addressed to improve their probability of success into clinical trials and ultimately as commercial products.

As discussed, TPEs, NPs, transporter targeting, and ingestible injectable devices have made significant progress, with recent commercial products utilizing TPEs. From an industrial pharmaceutical development perspective, there are still issues and open questions to be addressed. The drug delivery and CMC aspects are mainly related to late discovery and early development and are summarized in Table 1.5. Addressing these will be critical in increasing the number of studies progressing into the clinic with an increased chance of becoming a commercial product.

Acknowledgments

Thanks to Richard England, Advanced Drug Delivery, BioPharmaceuticals R&D Macclesfield, UK for help with chemical structures, Wikipedia, and https://smart.servier.com/ for the free clip art that formed the basis of many figures in this chapter.

Thanks to the AZ Oral Biologics Technology Team, Anand Subramony, Sivaneswary Genapathy, Iain Comley, Victoria Foster, Shawn Davis, Sergei Pechenov, Bertil Abrahamsson, Stanko Skrtic, Susanna Abrahmsén-Alami, Joe Grimsby, and Tina Rydén-Bergsten for many great discussions in this field.

References

Abeer, M. M., Rewatkar, P., Qu, Z., Talekar, M., Kleitz, F., Schmid, R., Lindén, M., Kumeria, T., & Popat, A. (2020). Silica nanoparticles: A promising platform for enhanced oral delivery of macromolecules. *Journal of Controlled Release, 326*, 544–555. Available from https://doi.org/10.1016/j.jconrel.2020.07.021.

Abramson, A., Caffarel-Salvador, E., Khang, M., Dellal, D., Silverstein, D., Gao, Y., Frederiksen, M. R., Vegge, A., Hubálek, F., Water, J. J., Friderichsen, A. V., Fels, J., Kirk, R. K., Cleveland, C., Collins, J., Tamang, S., Hayward, A., Landh, T., Buckley, S. T., & Traverso, G. (2019). An ingestible self-orienting system for oral delivery of macromolecules. *Science, 363*(6427), 611–615. Available from https://doi.org/10.1126/science.aau2277.

Agrawal, A. K., et al. (2014). Improved stability and antidiabetic potential of insulin containing folic acid functionalized polymer stabilized multilayered liposomes following oral administration. *Biomacromolecules, 15*(1), 350–360. Available from https://doi.org/10.1021/bm401580k.

Alame, M. M., Massaad, E., & Zaraket, H. (2016). Peramivir: A novel intravenous neuraminidase inhibitor for treatment of acute influenza infections. *Frontiers in Microbiology*, 7, 450.

Alexander, V. & Orna, G. (2011). Methods and compositions for oral administration of insulin. PCT/IL2009/000037. https://www.freepatentsonline.com/EP2254590B1.html

Alsenz, J., et al. (2000). Oral absorption of peptides through the cobalamin (Vitamin B12) pathway in the rat intestine. *Pharmaceutical Research, 17*(7), 825–832. Available from https://doi.org/10.1023/A:1007556108673.

Aran, K., Chooljian, M., Paredes, J., Rafi, M., Lee, K., Kim, A. Y., An, J., Yau, J. F., Chum, H., Conboy, I., Murthy, N., & Liepmann, D. (2017). An oral microjet vaccination system elicits antibody production in rabbits. *Science Translational Medicine, 9*(380). Available from https://doi.org/10.1126/scitranslmed.aaf6413, eaaf6413–eaaf6413.

Arrieta-Molero, J. F., et al. (1982). Orally administered liposome-entrapped insulin in diabetic animals. *Hormone Research in Paediatrics, 16*(4), 249–256. Available from https://doi.org/10.1159/000179509.

Augustijns, P. F., Brown, S. C., Willard, D. H., Consler, T. G., Annaert, P. P., Hendren, R. W., & Bradshaw, T. P. (2000). Hydration changes implicated in the remarkable temperature-dependent membrane permeation of cyclosporin A. *Biochemistry, 39*(25), 7621–7630. Available from https://doi.org/10.1021/bi9929709.

Azevedo, C., et al. (2020). Engineered albumin-functionalized nanoparticles for improved FcRn binding enhance oral delivery of insulin. *Journal of Controlled Release, 327*, 161–173. Available from https://doi.org/10.1016/j.jconrel.2020.08.005.

Bai, J. P., & Amidon, G. L. (1992). Structural specificity of mucosal-cell transport and metabolism of peptide drugs: implication for oral peptide drug delivery. *Pharmaceutical Research*, *9*(8), 969–978. Available from https://doi.org/10.1023/a:1015885823793.

Bai, J. P., Hu, M., Subramanian, P., Mosberg, H. I., & Amidon, G. L. (1992). Utilization of peptide carrier system to improve intestinal absorption: Targeting prolidase as a prodrug-converting enzyme. *Journal of Pharmaceutical Sciences*, *81*(2), 113–116. Available from https://doi.org/10.1002/jps.2600810202.

Bai, P. F., Subramanian, P., Mosberg, H. I., & Amidon, G. L. (1991). Structural requirements for the intestinal mucosal-cell peptide transporter: the need for N-terminal alpha-amino group. *Pharmaceutical Research*, *8*(5), 593–599. Available from https://doi.org/10.1023/a:1015848522228.

Banerjee, A., Ibsen, K., Brown, T., Chen, R., Agatemor, C., & Mitragotri, S. (2018). Ionic liquids for oral insulin delivery. *Proceedings of the National Academy of Sciences*, *115* (28), 7296–7301. Available from https://doi.org/10.1073/pnas.1722338115.

Banerjee, A., Qi, J., Gogoi, R., Wong, J., & Mitragotri, S. (2016). Role of nanoparticle size, shape and surface chemistry in oral drug delivery. *Journal of Controlled Release*, *238*, 176–185. Available from https://doi.org/10.1016/j.jconrel.2016.07.051.

Bao, X., Qian, K., & Yao, P. (2020). Oral delivery of exenatide-loaded hybrid zein nanoparticles for stable blood glucose control and β-cell repair of type 2 diabetes mice. *Journal of Nanobiotechnology*, *18*(1), 67. Available from https://doi.org/10.1186/s12951-020-00619-0.

Bernkop-Schnürch, A., Krauland, A. H., Leitner, V. M., & Palmberger, T. (2004). Thiomers: Potential excipients for non-invasive peptide delivery systems. *European Journal of Pharmaceutics and Biopharmaceutics*, *58*(2), 253–263. Available from https://doi.org/10.1016/j.ejpb.2004.03.032.

Bernkop-Schnürch, Andreas, Pinter, Y., Guggi, D., Kahlbacher, H., Schöffmann, G., Schuh, M., Schmerold, I., Del Curto, M. D., D'Antonio, M., Esposito, P., & Huck, C. (2005). The use of thiolated polymers as carrier matrix in oral peptide delivery—Proof of concept. *Journal of Controlled Release*, *106*(1), 26–33. Available from https://doi.org/10.1016/j.jconrel.2005.04.004.

Bernkop-Schnürch, Andreas (2018). Strategies to overcome the polycation dilemma in drug delivery. *Advanced Drug Delivery Reviews*, *136–137*, 62–72. Available from https://doi.org/10.1016/j.addr.2018.07.017.

Bhatnagar, S., Dave, K., & Venuganti, V. V. K. (2017). Microneedles in the clinic. *Journal of Controlled Release*, *260*, 164–182. Available from https://doi.org/10.1016/j.jconrel.2017.05.029.

Biograil Website (vol. 2020). (n.d.). Biograil Website. (accessed 2020); Available from: https://biograil.com/.

Bonengel, S., et al. (2018). Impact of different hydrophobic ion pairs of octreotide on its oral bioavailability in pigs. *Journal of Controlled Release*, *273*, 21–29. Available from https://doi.org/10.1016/j.jconrel.2018.01.012.

Brayden, D. J., & Alonso, M. J. (2016). Oral delivery of peptides: Opportunities and issues for translation. *Advanced Drug Delivery Reviews*, *106*, 193–195. Available from https://doi.org/10.1016/j.addr.2016.10.005.

Brayden, D. J., Hill, T. A., Fairlie, D. P., Maher, S., & Mrsny, R. J. (2020). Systemic delivery of peptides by the oral route: Formulation and medicinal chemistry approaches. *Advanced Drug Delivery Reviews*, *157*, 2–36. Available from https://doi.org/10.1016/j.addr.2020.05.007.

Brown, T. D., & Whitehead, K. A. (2020). Materials for oral delivery of proteins and peptides. *Nature Reviews Materials*, *5*, 127–148. Available from https://doi.org/10.1038/s41578-019-0156-6.

Buclin, T., et al. (2002). Bioavailability and biological efficacy of a new oral formulation of salmon calcitonin in healthy volunteers. *Journal of Bone Mineral Research*, *17*(8), 1478–1485.

Chalasani, K. B., et al. (2007). A novel vitamin B12-nanosphere conjugate carrier system for peroral delivery of insulin. *Journal of Controlled Release*, *117*(3), 421–429. Available from https://doi.org/10.1016/j.jconrel.2006.12.003.

Chen, M.-C., et al. (2011). A review of the prospects for polymeric nanoparticle platforms in oral insulin delivery. *Biomaterials*, *32*(36), 9826–9838. Available from https://doi.org/10.1016/j.biomaterials.2011.08.087.

Chiang, C.-M., & Weiner, N. (1987). Gastrointestinal uptake of liposomes. II. *In vivo studies*. *International Journal of Pharmaceutics*, *40*(1), 143–150. Available from https://doi.org/10.1016/0378-5173(87)90059-7.

Choudhari, K. B., Labhasetwar, V., & Dorle, A. K. (1994). Liposomes as a carrier for oral administration of insulin: Effect of formulation factors. *Journal of Microencapsulation*, *11* (3), 319–325. Available from https://doi.org/10.3109/02652049409040461.

Chuang, E.-Y., et al. (2013). Combination therapy via oral co-administration of insulin- and exendin-4-loaded nanoparticles to treat type 2 diabetic rats undergoing OGTT. *Biomaterials*, *34*(32), 7994–8001. Available from https://doi.org/10.1016/j.biomaterials.2013.07.021.

Clardy, S. M., Allis, D. G., Fairchild, T. J., & Doyle, R. P. (2011). Vitamin B12 in drug delivery: Breaking through the barriers to a B12 bioconjugate pharmaceutical. *Expert Opinion on Drug Delivery*, *8*(1), 127–140. Available from https://doi.org/10.1517/17425247.2011.539200.

Cox, A. B., et al. (2008). In vitro interactions between the oral absorption promoter, sodium caprate (C(10)) and S. typhimurium in rat intestinal ileal mucosae. *Pharmaceutics Research*, *25*(1), 114–122. Available from https://doi.org/10.1007/s11095-007-9354-9.

Craik, D. J., Fairlie, D. P., Liras, S., & Price, D. (2013). The future of peptide-based drugs. *Chemical Biology Drug Design*, *81*(1), 136–147. Available from https://doi.org/10.1111/cbdd.12055.

Cui, F.-d, et al. (2007). Preparation of insulin loaded PLGA-Hp55 nanoparticles for oral delivery. *Journal of Pharmaceutical Sciences*, *96*(2), 421–427. Available from https://doi.org/10.1002/jps.20750.

Custodio, J. M., Wu, C.-Y., & Benet, L. Z. (2008). Predicting drug disposition, absorption/elimination/transporter interplay and the role of food on drug absorption. *Advanced Drug Delivery Reviews*, *60*(6), 717–733. Available from https://doi.org/10.1016/j.addr.2007.08.043.

Damgé, C., et al. (1988). *New approach for oral administration of insulin with polyalkylcyanoacrylate nanocapsules as drug carrier*, *37*(2), 246–251, https://doi.org/10.2337/diab.37.2.246%JDiabetes.

Davis, S. S., Hardy, J. G., & Fara, J. W. (1986). Transit of pharmaceutical dosage forms through the small intestine. *Gut*, *27*(8), 886–892. Available from https://doi.org/10.1136/gut.27.8.886.

Desmopressin patient Leaflet 0.2 mg (2012). https://www.medicines.org.uk/emc/product/3795/smpc#gref.

Dumont, C., Bourgeois, S., Fessi, H., & Jannin, V. (2018). Lipid-based nanosuspensions for oral delivery of peptides, a critical review. *International Journal of Pharmaceutics*, *541* (1), 117–135. Available from https://doi.org/10.1016/j.ijpharm.2018.02.038.

Durán-Lobato, M., Niu, Z., & Alonso, M. J. (2019). Oral delivery of biologics for precision medicine. *Advanced Materials*, *1901935*, 1–27. Available from https://doi.org/10.1002/adma.201901935.

EMA guideline (2018). Guideline on the requirements for quality documentation concerning biological investigational medicinal products in clinical trials [cited 2020]. Available from: https://www.ema.europa.eu/en/documents/scientific-guideline/guideline-requirements-quality-documentation-concerning-biological-investigational-medicinal_en-0.pdf.

EMEA. (2007). Guideline on excipients in the dossier for application for marketing authorisation of a medicinial product. https://www.ema.europa.eu/en/documents/scientific-guideline/guideline-excipients-dossier-application-marketing-authorisation-medicinal-product-revision-2_en.pdf.

Estudante, M., et al. (2013). Intestinal drug transporters: An overview. *Advanced Drug Delivery Reviews*, *65*(10), 1340−1356.

Fan, W., Xia, D., Zhu, Q., Li, X., He, S., Zhu, C., Guo, S., Hovgaard, L., Yang, M., & Gan, Y. (2013). Functional nanoparticles exploit the bile acid pathway to overcome multiple barriers of the intestinal epithelium for oral insulin delivery. *Biomaterials*, *151*, 13−23. Available from https://doi.org/10.1016/j.biomaterials.2017.10.022.

Fernandez, S., Jannin, V., Chevrier, S., Chavant, Y., Demarne, F., & Carrière, F. (2013). In vitro digestion of the self-emulsifying lipid excipient Labrasol® by gastrointestinal lipases and influence of its colloidal structure on lipolysis rate. *Pharmaceutical Research*, *30*(12), 3077−3087. Available from https://doi.org/10.1007/s11095-013-1053-0.

Florence, A. T. (1997). The oral absorption of micro- and nanoparticulates: Neither exceptional nor unusual. *Pharmaceutical Research*, *14*(3), 259−266. Available from https://doi.org/10.1023/A:1012029517394.

Foley, D. W., Pathak, R. B., Phillips, T. R., Wilson, G. L., Bailey, P. D., Pieri, M., Senan, A., & Meredith, D. (2018). Thiodipeptides targeting the intestinal oligopeptide transporter as a general approach to improving oral drug delivery. *European Journal Medicinal Chemistry*, *156*, 180−189. Available from https://doi.org/10.1016/j.ejmech.2018.06.064.

Foley, D., Bailey, P., Pieri, M., & Meredith, D. (2009). Targeting ketone drugs towards transport by the intestinal peptide transporter, PepT1. *Organic Biomolecular Chemistry*, 7 (6), 1064−1067. Available from https://doi.org/10.1039/b818606e.

Foley, D. W., Rajamanickam, J., Bailey, P. D., & Meredith, D. (2010). Bioavailability through PepT1: the role of computer modelling in intelligent drug design. *Current Computer-Aided Drug Design*, *6*(1), 68−78.

Fonte, P., Nogueira, T., Gehm, C., Ferreira, D., & Sarmento, B. (2011). Chitosan-coated solid lipid nanoparticles enhance the oral absorption of insulin. *Drug Delivery and Translational Research*, *1*(4), 299−308. Available from https://doi.org/10.1007/s13346-011-0023-5.

Geho, W. B., & Lau, G. H., JR (2009). Hepatic-directed vesicle insulin: A review of formulation development and preclinical evaluation. *Journal of Diabetes Science Technology*, *3*(6), 1451−1459.

Gopalakrishnan, S., Pandey, N., Tamiz, A. P., Vere, J., Carrasco, R., Somerville, R., Tripathi, A., Ginski, M., Paterson, B. M., & Alkan, S. S. (2009). Mechanism of action of ZOT-derived peptide AT-1002, a tight junction regulator and absorption enhancer. *International Journal of Pharmaceutics*, *365*(1), 121−130. Available from https://doi.org/10.1016/j.ijpharm.2008.08.047.

Gourdon, B., Chemin, C., Moreau, A., Arnauld, T., Delbos, J. M., Péan, J. M., & Declèves, X. (2015). Influence of PLA-PEG nanoparticles manufacturing process on intestinal transporter PepT1 targeting and oxytocin transport. *Euroupean Journal of Pharmaceutics Biopharmaceutics*, *129*, 122−133. Available from https://doi.org/10.1016/j.ejpb.2018.05.022.

Gradauer, K., Nishiumi, A., Unrinin, K., Higashino, H., Kataoka, M., Pedersen, B. L., Buckley, S. T., & Yamashita, S. (2019). Interaction with mixed micelles in the intestine attenuates the permeation enhancing potential of Alkyl-Maltosides. *Molecular Pharmaceutics*, *12*(7), 2245–2253. Available from https://doi.org/10.1021/mp500776a.

Griesser, J., Hetényi, G., Moser, M., Demarne, F., Jannin, V., & Bernkop-Schnürch, A. (2017). Hydrophobic ion pairing: Key to highly payloaded self-emulsifying peptide drug delivery systems. *International Journal of Pharmacy*, *520*(1–2), 267–274. Available from https://doi.org/10.1016/j.ijpharm.2017.02.019.

Granhall, C., Donsmark, M., Blicher, T. M., Golor, G., Søndergaard, F. L., Thomsen, M., & Bækdal, T. A. (2019). Safety and Pharmacokinetics of Single and Multiple Ascending Doses of the Novel Oral Human GLP-1 Analogue, Oral Semaglutide, in Healthy Subjects and Subjects with Type 2 Diabetes. *Clinical Pharmacokinetics*, *58*(6), 781–791.

Griesser, Janine, Hetényi, G., Kadas, H., Demarne, F., Jannin, V., & Bernkop-Schnürch, A. (2018). Self-emulsifying peptide drug delivery systems: How to make them highly mucus permeating. *International Journal of Pharmaceutics*, *538*(1), 159–166. Available from https://doi.org/10.1016/j.ijpharm.2018.01.018.

Griffin, B. T., Guo, J., Presas, E., Donovan, M. D., Alonso, M. J., & O'Driscoll, C. M. (2016). Pharmacokinetic, pharmacodynamic and biodistribution following oral administration of nanocarriers containing peptide and protein drugs. *Advanced Drug Delivery Reviews*, *106*, 367–380. Available from https://doi.org/10.1016/j.addr.2016.06.006.

Halberg, I. B., Lyby, K., Wassermann, K., Heise, T., Plum-Mörschel, L., & Zijlstra, E. (2019). The effect of food intake on the pharmacokinetics of oral basal insulin: A randomised crossover trial in healthy male subjects. *Clinical Pharmacokinetics*, *58*(11), 1497–1504. Available from https://doi.org/10.1007/s40262-019-00772-2.

He, H., Lu, Y., Qi, J., Zhu, Q., Chen, Z., & Wu, W. (2019). Adapting liposomes for oral drug delivery. *Acta Pharmaceutica Sinica B*, *9*(1), 36–48. Available from https://doi.org/10.1016/j.apsb.2018.06.005.

Henry, S., McAllister, D. V., Allen, M. G., & Prausnitz, M. R. (1998). Microfabricated microneedles: A novel approach to transdermal drug delivery. *Journal of Pharmaceutical Sciences*, *87*(8), 922–925. Available from https://doi.org/10.1021/js980042+.

Hetényi, G., Griesser, J., Moser, M., Demarne, F., Jannin, V., & Bernkop-Schnürch, A. (2017). Comparison of the protective effect of self-emulsifying peptide drug delivery systems towards intestinal proteases and glutathione. *International Journal of Pharmaceutics*, *523*(1), 357–365. Available from https://doi.org/10.1016/j.ijpharm.2017.03.027.

Hristov, D., McCartney, F., Beirne, J., Mahon, E., Reid, S., Bhattacharjee, S., Penarier, G., Werner, U., Bazile, D., & Brayden, D. J. (2020). Silica-coated nanoparticles with a core of zinc, l-Arginine, and a peptide designed for oral delivery. *ACS Applied Materials & Interfaces*, *12*(1), 1257–1269. Available from https://doi.org/10.1021/acsami.9b16104.

Jain, A., & Jain, S. K. (2015). L-Valine appended PLGA nanoparticles for oral insulin delivery. *Acta Diabetology*, *52*(4), 663–676. Available from https://doi.org/10.1007/s00592-015-0714-3.

Jeong, H.-R., Lee, H.-S., Choi, I.-J., & Park, J.-H. (2017). Considerations in the use of microneedles: Pain, convenience, anxiety and safety. *Journal of Drug Targeting*, *25*(1), 29–40. Available from https://doi.org/10.1080/1061186X.2016.1200589.

Karsdal, M. A., Byrjalsen, I., Azria, M., Arnold, M., Choi, L., Riis, B. J., & Christiansen, C. (2009). Influence of food intake on the bioavailability and efficacy of oral calcitonin. *British Journal of Clinical Pharmacology*, *67*(4), 413–420. Available from https://doi.org/10.1111/j.1365-2125.2009.03371.x.

Kawashima, Y., Yamamoto, H., Takeuchi, H., & Kuno, Y. (2000). Mucoadhesive DL-lactide/glycolide copolymer nanospheres coated with chitosan to improve oral delivery of elcatonin. *Pharmaceutical Development and Technology*, *5*(1), 77–85. Available from https://doi.org/10.1081/PDT-100100522.

Knütter, I., Wollesky, C., Kottra, G., Hahn, M. G., Fischer, W., Zebisch, K., Neubert, R. H. H., Daniel, H., & Brandsch, M. (2008). Transport of angiotensin-converting enzyme inhibitors by H⁺/peptide transporters revisited. *Journal of Pharmacology and Experimental Therapeutics*, *327*(2), 432. Available from https://doi.org/10.1124/jpet.108.143339.

Lai, S. K., Wang, Y.-Y., & Hanes, J. (2009). Mucus-penetrating nanoparticles for drug and gene delivery to mucosal tissues. *Advanced Drug Delivery Reviews*, *61*(2), 158–171. Available from https://doi.org/10.1016/j.addr.2008.11.002.

Lambert, W. J. (2010). Considerations in developing a target product profile for parenteral pharmaceutical products. *AAPS PharmSciTech*, *11*(3), 1476–1481. Available from https://doi.org/10.1208/s12249-010-9521-x.

Lamson, N. G., Berger, A., Fein, K. C., & Whitehead, K. A. (2020a). Anionic nanoparticles enable the oral delivery of proteins by enhancing intestinal permeability. *Nature Biomedical Engineering*, *4*(1), 84–96. Available from https://doi.org/10.1038/s41551-019-0465-5.

Lamson, N. G., Fein, K. C., Gleeson, J. P., Xian, S., Newby, A., Chaudhary, N., Melamed, J. R., Cochran, K., Ball, R. L., Suri, K., Ahuja, V., Zhang, A., Berger, A., Kolodieznyi, D., Schmidt, B. F., Silva, G. L., & Whitehead, K. A. (2020b). From farm to pharmacy: Strawberry-enabled oral delivery of protein drugs. *bioRxiv*. Available from https://doi.org/10.1101/2020.03.11.987461, 2020.03.11.987461.

Lehr, C.-M., Poelma, F. G. J., Junginger, H. E., & Tukker, J. J. (1991). An estimate of turnover time of intestinal mucus gel layer in the rat in situ loop. *International Journal of Pharmaceutics*, *70*(3), 235–240. Available from https://doi.org/10.1016/0378-5173(91)90287-X.

Leonard, T. W., et al. (2006). Promoting absorption of drugs in humans using medium-chain fatty acid-based solid dosage forms: GIPET. *Expert Opinion Drug Delivery*, *3*(5), 685–692.

Li, J., Wang, Y., Han, L., Sun, X., Yu, H., & Yu, Y. (2012). Time-action profile of an oral enteric insulin formulation in healthy Chinese volunteers. *Clinical Therapeutics*, *34*(12), 2333–2338. Available from https://doi.org/10.1016/j.clinthera.2012.11.004.

Lipinski, C. A., Lombardo, F., Dominy, B. W., & Feeney, P. J. (1997). Experimental and computational approaches to estimate solubility and permeability in drug discovery and development settings. *Advanced Drug Delivery Reviews*, *23*(1), 3–25. Available from https://doi.org/10.1016/S0169-409X(96)00423-1.

Lundquist, P., & Artursson, P. (2016). Oral absorption of peptides and nanoparticles across the human intestine: Opportunities, limitations and studies in human tissues. *Advanced Drug Delivery Reviews*, *106*, 256–276. Available from https://doi.org/10.1016/j.addr.2016.07.007.

Mahato, R. I., Narang, A. S., Thoma, L., & Miller, D. D. (2003). *Emerging Trends in Oral Delivery of Peptide and Protein Drugs*, *20*(2 & 3), 62. Available from https://doi.org/10.1615/CritRevTherDrugCarrierSyst.v20.i23.30.

Maher, S., Mrsny, R. J., & Brayden, D. J. (2016). Intestinal permeation enhancers for oral peptide delivery. *Advanced Drug Delivery Reviews*, *106*, 277–319. Available from https://doi.org/10.1016/j.addr.2016.06.005.

Matsson, P., Doak, B. C., Over, B., & Kihlberg, J. (2016). Cell permeability beyond the rule of 5. *Advanced Drug Delivery Reviews*, *101*, 42–61. Available from https://doi.org/10.1016/j.addr.2016.03.013.

Medicine, U. S. N. L.of. (2020). ClinicalTrials.gov. https://www.clinicaltrials.gov/.

Menzel, C., Holzeisen, T., Laffleur, F., Zaichik, S., Abdulkarim, M., Gumbleton, M., & Bernkop-Schnürch, A. (2018). In vivo evaluation of an oral self-emulsifying drug delivery system (SEDDS) for exenatide. *Journal of Control Release, 277*, 165–172. Available from https://doi.org/10.1016/j.jconrel.2018.03.018.

Michel, C., Aprahamian, M., Defontaine, L., Couvreur, P., & Damage, C. (1991). The effect of site of administration in the gastrointestinal tract on the absorption of insulin from nanocapsules in diabetic rats. *Journal of Pharmacy and Pharmacology, 43*(1), 1–5. Available from https://doi.org/10.1111/j.2042-7158.1991.tb05437.x.

Mohammadi-Samani, S., & Taghipour, B. (2015). PLGA micro and nanoparticles in delivery of peptides and proteins; problems and approaches. *Pharmaceutical Development Technology, 20*(4), 385–393. Available from https://doi.org/10.3109/10837450.2014.882940.

Mohammed, M. A. S., Syeda, J. T. M., Wasan, K. M., & Wasan, E. K. (2017). An overview of chitosan nanoparticles and its application in non-parenteral drug delivery. *Pharmaceutics, 9*(4), 53.

Morales, J. O., Fathe, K. R., Brunaugh, A., Ferrati, S., Li, S., Montenegro-Nicolini, M., Mousavikhamene, Z., McConville, J. T., Prausnitz, M. R., & Smyth, H. D. C. (2017). Challenges and future prospects for the delivery of biologics: Oral mucosal, pulmonary, and transdermal routes. *The AAPS Journal, 19*(3), 652–668. Available from https://doi.org/10.1208/s12248-017-0054-z.

Müller, J., Keiser, M., Drozdzik, M., & Oswald, S. (2017). Expression, regulation and function of intestinal drug transporters: An update. *Biological Chemistry, 398*(2), 175–192. Available from https://doi.org/10.1515/hsz-2016-0259.

Murakami, T., & Takano, M. (2008). Intestinal efflux transporters and drug absorption. *Expert Opinion Drug Metabolism Toxicology, 4*(7), 923–939.

Mycapssa Prescribing Information (2020). https://www.accessdata.fda.gov/drugsatfda_-docs/label/2020/208232s000lbl.pdf.

Nava, P., López, S., Arias, C. F., Islas, S., & González-Mariscal, L. (2004). The rotavirus surface protein VP8 modulates the gate and fence function of tight junctions in epithelial cells. *Journal of Cell Science, 117*(23), 5509–5519. Available from https://doi.org/10.1242/jcs.01425.

Nazir, I., Shahzadi, I., Jalil, A., & Bernkop-Schnürch, A. (2020). Hydrophobic H-bond pairing: A novel approach to improve membrane permeability. *International Journal of Pharmaceutics, 573*, 118863. Available from https://doi.org/10.1016/j.ijpharm.2019.118863.

Neoral Soft Gelatin Capsules SmPC (2020). https://www.medicines.org.uk/emc/medicine/1307.

Newstead, S. (2011). Towards a structural understanding of drug and peptide transport within the proton-dependent oligopeptide transporter (POT) family. *Biochemical Society Transactions, 39*(5), 1353–1358. Available from https://doi.org/10.1042/bst0391353.

Newstead, S. (2017). Recent advances in understanding proton coupled peptide transport via the POT family. *Current Opinion Structural Biology, 45*, 17–24. Available from https://doi.org/10.1016/j.sbi.2016.10.018.

Newswire, P. (2020). Rani Therapeutics announces positive phase I study results of oral cctreotide using RaniPillTM (vol. 2020). https://www.prnewswire.com/news-releases/rani-therapeutics-announces-positive-phase-i-study-results-of-oral-octreotide-using-ranipill-300992818.html.

Niu, Z. (2016). Lipid-based nanocarriers for oral peptide delivery. *Advanced Drug Delivery Reviews, 106*, 337–354.

Niu, M., Tan, Y., Guan, P., Hovgaard, L., Lu, Y., Qi, J., Lian, R., Li, X., & Wu, W. (2016). Enhanced oral absorption of insulin-loaded liposomes containing bile salts: A mechanistic study. *International Journal of Pharmacy, 460*(1–2), 119–130. Available from https://doi.org/10.1016/j.ijpharm.2013.11.028.

Okada, J., Cohen, S., & Langer, R. S. (1998). Oral delivery of vaccines using polymerized liposomes. In https://www.freepatentsonline.com/5762904.html (vol. 5762904). Massachusetts Institute of Technology (Cambridge, MA). https://www.freepatentsonline.com/5762904.html.

Parmentier, J., Thewes, B., Gropp, F., & Fricker, G. (2011). Oral peptide delivery by tetraether lipid liposomes. *International Journal of Pharmaceutics*, *415*(1), 150–157. Available from https://doi.org/10.1016/j.ijpharm.2011.05.066.

Petrus, A. K., Vortherms, A. R., Fairchild, T. J., & Doyle, R. P. (2007). Vitamin B12 as a carrier for the oral delivery of insulin. *ChemMedChem*, *2*(12), 1717–1721. Available from https://doi.org/10.1002/cmdc.200700239.

Plapied, L., Duhem, N., des Rieux, A., & Préat, V. (2011). Fate of polymeric nanocarriers for oral drug delivery. *Current Opinion in Colloid & Interface Science*, *16*(3), 228–237. Available from https://doi.org/10.1016/j.cocis.2010.12.005.

Prego, C., Torres, D., & Alonso, M. J. (2005). The potential of chitosan for the oral administration of peptides. *Expert Opinion on Drug Delivery*, *2*(5), 843–854. Available from https://doi.org/10.1517/17425247.2.5.843.

Prego, C., García, M., Torres, D., & Alonso, M. J. (2005). Transmucosal macromolecular drug delivery. *Journal of Controlled Release*, *101*(1), 151–162. Available from https://doi.org/10.1016/j.jconrel.2004.07.030.

Pridgen, E. M., Alexis, F., Kuo, T. T., Levy-Nissenbaum, E., Karnik, R., Blumberg, R. S., Langer, R., & Farokhzad, O. C. (2013). Transepithelial transport of Fc-targeted aanoparticles by the neonatal Fc receptor for oral delivery. *Science Translational Medicine*, *5*(213), 213ra167. Available from https://doi.org/10.1126/scitranslmed.3007049.

Rachmiel, M., Barash, G., Leshem, A., Sagi, R., doenyas-barak, K., & Koren, S. (2019). OR14-1 pharmacodynamics, safety, tolerability, and efficacy of oral insulin formulation (*Oshadi Icp*) among young adults with type 1 diabetes: A summary of clinical studies phases I, Ib, and Ii. *Journal of the Endocrine Society*, *3*(Supplement_1). Available from https://doi.org/10.1210/js.2019-OR14-1.

Raoof, A. A., Ramtoola, Z., McKenna, B., Yu, R. Z., Hardee, G., & Geary, R. S. (2008). Effect of sodium caprate on the intestinal absorption of two modified antisense oligonucleotides in pigs. *European Journal of Pharmaceutical Science*, *17*(3), 131–138. Available from https://doi.org/10.1016/s0928-0987(02)00162-8.

Rawat, M., Singh, D., Saraf, S., & Saraf, S. (2006). Nanocarriers: Promising vehicle for bioactive drugs. *Biological and Pharmaceutical Bulletin*, *29*(9), 1790–1798. Available from https://doi.org/10.1248/bpb.29.1790.

Rossi Sebastiano, M., Doak, B. C., Backlund, M., Poongavanam, V., Over, B., Ermondi, G., Caron, G., Matsson, P., & Kihlberg, J. (2018). Impact of dynamically exposed polarity on permeability and solubility of chameleonic drugs beyond the rule of 5. *Journal of Medicinal Chemistry*, *61*(9), 4189–4202. Available from https://doi.org/10.1021/acs.jmedchem.8b00347.

Rybelsus Prescribing Information (2021). https://www.novo-pi.com/rybelsus.pdf.

Sarmento, B., Martins, S., Ferreira, D., & Souto, E. B. (2007). Oral insulin delivery by means of solid lipid nanoparticles. *International Journal of Nanomedicine*, *2*(4), 743–749. Available from https://pubmed.ncbi.nlm.nih.gov/18203440.

Shan, W., Zhu, X., Liu, M., Li, L., Zhong, J., Sun, W., Zhang, Z., & Huang, Y. (2015). Overcoming the diffusion barrier of mucus and absorption barrier of epithelium by self-assembled nanoparticles for oral delivery of insulin. *ACS Nano*, *9*(3), 2345–2356. Available from https://doi.org/10.1021/acsnano.5b00028.

Spanier, B. (2014). Transcriptional and functional regulation of the intestinal peptide transporter PEPT1. *Journal of Physiology*, *592*(5), 871–879. Available from https://doi.org/10.1113/jphysiol.2013.258889.

Sturmer, A., et al. (2013). Pharmacokinetics of oral recombinant human parathyroid hormone [rhPTH(1–31)NH2] in postmenopausal women with osteoporosis. *Clinical Pharmacokinetics*, *52*(11), 995–1004.

Su, H., Wang, Y., Liu, S., Wang, Y., Liu, Q., Liu, G., & Chen, Q. (2019). Emerging transporter-targeted nanoparticulate drug delivery systems. *Acta Pharmaceutica Sinica B*, *9*(1), 49–58. Available from https://doi.org/10.1016/j.apsb.2018.10.005.

Sun, Y., Gan, W., Lei, M., Jiang, W., Cheng, M., He, J., Sun, Q., Liu, W., Hu, L., & Jin, Y. (2018). PEPT1-mediated prodrug strategy for oral delivery of peramivir. *Asian Journal Pharmaceutical Sciences*, *13*(6), 555–565. Available from https://doi.org/10.1016/j.ajps.2018.05.008.

Tamai, I. (2012). Oral drug delivery utilizing intestinal OATP transporters. *Advanced Drug Delivery Review*, *64*(6), 508–514. Available from https://doi.org/10.1016/j.addr.2011.07.007.

Terada, T., & Inui, K. (2004). Peptide transporters: Structure, function, regulation and application for drug delivery. *Current Drug Metabolism*, *5*(1), 85–94.

Therapeutics, R. (n.d.). Rani Therapeutics Website (accessed 2020); Available from https://www.ranitherapeutics.com/applications/.

Thorn, K., Andersen, R., Christensen, J., Jakobsen, P., Nielsen, C. U., Steffansen, B., & Begtrup, M. (2007). Design, synthesis, and evaluation of tripeptidic promoieties targeting the intestinal peptide transporter hPEPT1. *ChemMedChem*, *2*(4), 479–487. Available from https://doi.org/10.1002/cmdc.200600200.

Tillman, L. G., Geary, R. S., & Hardee, G. E. (2008). Oral delivery of antisense oligonucleotides in man. *Journal of Pharmaceutical Sciences*, *97*(1), 225–236. Available from https://doi.org/10.1002/jps.21084.

Tjia, J. F., Webber, I. R., & Back, D. J. (1991). Cyclosporin metabolism by the gastrointestinal mucosa. *British Journal of Clinical Pharmacology*, *31*(3), 344–346. Available from https://doi.org/10.1111/j.1365-2125.1991.tb05540.x.

Traverso, G., Schoellhammer, C. M., Schroeder, A., Maa, R., Lauwers, G. Y., Polat, B. E., Anderson, D. G., Blankschtein, D., & Langer, R. (2015). Microneedles for drug delivery via the gastrointestinal tract. *Journal of Pharmaceutical Sciences*, *104*(2), 362–367. Available from https://doi.org/10.1002/JPS.24182.

Tuvia, S., Atsmon, J., Teichman, S. L., Katz, S., Salama, P., Pelled, D., Landau, I., Karmeli, I., Bidlingmaier, M., Strasburger, C. J., Kleinberg, D. L., Melmed, S., & Mamluk, R. (2012). Oral octreotide absorption in human subjects: Comparable pharmacokinetics to parenteral octreotide and effective growth hormone suppression. *The Journal of Clinical Endocrinology & Metabolism*, *97*(7), 2362–2369. Available from https://doi.org/10.1210/jc.2012-1179.

Tuvia, S., Pelled, D., Marom, K., Salama, P., Levin-Arama, M., Karmeli, I., Idelson, G. H., Landau, I., & Mamluk, R. (2014). A novel suspension formulation enhances intestinal absorption of macromolecules via transient and reversible transport mechanisms. *Pharmaceutical Research*, *31*(8), 2010–2021. Available from https://doi.org/10.1007/s11095-014-1303-9.

Wang, C. Y., Liu, S., Xie, X. N., & Tan, Z. R. (2017). Regulation profile of the intestinal peptide transporter 1 (PepT1). *Drug Design Development and Therapy*, *11*, 3511–3517. Available from https://doi.org/10.2147/dddt.S151725.

Wang, G., Zhao, L., Jiang, Q., Sun, Y., Zhao, D., Sun, M., He, Z., Sun, J., & Wang, Y. (2020). Intestinal OCTN2- and MCT1-targeted drug delivery to improve oral bioavailability. *Asian Journal of Pharmaceutical Science*, *15*(2), 158–173. Available from https://doi.org/10.1016/j.ajps.2020.02.002.

Wang, Y., Cui, Y., Zhao, Y., Zhao, Q., He, B., Zhang, Q., & Wang, S. (2018). Effects of surface modification and size on oral drug delivery of mesoporous silica formulation. *Journal of Colloid and Interface Science*, *513*, 736–747. Available from https://doi.org/10.1016/j.jcis.2017.11.065.

Wong, C. Y., Al-Salami, H., & Dass, C. R. (2017). Potential of insulin nanoparticle formulations for oral delivery and diabetes treatment. *Journal of Control Release, 264*, 247–275. Available from https://doi.org/10.1016/j.jconrel.2017.09.003.

Wright, S., Yelland, M., Heathcote, K., Ng, S. K., & Wright, G. (2009). Fear of needles—nature and prevalence in general practice. *Australian Family Physician, 38*(3), 172–176.

Wu, Y., Sun, M., Wang, D., Li, G., Huang, J., Tan, S., Bao, L., Li, Q., Li, G., & Si, L. (2019). A PepT1 mediated medicinal nano-system for targeted delivery of cyclosporine A to alleviate acute severe ulcerative colitis. *Biomaterials Science*, 7(10), 4299–4309. Available from https://doi.org/10.1039/c9bm00925f.

Xiong, X. Y., Li, Y. P., Li, Z. L., Zhou, C. L., Tam, K. C., Liu, Z. Y., & Xie, G. X. (2007). Vesicles from Pluronic/poly(lactic acid) block copolymers as new carriers for oral insulin delivery. *Journal of Controlled Release, 120*(1–2), 11–17. Available from https://doi.org/10.1016/j.jconrel.2007.04.004.

Xu, Y., Shrestha, N., Préat, V., & Beloqui, A. (2020). Overcoming the intestinal barrier: A look into targeting approaches for improved oral drug delivery systems. *Journal of Control Release, 322*, 486–508. Available from https://doi.org/10.1016/j.jconrel.2020.04.006.

Xu, Yining, Keersmaecker, H., Braeckmans, K., De Smedt, S., Cani, P., Preat, V., & Beloqui, A. (2020). Targeted nanoparticles towards increased L cell stimulation as a strategy to improve oral peptide delivery in incretin-based diabetes treatment. *Biomaterials, 255*, 120209. Available from https://doi.org/10.1016/j.biomaterials.2020.120209.

Yang, C. Y., Dantzig, A. H., & Pidgeon, C. (1999). Intestinal peptide transport systems and oral drug availability. *Pharmaceutical Research, 16*(9), 1331–1343.

Zhang, E. Y., Phelps, M. A., Cheng, C., Ekins, S., & Swaan, P. W. (2006). Modeling of active transport systems. *Advanced Drug Delivery Reviews, 54*(3), 329–354. Available from https://doi.org/10.1016/S0169-409X(02)00007-8.

Zhang, N., Ping, Q. N., Huang, G. H., & Xu, W. F. (2005). Investigation of lectin-modified insulin liposomes as carriers for oral administration. *International Journal of Pharmaceutics, 294*(1), 247–259. Available from https://doi.org/10.1016/j.ijpharm.2005.01.018.

Zhang, N., Ping, Q., Huang, G., Xu, W., Cheng, Y., & Han, X. (2006). Lectin-modified solid lipid nanoparticles as carriers for oral administration of insulin. *International Journal of Pharmaceutics, 327*(1), 153–159. Available from https://doi.org/10.1016/j.ijpharm.2006.07.026.

Zhu, T., Chen, X.-Z., Steel, A., Hediger, M. A., & Smith, D. E. (2000). Differential recognition of ACE inhibitors in *Xenopus Laevis* oocytes expressing rat PEPT1 and PEPT2. *Pharmaceutical Research, 17*(5), 526–532. Available from https://doi.org/10.1023/A:1007556630189.

Zijlstra, E., Heinemann, L., & Plum-Mörschel, L. (2014). Oral insulin reloaded: A structured approach. *Journal of Diabetes Science and Technology, 8*(3), 458–465. Available from https://doi.org/10.1177/1932296814529988.

CHAPTER TWO

Design and synthesis of chemically modified peptides and proteins —critical considerations for oral delivery

Jefferson D. Revell
AstraZeneca R&D, Discovery Sciences, Cambridge, United Kingdom

2.1 Introduction

The peptide and protein therapeutics sector is a rapidly growing billion-dollar market in the pharmaceutical industry. Although there are hundreds of peptide-based therapeutics in clinical trials worldwide, only around 60 peptides are currently approved for use as drugs (Kong et al., 2020; Moroz et al., 2016); still fewer are approved for oral administration, highlighting the difficulty in achieving this desirable route of administration (Brayden et al., 2020). Several notable examples of peptide therapeutics which are approved as formulations for oral delivery include Romidepsin (Istodax), octreotide (Mycapssa), taltirelin (Ceredist), salmon calcitonin (Miacalcin), desmopressin (Minrin), plecanatide (Trulance), linaclotide (Linzess) and cyclosporin A (Neoral, Sandimmune), with oral semaglutide (Rybelsus) being the most recently approved by the FDA in 2019 (Fig. 2.1).

As a natural class peptides offer a set of well-defined biophysical properties, with both advantages and disadvantages unique to their modality in comparison to other chemical classes such a small molecule. Naturally occurring peptides tend to be highly potent, selective and target specific, with low associated toxicity; making them an appealing class for therapeutic development. Conversely, natural peptides are liable to proteolytic degradation, elimination by renal clearance, and show poor membrane permeability. Whilst the vast majority of small molecule drugs can readily be formulated for oral administration using well recognized and validated approaches, the same cannot be said for the majority of naturally occurring peptides, which

Oral Delivery of Therapeutic Peptides and Proteins.
DOI: https://doi.org/10.1016/B978-0-12-821061-1.00003-4

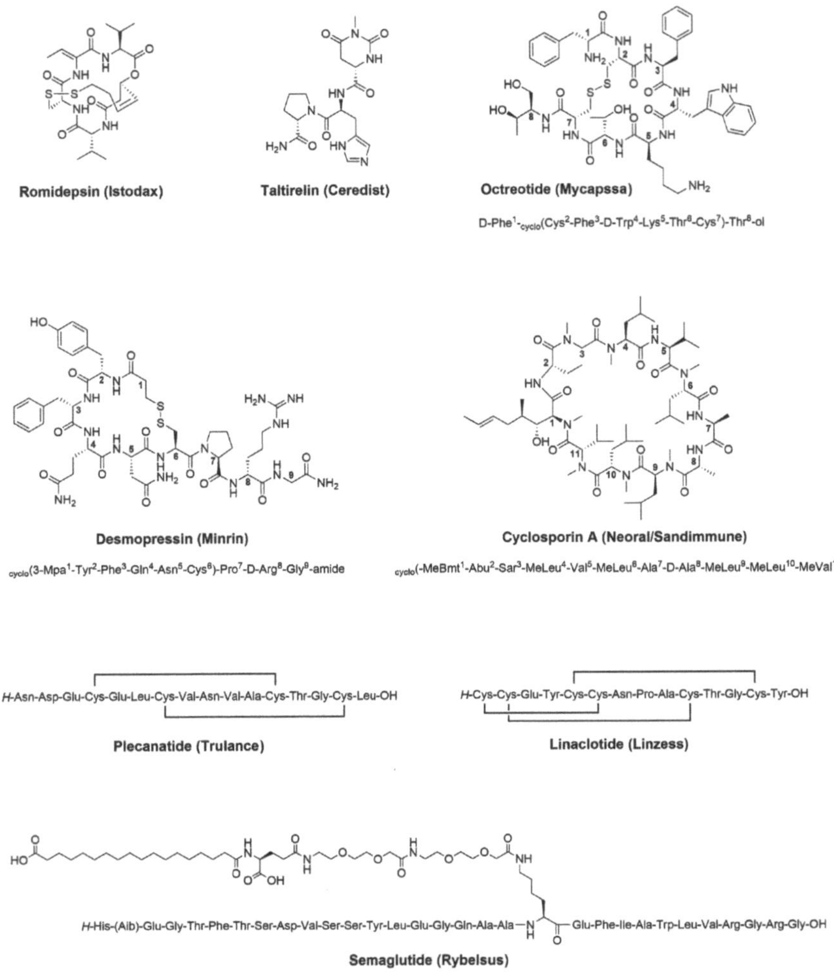

Figure 2.1 Approved peptide therapeutics for oral delivery.

until recently could generally be given only by invasive dosing routes such as intravenous or subcutaneous injection. With a significant numbers of patients reporting needle phobia (Nir et al., 2003) and concerns over injection site scarring (Fu et al., 2009), there has been an increased drive to move from parenteral to enteral dosing technologies. Recognizing the multiple advantages of oral delivery including generally higher patient compliance, dosing simplicity and lower production cost of goods, researchers have sought for decades to overcome the manifold issues facing safe and effective oral delivery of peptide drugs. The development of long-acting peptide therapeutics specifically targeting the oral route is however still generally hampered by factors such as short plasma half-life and poor oral bioavailability, largely a result of the natural susceptibility of native peptides to enzymatic degradation and their inherently polar character. The majority of proteolytic functions are crucial in regulating essential biomolecular processes such as turning off specific peptide signaling events at cell surfaces, or the gastric breakdown of proteins and peptides during digestion.

Thus the activity of the responsible proteases cannot simply be inhibited without, in many cases, causing other metabolic disturbances. Increasing the enzymatic resistance of a peptide of interest is therefore a more desirable solution. Generally, two methods are utilized to increase enzymatic resistance: sequence specific modifications, for example, those affecting the primary structure of the peptide itself; and global modifications, for example, those which modify the overall physicochemical characteristics of the peptide. Introduced strategically, such modifications can safely reduce the effects of natural physiological processes which would otherwise eliminate or inactivate a peptide whose action is desired, for example, enzymatic degradation and/or clearance by renal ultrafiltration.

Sequence specific modifications include incorporation of proteolysis-resistant noncanonical amino acids at particularly vulnerable site, or more involved modifications including cyclization between naturally occurring side-chain functions, for example, disulfide formation (Cys—Cys), or lactamization (Lys—Glu or Lys—Asp). Additional modifications include cyclization between noncanonical amino acid surrogates within the peptide backbone, for example, olefin metathesis stapling (Bird et al., 2010).

Global modifications include processes such as peptide lipidation, for example, palmitoylation, PEGylation or derivatization with polysarcosine. Lipidation has the effect of creating a circulating reservoir of peptide which reversibly associates with naturally abundant albumin in blood serum. Peptide associated with albumin effectively escapes renal ultrafiltration since the size of the associated complex is above the glomerular filtration cut-off. As the peptide dissociates from the surface of the albumin, it is again free to interact with endogenous receptors. Derivatization with polyethylene glycols (PEG) or polysarcosines has the effect of physically shielding the peptide from proteolysis and imparts significant hydrophilicity which upon hydration greatly increases the hydrodynamic radius of the therapeutic molecule to overcome renal clearance (D'souza & Shegokar, 2016; Hu et al., 2018). Whilst these technologies can be broadly applicable to therapeutic peptides in general and to an extent are able to extend circulatory half-life, a need still exists for methods of increasing stability of peptides and proteins to enzymatic degradation, particularly in light of the desire to produce peptides suitable for oral administration. There have been a great many approaches undertaken to address the manifold roadblocks imposed by a healthy digestive and circulatory system, often individually however. Early attempts generally consisted of coformulation of unmodified peptides with various combinations of transient permeation enhancers, enzyme inhibitors and/or pH modifiers (Zupančič & Bernkop-Schnürch, 2017). These approaches generally met with very little success if any, delivering extremely low and variable oral bioavailabilities of <1% of the dosed API (Active Pharmaceutical Ingredient). Given then that >99% of the peptide is essentially wasted through either degradation or excretion, only the safest, cheapest and most potent peptides could be considered viable candidates for this approach. As high value substances, peptide pharmaceutics are relatively costly to manufacture; hence, there are very few unmodified peptides that fulfill these criteria. This chapter discusses the major digestive and circulatory proteases most likely to destroy peptide and protein function before reaching their respective targets. We discuss strategies to circumvent degradation as well as the molecular design strategies and chemical modifications of peptide drugs which can rendering them more suitable for oral delivery (Drucker, 2020).

2.2 Enzymatic barriers to absorption of intact peptides and proteins

With very few exceptions animals have evolved highly efficient digestive mechanisms with which to extract and absorb nutrition from the food they consume. Essential life-supporting nutrients including water, carbohydrates, lipids, proteins, vitamins and minerals must all be ingested and processed from more complex forms before absorption and utilization. An ability to digest and absorb this broad range of substances which encompasses small molecules, fats and proteinaceous macromolecules from an extremely diverse 'omnivorous' diet has allowed humans and many animals to successfully populate all continents. However, the human gastrointestinal (GI) tract is not only a highly efficient and specialized machine for intaking, processing and absorbing nutrients, it is also a protective and selectively permeable physical barrier against many potentially harmful entities including bacteria, viruses and parasites which may be consumed inadvertently alongside the food we eat (Worthington et al., 2018). To accomplish its digestive, absorptive and protective functions, the GI tract has evolved specialized tissues and compartments with each section hosting a unique bacterial microflora, maintaining a well-defined pH range, and secreting specific enzymes along its length. Additionally, the epithelial lumen continuously secretes a protective mucus lining to prevent digestive damage itself. Selective permeability of the epithelial lumen is maintained by a large number of influx and efflux pumps, specialized receptors and transporters which together make the GI tract a formidable barrier through which to accomplish safe and effective delivery of intact naturally occurring macromolecules (New, 2021). Of the major digestive enzymes broadly classified as proteases, lipases or amylases, it is the proteases which present the biggest obstacle to oral absorption of intact peptides given their proteinaceous nature. The four major digestive endoproteases, those which target and hydrolyze specific motifs within the peptide backbone, are secreted in abundance early on in the digestive process at various sites in the GI tract. These include pepsin A, trypsin, chymotrypsin and elastase, each specific and complimentary in their hydrolytic selectivity, reducing the vast majority of ingested natural peptides and proteins to a soup of dipeptides, tripeptides and short oligopeptides. Following the gross division of proteins and peptides into fragments, these are either absorbed directly or processed further into still smaller single amino acids, di- and tripeptides by various carboxy- and amino exopeptidases; enzymes which hydrolyze peptides from the C- or N-termini of peptides, respectively. Following this rapid digestive assault, peptides and proteins which are now fully disassembled into their constituent amino acids are then quickly and readily absorbed through the intestinal lumen by a multitude of luminal transporters, for example, PEPT1 and PEPT2 (Matthews & Adibi, 1976; Shen et al., 2001) and organic anion transporting polypeptides (Oswald, 2019; Tamai, 2012). Pepsin A [EC 3.4.23.1] is the major active form of the aspartic endopeptidase produced by and secreted in abundance as its zymogen precursor pepsinogen by the chief cells of the gastric lumen. Pepsinogen is converted to pepsin either by activation at low pH by hydrochloric acid secreted into the gastric fluid, or autocatalytically by pepsin itself. Activation begins below pH 6 through dissociation of a small basic peptide from the precursor zymogen as the pH of the stomach drops in response to food intake. As the pH of the stomach continues to fall, this peptide, which is itself an inhibit of further activation is completely digested by the activated pepsin. For this reason, the aspartic peptidases are collectively termed acidic peptidases given their high stability and activity at low pH. However, pepsin is quickly and irreversibly denatured as the pH of the digestive tract increases past pH 7 on going from the stomach to the duodenum (Neurath et al., 1967; Tang, 2013). Pepsin recognizes primarily large aromatic and hydrophobic residues such as phenylalanine, tyrosine, leucine, and to a lesser extent tryptophan and isoleucine, hydrolyzing them on the amino side (Tang, 2013). The gastric output from the digestion of proteins and peptides is a highly acidic mixture of polypeptides and oligopeptides, suitably prepared for further disassembly by the remaining digestive proteases secreted by the exocrine pancreas into the duodenum. Furthermore, the highly acidic gastric juice of the stomach serves to rapidly denature the secondary and tertiary structures of proteins and peptides, as well as partially deamidating the amide side chains in both asparagine and glutamine residues. Trypsin (EC 3.4.21.4), chymotrypsin (EC 3.4.21.1) and elastase (EC 3.4.21.36) are all members of the chymotrypsin superfamily of serine endopeptidases produced by and stored in the acinar cells of the pancreas as their respective inactive zymogens. Active trypsin is produced upon cleavage of a short N-terminal peptide from trypsinogen, either slowly by trypsin itself or much more rapidly by enterokinase, a large protein expressed on the luminal brush border of the upper intestine. Trypsin specifically recognizes the unprotonated side-chains of the basic residues arginine and lysine, at an optimal pH between 7.5 and 8.5, cleaving them on the carboxyl side unless either residue is followed by proline (Huber & Bode, 1978). The hydrolysis of trypsinogen to trypsin triggers an activation cascade initiating hydrolysis of all other pancreatic zymogens to their respective active forms, including the precursors of two important exopeptidases, namely carboxypeptidase A and B (discussed later). Chymotrypsin recognizes and cleaves large aromatic and hydrophobic residues phenylalanine, tyrosine, tryptophan and leucine on their carboxyl side, complimenting pepsin in its action at their amino side. Elastase recognizes the small hydrophobic residues glycine, alanine and valine, hydrolyzing them on the carboxyl side. During normal digestion, elastase is required to aid the breakdown of elastin and other connective proteins such as collagen. The actions of elastase are not restricted to connective tissues however, readily hydrolyzing any soluble protein or peptide presenting the appropriate, surface-exposed amino acid motifs

(Narayanan & Anwar, 1969). The exopeptidase carboxypeptidase A (3.4.17.1) also known as CPA1 hydrolyzes residues with aromatic and aliphatic side chains from the C-terminus of polypeptides; whilst carboxypeptidase B (3.4.17.2) otherwise known as CPA2 hydrolyzes lysine and arginine residues, recognizing their basic side chain functions (McKay et al., 1979) (Table 2.1).

Once in circulation, peptides and proteins remain continually subject to potential proteolysis by an extremely large number of both membrane-bound and soluble enzymes, limiting the plasma half-life of unmodified species. Of the most studied and relevant to the development of both orally and parenterally delivered peptides are Neprilysin (NEP 24.11) (Hupe-Sodmann et al., 1997), dipeptidyl peptidase-IV (DPP-IV) (Mentlein, 1999) and angiotensin-converting enzyme-2 (ACE-2) (Vickers et al., 2002) (Table 2.2).

Table 2.1 Major digestive proteases relevant to development of peptides for oral delivery.

Protease	Distribution	Family	Specificity
Pepsin	Stomach	Aspartate endoprotease	Amino side: Y, F, W, L
Trypsin A	Duodenum	Serine endoprotease	Carboxyl side: R, K
Chymotrypsin	Duodenum	Serine endoprotease	Carboxyl side: Y, F, W, L, M
Elastase	Duodenum	Serine endoprotease	Carboxyl side: G, A, V
Carboxypeptidase A	Duodenum	Serine endoprotease	C-terminal: Y, F, W, L, I, V
Carboxypeptidase B	Duodenum	Serine endoprotease	C-terminal: K, R

Table 2.2 Major circulatory proteases relevant to development of peptides therapeutics.

Protease	Distribution	Family	Specificity
Neprilysin (NEP 24.11)	Kidney brush border and circulation	Zinc endopeptidase	Amino side: Y, F, W, L
Dipeptidyl peptidase-IV (DPP-IV)	Kidney brush border and circulation	Serine	aminodipeptidase
N-terminal dipeptides			
Angiotensin-converting enzyme-2 (ACE-2)	Lung, intestinal lumen and circulation	Zinc carboxypeptidase	C-terminal amino acids, particularly F, Y, L

2.3 Overcoming renal clearance of unmodified peptides

Once in circulation, the majority of unmodified peptides and small proteins exhibit very short plasma half-life durations, severely limiting their therapeutic potential through a lack of receptor engagement. Aside from proteolytic instabilities, renal clearance by kidney glomerular filtration efficiently removes lower molecular weight substances from circulation whilst preventing the loss of large proteins. Renal filtration occurs across the glomerular capillary wall, a size-selective barrier which allows the excretion of substances below a cut-off threshold of approximately 30–50 kDa into the urine for disposal. The main approaches undertaken to extend the circulatory half-life of peptide and protein drugs have therefore focussed on increasing the apparent size of these entities, such that they are recovered into the systemic circulation. It is estimated that renal ultrafiltration accounts for up to 80% of the total metabolism of unmodified peptides and proteins below the filtration cut-off size (Rabkin & Dahl, David, 1993). There are two main approaches to increase the size of peptides and proteins presented to the glomerulus, chemical derivatization with high molecular weight polymers, and reversible association with large circulatory proteins such as serum albumin. The first method increases the physical size and hydrodynamic radius of the molecule of interest through derivatization with inert polymers, for example, PEG or polysarcosine (pSar). The second approach relies on strong but reversible plasma protein binding, usually to albumin, the most abundant serum protein, to decrease plasma clearance. Filtration of serum proteins and the size of albumin (~60 kDa) are severally limited by the renal vascular barrie; hence, molecules associated strongly with these proteins will not be appreciably removed from circulation (H. Wu & Huang, 2018). With regard to oral delivery of peptides and proteins, an important fundamental difference exists between these approaches, precluding the first from consideration. Any modification which increases the molecular weight of a substance strongly acts to reduce potential absorption through the GI tract, thereby further hindering access to the circulation. For this reason, we will focus here on successful methods through which therapeutic peptides can be derivatized with small albumin-binding lipids which allow both absorption from the GI tract and escape from renal filtration.

2.4 Peptide design considerations to improve oral bioavailability

Alongside the considerable efforts and recent advances in both API formulation and smart device development (Ibrahim et al., 2020; Mandsberg et al., 2020; Mitragotri et al., 2014), synthetic chemists have turned to designing peptide therapeutics which themselves overcome in part the barriers preventing their precursors from successful oral dosing. Although such strategies have been well known for many years, again, pricing pressure has dominated widespread acceptance and adoption of these often costly modifications in all but the most critical medical areas. Only nowadays is the cost of unnatural amino acid derivatives acceptably low enough that pharmaceutical companies are investing in peptidomimetic development. Design considerations for peptide candidates intended for oral delivery should take into account the following important points:

Safety: Where the oral bioavailability of a drug substance is generally very low (1%–5%), there may exist significant interpatient variability, with some patients absorbing the drug substance in the lower range and others in the higher range. Where drug substances have a relatively narrow therapeutic range, for example, insulin (Kalra et al., 2010) or warfarin (Sridharan et al., 2020), this can result in either significant underdosing or overdosing, both of which have a negative outcome for the patient. Hence, the desirable oral bioavailability of therapeutic peptides should be >30%, and the active substance should have a high therapeutic index to limit such interpatient variability and its associated negative effects.

Proteolytic stability: Orally dosed peptides and proteins will encounter an extremely broad range of peptidases across at a variety of pH values before reaching their target receptors (Lundquist & Artursson, 2016). Whilst protease inhibitors and pH modifiers may be coformulated and combined with the site-specific release of the peptide API at an optimal site for absorption across the GI tract, some in-built protease resistance and tolerance of different pHs have proven effective at increasing bioavailability.

Optimal lipophilicity: By their very nature, most unmodified endogenous peptides are highly polar, a property which positively impacts their water solubility, whilst at the same time unfortunately precluding their free passage across cellular membranes. The polar character of natural peptides stems from extensive hydrogen bond networks formed not only through the polyamide backbone but also between ionizable side-chain functions such as glutamate, aspartate, lysine and arginine, giving rise to zwitterionic species at physiological pH. Synthetic modifications which offset and reduce the overall polarity of therapeutic peptides without compromising water solubility result in generally enhanced oral bioavailability by increasing membrane permeability (Di, 2015).

Optimized circulatory half-life: Glomerular filtration is a process whereby the kidneys remove small peptides, proteins and metabolic waste from the circulation in a size-specific manner. Unmodified peptides and proteins below $\sim$60 kDa in size are susceptible to renal clearance, rapidly eliminating them from circulation and strongly reducing systemic exposure. Modifications such as lipidation, which allow reversible association with serum proteins increase the apparent size and hydrodynamic radius of these molecules, overcoming clearance without largely impacting the true molecular weight of the therapeutic (Zorzi et al., 2019).

2.4.1 Improving stability of peptide drugs in the GI tract

Realizing that digestive and circulatory enzymes recognize and rapidly inactivate natural peptides and proteins, researchers have sought ingenious ways to substitute natural residues with unrecognized nonnatural surrogates; fulfilling the required biochemical role of the native residue whilst circumventing proteolysis around the point of substitution. There are numerous adopted strategies, many borrowed from small molecule development, which may be combined to generate peptidomimetic drugs, offering superior metabolic stability, enhanced membrane permeability and extended circulatory half-life. In optimizing any peptide ligand for oral delivery, there can be many starting points dependent upon the peptide in question; however, there are several general approaches described below to explore the most common options (Adessi & Soto, 2002; Gentilucci et al., 2010).

2.4.1.1 Substitution of sequence liabilities

As we have described, there are very few natural α-amino acid residues and motifs which are not specifically recognized and hydrolyzed by either digestive, brush boarder (either kidney or intestinal lumen) or soluble circulatory enzymes. In designing metabolically stabilized therapeutic peptides it is important to consider both the structure and activity of the native ligand, taking into account the role of secondary motifs such as helices, sheets and turns. Significant deviations in peptide conformational structure resulting from the incorporation of incompatible residues are likely to result in decreased receptor interaction and loss of function.

Initial structure–activity assessment frequently starts with alanine scanning of an L-peptide of interest and provides a wealth of information about the role of each native amino acid side chain in the function of the peptide (London et al., 2010). Alanine scanning is frequently successful in identifying the core pharmacophore of a peptide ligand. Identifying the residues critical to function over those which are less important provides a direction to enable potential simplification, substitution, truncation and removal of proteolytic liabilities. The systematic incorporation of D-amino acids in place of native L-residues provides information complimentary to alanine scanning. Rather than removing the original side chain function entirely, as is the case with alanine scanning, the D-amino acid residue incorporates the function albeit presented in its inverse configuration, relaying information as to the structural requirements of the ligand–receptor interaction (Fig. 2.2).

Importantly, those positions which tolerate incorporation of D-amino acids without adversely affecting potency or selectivity provide an opportunity to overcome proteolysis in this position as D-amino acids are not recognized by proteases. α-Methyl amino acids are structurally similar to their natural counterparts but differ in that the α-proton is replaced by a methyl group which is therefore presented in the D-position relative to the L-side-chain function. Aminoisobutyric acid (Aib) is the simplest member of the α-methyl amino acids, a relatively new class of synthetic residues belonging to the α,α-disubstituted family (e.g. cycloleucine). Aib is now inexpensive and easily accessible due not only to its widespread use in peptidomimetic research but also its low manufacturing cost, which does not require a chiral resolution stage since Aib is nonchiral (Islam et al., 2020). Chiral α-methyl amino acids have only recently become commercially available at reasonable cost and have therefore been utilized in peptide therapeutics to a far lesser extent (Aravinda et al., 2008) (Fig. 2.3).

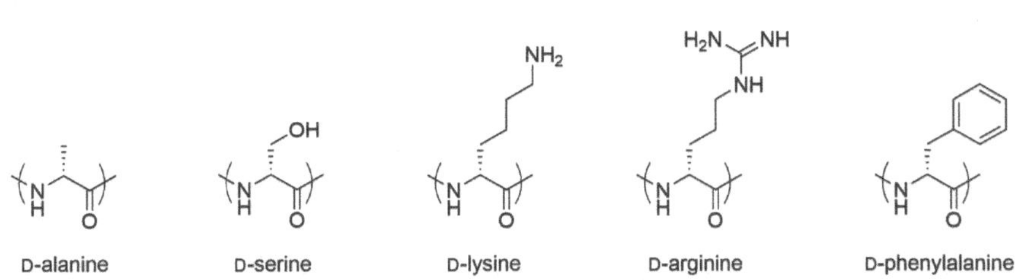

Figure 2.2 Examples of commercially available D-amino acids.

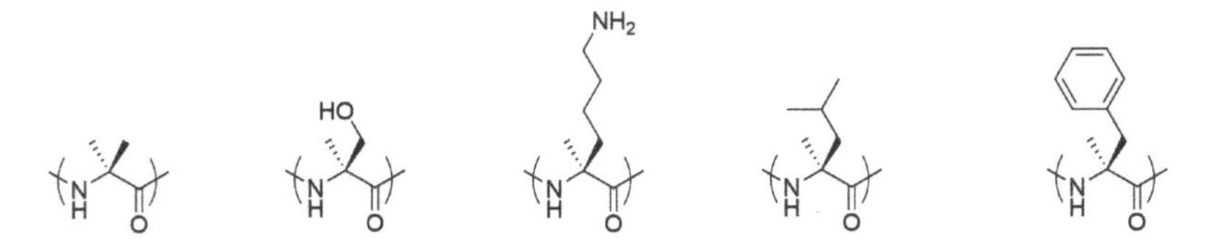

Figure 2.3 Examples of commercial α-methyl-L-amino acids.

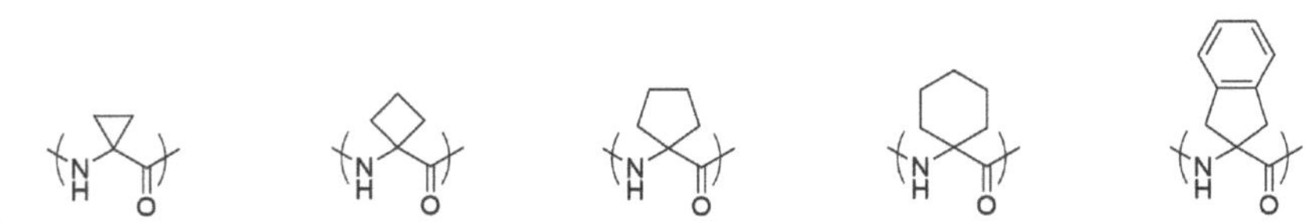

Figure 2.4 Examples of commercially available α,α-disubstituted-amino acids.

The absence of an α-proton renders α-methyl amino acids entirely stable towards both racemization and epimerization during solid-phase peptide synthesis reducing side reactions and simplifying purification. Regarded as conformationally restricted, the presence of the geminal methyl group on the α-carbon strongly influences the secondary structure of peptides incorporating α-methyl residues, limiting rotational freedom around the $N-C^{\alpha}$ and $C^{\alpha}-C'$ bonds at the site of substitution. With fewer low-energy conformations available, such substitutions favor intramolecular hydrogen bonding between the amino −NH group and carbonyl function in every fourth (i to $i+4$) or third (i to $i+3$) residue along the peptide backbone, result in a higher propensity for α-helix and 3_{10} helix formation, respectively. Where the side chain function is cyclized at the α-carbon atom a class of important nonnatural residues referred to as α,α-disubstituted amino acids arises (Fig. 2.4). The increased rigidity of this unit exerts a profound steric constraint around neighboring residues, strongly influencing the conformation of peptides bearing these achiral units in the same manner as with the α-methyl amino acids.

Exceptionally resistant to proteolysis, compact in size and hydrophobic in nature, α,α-disubstituted cycloalkanes have been used in the synthesis of many metabolically stable peptidomimetics (Formaggio et al., 2004).

The *N*-alkylation of amide bond nitrogen atoms results in tertiary amides with significantly different physicochemical properties to the parent bond. Loss of the amide proton by alkylation precludes hydrogen bond formation at this point in the peptide backbone, altering significantly structural characteristics which may be required for receptor interaction. Additionally, in preventing keto-enol tautomerization of the parent amide bond which reduces free rotation, alkylation promotes *cis/trans* isomerization and significantly greater local flexibility and hydrophobicity which may aid with membrane permeability.

Ribosomal proteases do not effectively recognize or hydrolyze alkylated amide bonds within peptides, rendering this modification useful for increasing metabolic stability (Biron et al., 2008). Given their wide commercially availability, the most commonly applied alkyl amide amino acids are *N*-methylated, again providing the opportunity to scan a peptide of interest along its length to determine the biochemical and physicochemical effect of the substitution (Fig. 2.5).

Insertion of an additional methylene ($-CH_2-$) group within the side chain of a standard α-amino acid extends the distance between the α-carbon atom and the side chain function, giving rise to the homo-amino acid analog (Fig. 2.6). As close analogs of their standard counterparts, these residues offer greater side-chain flexibility and the ability to explore interactions with more distant receptor pockets and surfaces.

With reduced recognition by proteases, well-tolerated homo-amino acid substitutions offer a further opportunity to incorporate a degree of proteolytic resistance into a peptide ligand of interest.

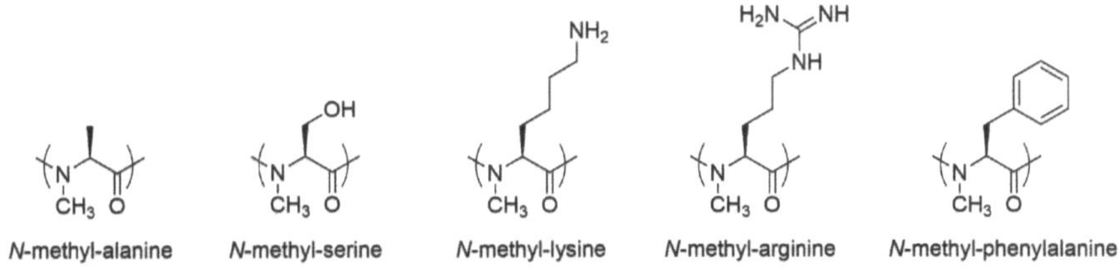

Figure 2.5 Examples of commercially available *N*-methyl-L-amino acids.

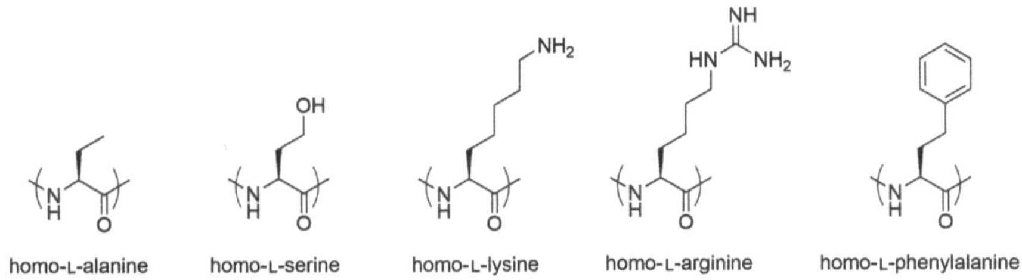

Figure 2.6 Examples of commercially available homo-L-amino acids.

The most commonly incorporated β-amino acids can be classified as 2-(α-) or 3-(β-) depending on the relative position of the amino function to the carboxyl group. β-amino acids as a class offer the chemist an ability to modulate the conformation of peptides through alternate positioning of side-chain functions coupled with improved metabolic stability. Two general classes of linear β-amino acids are well known, namely the β2- and β3-family members, many side-chain (-R) variants of which are commercially available (Fig. 2.7) and straightforward to substitute into synthetic peptides using standard chemical protocols.

Complimenting the linear family of β-amino acids are the sterically constrained cyclic β-amino acids (Fig. 2.8).

There are now a number of commercially available nonnatural aromatic amino acids (Fig. 2.9) which can readily be substituted through chemical synthesis for natural residues in a peptide of interest.

Given the sterically demanding characteristics of many of these aromatic derivatives, they offer not only proteolytic stability and increased hydrophobicity, but also alternative space-filling properties which may strongly influence binding affinity and thereby receptor potency and selectivity.

Although the numerous nonnatural amino acid classes described earlier are by no means exhaustive, these are some of the most commonly applied substitutions used to guide potentially more elaborate future modifications through iterative design.

Figure 2.7 Structural topology of linear β2- and β3-amino acids.

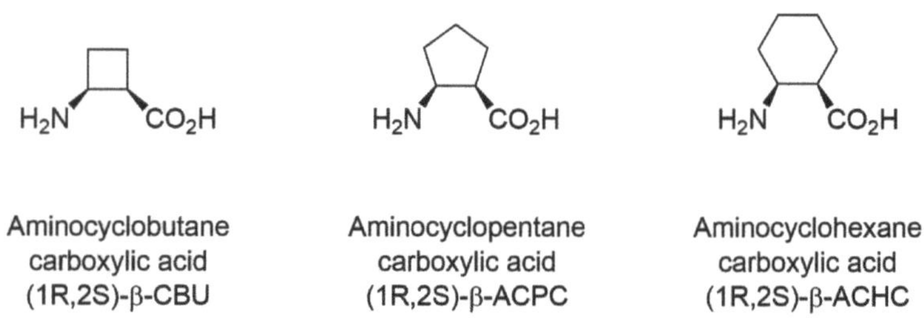

Figure 2.8 Examples of commercially available cyclic β-amino acids.

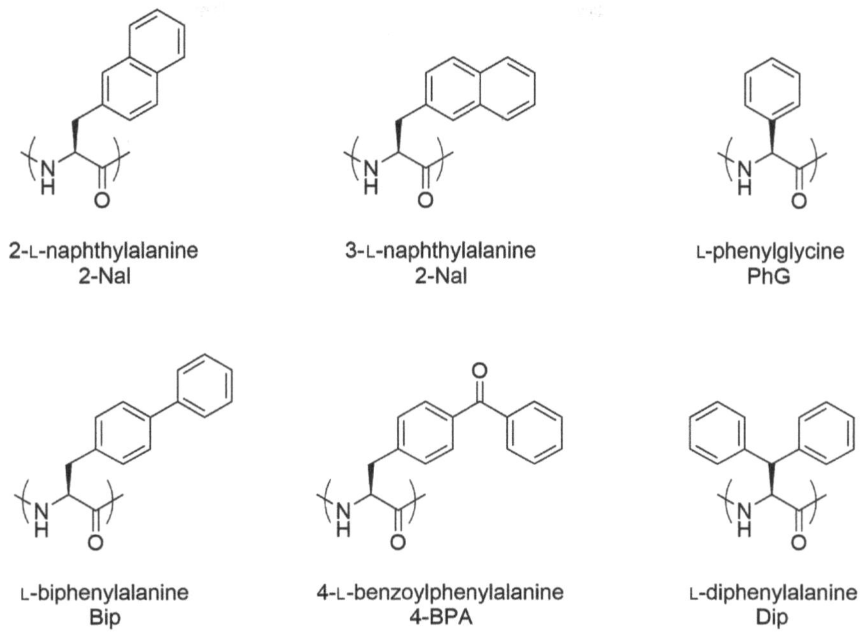

Figure 2.9 Examples of commercially available nonnatural aromatic amino acids.

2.4.1.2 Global modifications to improve metabolic stability

Aside from making point mutations within the backbone of the peptide of interest to remove protease liabilities, sites prone to hydrolysis or deamidation, etc., there are also a number of global modifications which can readily be applied to synthetic peptides and some proteins, to stabilize secondary structural elements, improve metabolic stability, increase membrane permeability and reduce clearance from circulation (Zhang et al., 2018). Where permitted by the ligand–receptor interaction, the N- and/or C- termini may be derivatized, truncated or connected (generating the cyclic peptide) using appropriate chemistries. Many such modifications have a profound effect on the physicochemical properties of the parent structure and can greatly enhance or ameliorate activity. Cyclization can take basically two forms, macrocyclization, the covalent connection of the peptide termini, or stapling, the connection of side chains between different residues within the backbone of the peptide (Fig. 2.10).

Macrocyclization can occur during ribosomal synthesis as a posttranslational modification, rendering peptides more resistant to proteolytic degradation and imparting conformational constraint due to reduced flexibility of the circular structure, particularly in smaller cycles of 10 or

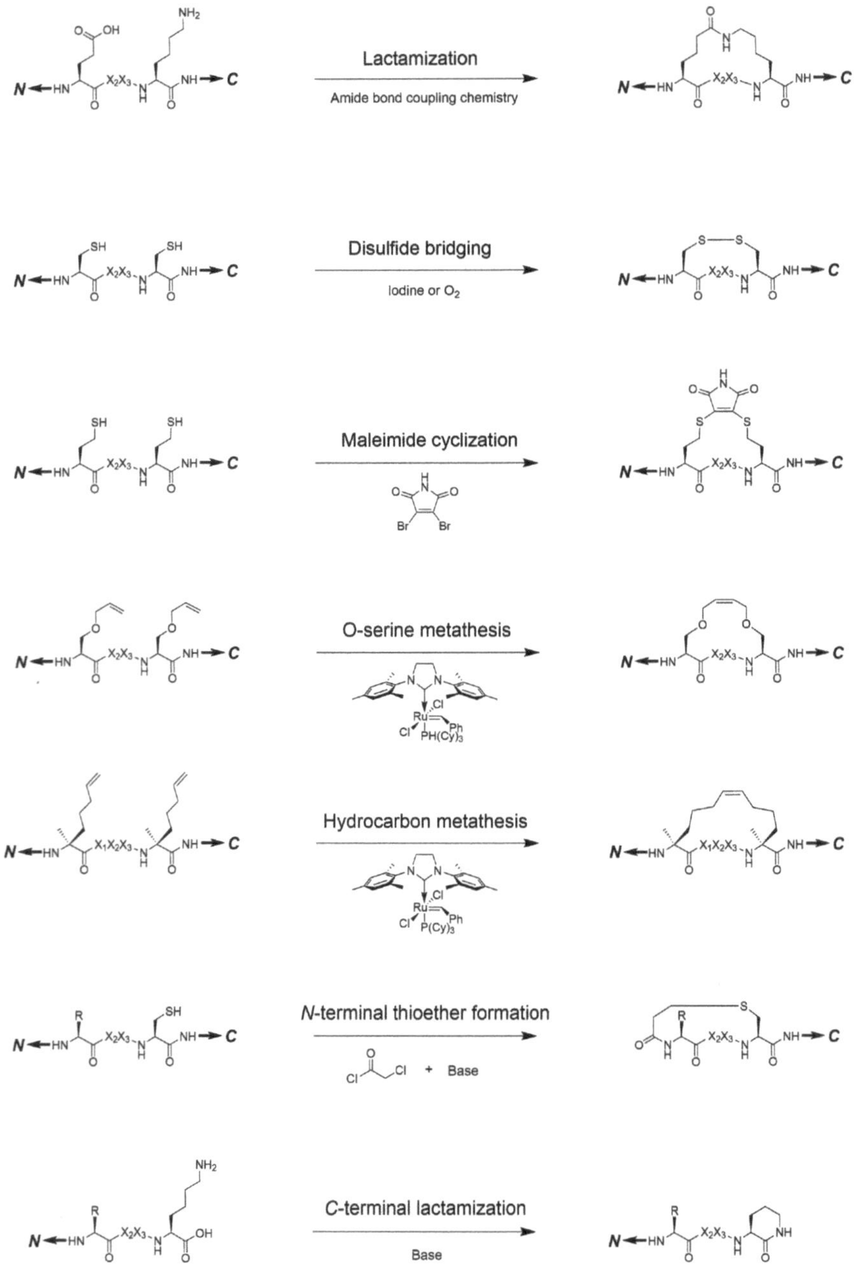

Figure 2.10 Global modifications resulting in peptide cyclization.

fewer residues. During proteolysis, the peptide substrate must adopt a specific conformation to engage the active site before hydrolysis at the scissile bond. If this cannot occur due to conformational restraints in the peptide substrate itself, then the rate of hydrolysis will be greatly reduced or prevented entirely.

Peptide stapling is usually a synthetic process through which specific side-chain functions are reacted and covalently connected to introduce a stable intramolecular cycle. This may take several common forms, including disulfide formation between cysteine residues (Ali et al., 2019), lactamization between side-chain functions of either glutamate or aspartate and lysine (Lambert et al., 2001), or hydrocarbon stapling by metathesis of nonnatural residues bearing terminal alkenes (Stone et al., 2018). In addition, either terminus of a peptide may be cyclized chemically to protect this region prone to exopeptidase degradation, reduce polarity and modify biochemical activity (Fischer et al., 2021; Otero-Ramirez et al., 2018).

As with linear peptides which may naturally fold into helices, sheets and turns, the allowed conformations of cyclic peptides will be determined largely by intramolecular hydrogen bonding, ionic interactions and hydrophobic interactions predetermined within the peptide primary sequence. This sequence can be modified as we have seen to enhance certain structural features within the peptide, and an understanding of not only the intramolecular contacts but also the ligand–receptor interactions comes into play during development. The ability to preorganize the intra-annular contacts within small cyclic peptides has led to the design of many structures which preferentially bury polar residues within the cycle, exposing their more hydrophobic residues for interaction with lipid membranes and thereby increasing passive permeability through lipid membranes (Nielsen et al., 2017). However, despite these advances in our understanding and ability to manipulate cyclic peptide properties there are still only a handful of cyclic peptides which are approved for oral delivery. As with almost all linear peptides, the majority of cyclic peptides have very low membrane permeability and still fewer are effectively absorbed through the digestive tract.

2.4.2 Optimizing membrane permeability

Peptide membrane permeability can be quantitatively measured and reported as a permeability coefficient, typically expressed in units of cm/s, a measurement of how rapidly a given peptide moves across the membrane. In vitro cultures of Caco-2 (Colorectal carcinoma) or MDCK (Madin Darby Canine Kidney) cell lines are the most commonly studied artificial lipid bilayers and are widely used as models to study permeability. Caco-2 models allow assessment of passive permeation only, providing information on the physical ability of the peptide to translocate from the top face (apical) to the bottom face (basolateral) of the monolayer. In contrast, MDCK models provide a tighter junction monolayer which expresses numerous active transport proteins, potentially offering a more realistic model of in vivo permeability. In contrast to small molecules, the unaided passage of peptides or proteins across such model membranes is remarkably low and has rarely been reported. Given that the plasma membrane is virtually impermeable to larger uncharged polar peptides and excludes charged peptides entirely, the number of natural molecules known to passively diffuse across cell membranes is very low. Of note is the natural cyclic peptide cyclosporin A (CsA), which shows appreciable permeability across cell membranes despite its relatively large size and polarity (Bockus et al., 2013). Even so, the reported permeability coefficient of cyclosporin A across artificial membranes is low in comparison to that of a typical highly permeable small molecule (2.5×10^{-7} cm/s vs 10^{-5} cm/s, respectively) (Augustijns et al., 1993). Cyclosporin A is a calcineurin inhibitor, used as an immunosuppressant during organ transplants operations to prevent rejection, and in the treatment of certain autoimmune diseases. CsA is a nonribosomal peptide first isolated in 1971 from the fungus *Tolypocladium inflatum*, receiving approval as both an orally administered solution and as soft gelatine capsules (Sandimmune, Sandoz) in 1983 (Kolata, 1983; Stähelin, 1996; Wu & Kuca, 2019). Following administration as a liquid solution, rapid absorption of CsA occurs through the digestive tract and peak plasma concentrations are achieved at 1.5−2 h after dosing. CsA is not degraded by the major digestive or circulatory proteases to any appreciable extent, rather it is metabolized in the small intestine, kidneys and liver by CYP3A4, a monooxygenase enzyme of the cytochrome P450 superfamily (Kolata, 1983).

Cyclosporin A is a head-to-tail cyclic 11-mer peptide containing seven *N*-methylated residues and a single rarely encountered D-alanine (Fig. 2.11). The high proportion of *N*-methylated residues greatly reduces the number of hydrogen bonds made by CsA and as a result it is appreciably hydrophobic, enabling its passive translocation across the GI tract. At ~1200 Da, CsA is also a well-recognized contradiction to Lipinski's rules, which state that molecules over ~500 Da, bearing more than 5 hydrogen-bond donors or

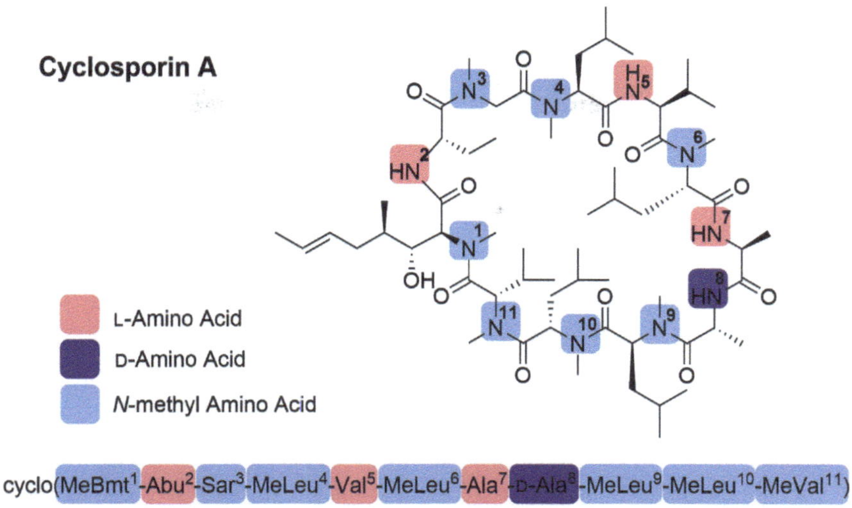

Figure 2.11 The structure of 11-mer nonribosomal peptide Cyclosporin A.

10 hydrogen-bond acceptors are likely to be poorly absorbed. The unusual physicochemical properties of CsA stem from its conformation-dependent hydrophobicity, switching of which allows membrane permeation with unexpected ease. However, despite enormous efforts in engineering novel *N*-methylated cyclic peptide therapeutics akin to CsA, very little success has been achieved to date, and CsA remains one of a kind when it comes to oral bioavailability. Nevertheless, there have been many learnings obtained from the structural features of CsA which may guide design and aid facilitated oral delivery in combination with other approaches, for example, the use of transient permeation enhancers. As a general observation, small, conformationally flexible, metabolically stable macrocycles exhibiting minimal polarity and charge will have improved membrane permeability over large, linear, polar peptides (Yang & Hinner, 2015).

2.4.3 Disulfide bond surrogates

With typical dissociation energies of around 60 kcal/mol disulfides are relatively strong covalent bonds formed across the two sulfhydryl functions of proximal cysteine residues and contribute significantly to the secondary and tertiary structures of peptides and proteins. Naturally occurring disulfides stabilize motifs and define conformational attributes which may drive crucial ligand interactions and structural functions. Disulfides may form spontaneously under mild oxidizing conditions, although bond formation is usually reversible and many disulfide-containing peptides and proteins are unstable towards the reducing environment of the cell cytosol (Patil et al., 2015). Given the integral nature of disulfide bonds to biochemical function, their inherent instability has prompted researchers to develop more stable surrogates, with improved properties (Góngora-Benítez et al., 2014; Thombare & Hutton, 2018). Whilst there have been many approaches to replacing disulfide bonds, in practice very few of these preserve the favorable characteristics of the native moiety which is frequently a crucial determinant of recognition and structural integrity, core to the arrangement of the pharmacophore (Jing & Jin, 2020). Amongst others, thioethers (Otero-Ramirez et al., 2018), diselenides (Medini et al., 2016; Metanis et al., 2006), lactams (Taylor, 2002; Vasco et al., 2020), triazoles (Li et al., 2013; White et al., 2020), ethers (Li et al., 2013; Zhao et al., 2020) and hydrocarbon staples (Watson et al., 2021; Yuen et al., 2019) have all been evaluated as disulfide surrogates in many different peptides of interest; however, the resulting analogs have not always proven sufficiently useful for further development (Fig. 2.12).

Of several notable examples, one recent innovation utilizes a novel ether mimetic, demonstrating excellent stability under both reducing and

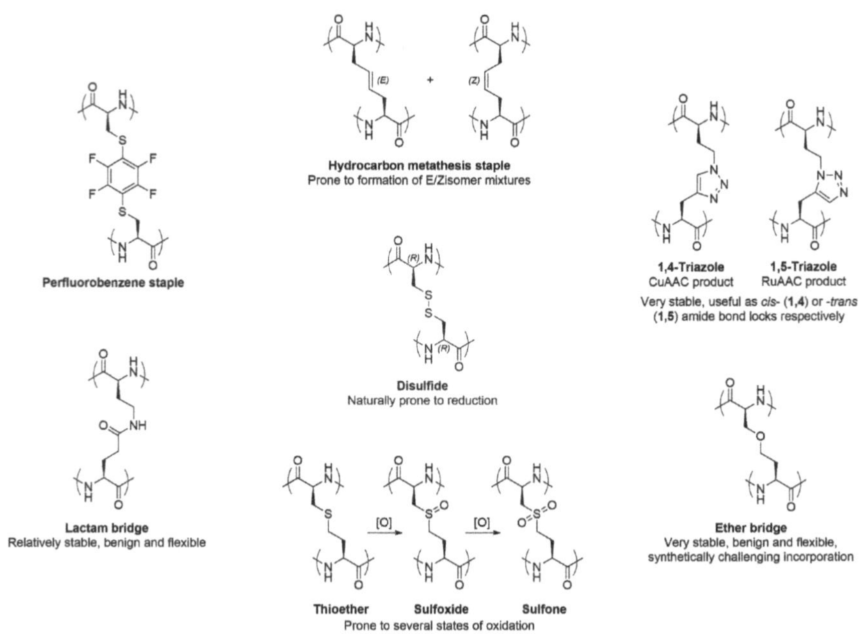

Figure 2.12 Commonly proposed disulfide bond surrogates.

oxidizing conditions and maintaining fully the activity of the original peptides into which it was incorporated (Fig. 2.13; Zhao et al., 2020).

Although this derivative is not commercially available and likely to be prohibitive in its manufacturing cost of goods, it demonstrates a successful and innovative strategy in overcoming the instabilities associated with disulfide-bridged therapeutics.

Incorporation of this diaminodiacid into the cyclic nonapeptide hormone oxytocin as a disulfide mimetic resulted in an ether-bridged analog which was shown to maintain the potency and selectivity of the native hormone whilst significantly improving metabolic stability (Fig. 2.14).

Modification of the related pituitary vasoconstrictor hormone vasopressin has also been the subject of decades of intense research, its potent antidiuretic effects making it a promising treatment for central diabetes insipidus. However, endogenous vasopressin has a very short systemic half-life of several minutes due to two major liabilities (*N*-terminal cysteine and L-arginine) which limit its utility as a therapeutic. Furthermore, vasopressin exhibits a broad pharmacology, acting systemically at the $V1_a$ receptor widely expressed throughout vascular tissues, resulting in

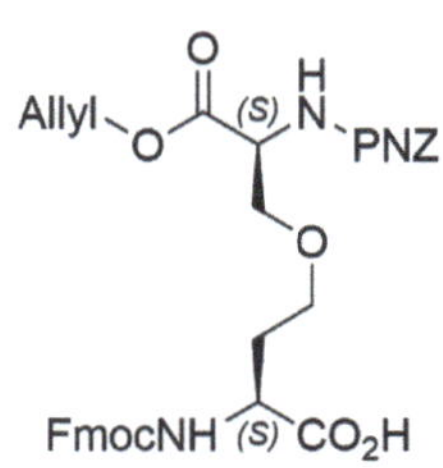

Figure 2.13 An orthogonally protected diaminodiacid ether mimetic for disulfide replacement.

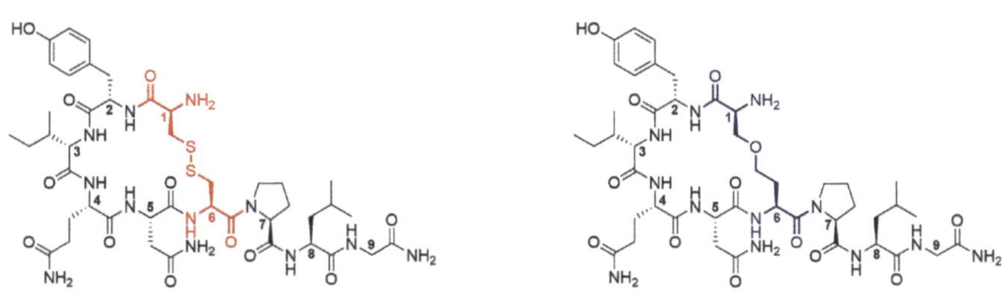

Oxytocin (Pitocin)

cyclo(Cys[1]-Tyr[2]-Ile[3]-Gln[4]-Asn[5]-Cys[6])-Pro[7]-Leu[8]-Gly[9]-amide

Ether-bridged oxytocin

O-cyclo(Ser[1]-Tyr[2]-Ile[3]-Gln[4]-Asn[5]-hSer[6])-Pro[7]-Leu[8]-Gly[9]-amide

Figure 2.14 (left) Natural oxytocin (Pitocin) and (right) ether-bridged oxytocin analog.

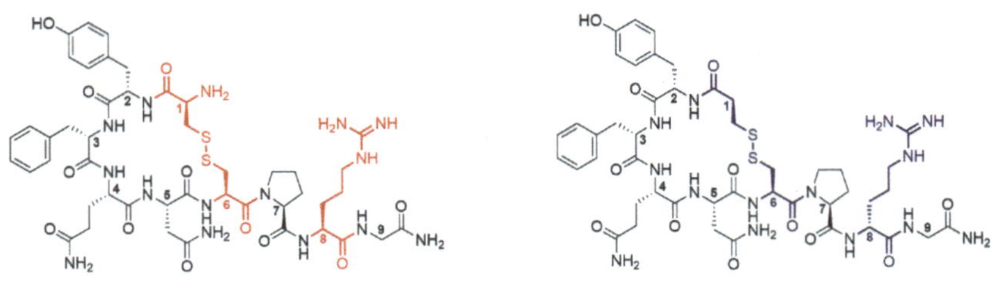

Vasopressin

$_{cyclo}$(Cys^1-Tyr^2-Phe^3-Gln^4-Asn^5-Cys^6)-Pro^7-Arg^8-Gly^9-amide

Desmopressin (Minrin)

$_{cyclo}$(3-Mpa^1-Tyr^2-Phe^3-Gln^4-Asn^5-Cys^6)-Pro^7-D-Arg^8-Gly^9-amide

Figure 2.15 (left) Natural vasopressin and (right) desmopressin (Minrin).

increased blood pressure. Vasopressin also acts centrally at the $V1_b$ receptor, increasing contraction of intestinal smooth muscle which causes painful colic like symptoms in some patients. Deamination of the N-terminus and substitution of L-arginine^8 for D-arginine entirely remove both pressor activity at $V1_a$ and central availability to $V1_b$, whilst increasing diuretic activity mediated by the V2 receptor (Fig. 2.15; Koshimizu et al., 2012). V2 is expressed in renal distal tubules and collecting ducts, stimulating water reabsorption in the kidney, making this a highly selective antidiuretic (Hans, 1990). The resulting commercialized analog desmopressin (Minrin) has substantially improved metabolic stability, resulting in a systemic half-life of several hours and allowing desmopressin to be administered orally, albeit with an effective bioavailability of between 0.08%–0.16%.

Although oral bioavailability of desmopressin is exceptionally low, its high potency and wide therapeutic index make this level of uptake both safe and effective across a broad patient population (Rembratt et al., 2004).

2.4.4 Reducing elimination of peptide drugs in the circulation

Lipidation is the simplest, cheapest and therefore most common strategy used to extend the circulatory half-life of peptide therapeutics by promoting reversible noncovalent interaction with serum albumin in the circulatory system. In addition, lipidation is a clinically validated approach also shown to reduce immunogenicity and enable intracellular uptake and delivery across epithelia (Menacho-Melgar et al., 2019). Albumin is the most abundant protein in the bloodstream and displays a remarkable ability to strongly but reversibly bind a very broad range of chemical moieties. Albumin binds all manner of lipids with affinities generally increasing with lipid length (Madsen et al., 2007), thereby allowing lipidated peptides to be recovered from renal ultrafiltration and also shielding the peptide from proteolytic degradation to some extent (Bech et al., 2018). Lipidation of peptide therapeutics has been used mainly in the context of injectable therapeutics and has allowed for impressive increases in plasma half-life and drug exposure. Liraglutide (Victoza, Novo Nordisk), a once-daily subcutaneously delivered lipidated GLP-1 analog indicated for type-2 diabetes and obesity, incorporates a saturated C_{16} palmitoyl lipid which extends plasma half-life from just several minutes in the unfunctionalized peptide out to around 13 h in the derivatized therapeutic (Guryanov et al., 2016; Knudsen et al., 2000). Semaglutide (Ozempic, Novo Nordisk), a next-generation lipidated GLP-1 analog of GLP-1, shows an even more impressive protraction of pharmacokinetics with a plasma half-life of around 168 h in man, permitting once-weekly subcutaneous administration (Jensen et al., 2017). Another recently approved peptide therapeutic demonstrating the utility of lipidation is the long-acting human growth hormone analog somapacitan (Sogroya, Novo Nordisk) indicated for the treatment of growth hormone deficiencies in adults (Ramírez-Andersen et al., 2018). The plasma half-life of human growth hormone is significantly extended in somapacitan in which a long linker region separates the protein from an albumin-binding tetrazolyl functionalized C_{17} lipid moiety (Madsen et al., 2007) (Fig. 2.16). With an albumin binding fraction in excess of 99%, somapacitan exhibits around 40 h of half-life in circulation, permitting once-weekly administration (Johannsson et al., 2018).

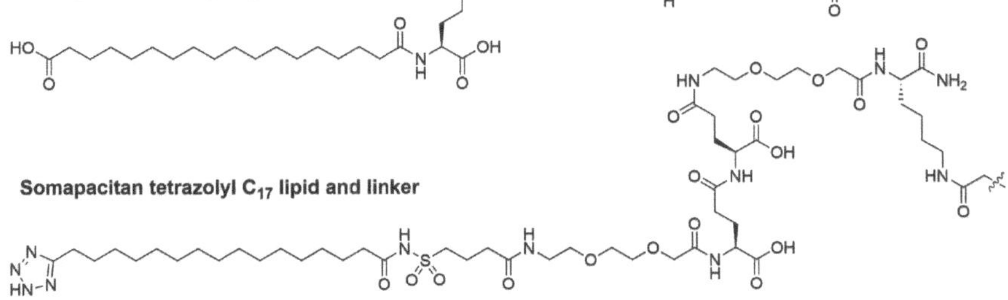

Figure 2.16 Examples of albumin-binding lipids used to extend peptide plasma half-life.

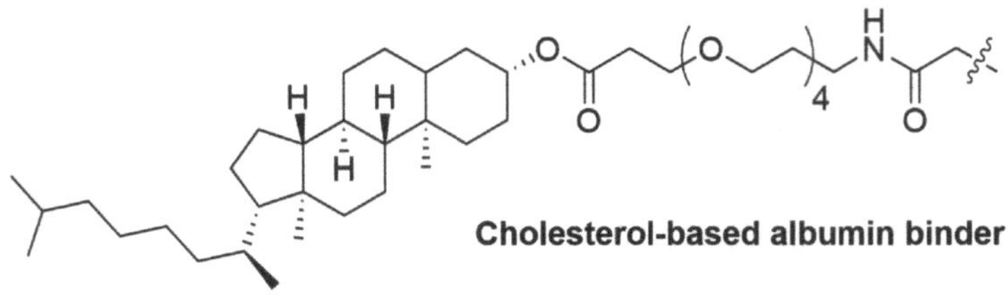

Figure 2.17 Example of a cholesterol albumin-binding moiety used to extend peptide plasma half-life.

Another compact and potentially useful alternative to the incorporation of linear saturated lipids is peptide derivatization using cholesterol as an albumin binder, which may extend plasma half-life and also increase receptor potency of certain peptide drugs (Fig. 2.17).

Cholesterol derivatized peptides are thought to associate more strongly with cellular membranes than linear lipids, potentiating their greater interaction with cell surface receptors (Santoprete et al., 2011).

2.5 Rybelsus, an oral semaglutide formulation, a short case study

In 2020 the US Food and Drug Administration (FDA) approved Rybelsus (Semaglutide; Novo Nordisk) oral tablets for the treatment of adults with insufficiently controlled T2DM. The active principal peptide semaglutide was previously approved as Ozempic in 2017 for once-weekly injection, demonstrating improved glycaemic control, decreased HbA1c, and reduced bodyweight in conjunction with a healthy diet and exercise. Rybelsus is notably the first GLP-1 receptor agonist (GLP-1RA) approved for use that does not need to be injected; an appealing alternative in patients with needle-phobia (Fu et al., 2009; Nir et al., 2003). Semaglutide is a long-acting monolipidated GLP-1 (7–37) in which the side chain ε-amino group of Lysine^{26} is acylated and the N-terminus protected against proteolytic digestion by DPP-IV though substitution of Ala^{8} to Aib^{8}. Substitution of native lysine^{34} for arginine^{34} in semaglutide leaves a singular lysine^{26} for site-specific lipidation, permitting simplified semi-synthesis through a combination of both synthetic and recombinant methodologies (Knudsen & Lau, 2019). In all other respects, semaglutide is identical in sequence to the native (7–37) endogenous ligand (Pratley et al., 2019), a potentially important factor in maintaining the desired pharmacology, tolerability and minimizing the potential for immunogenicity. With a molecular weight >4kD, Rybelsus requires additional formulation from the approved once-weekly injection of semaglutide (Ozempic). The truly remarkable achievement of reassigning semaglutide from injectable to oral tablet relies largely on the use of the transient permeation enhancer SNAC (sodium *N*-(8-[2-hydroxybenzoyl] amino)caprylate) which acts to open tight junctions in the gastric mucosa, permitting the absorption of the peptide in sufficiently large and uniform amounts to achieve safe and efficacious plasma levels in a broad patient population (Kalra & Sahay, 2020). However, semaglutide is a rare example of a beautifully engineered therapeutic exhibiting all of the necessary physicochemical and pharmacological properties required for successful oral delivery; primarily wide therapeutic index, extended pharmacokinetics and high receptor potency. GLP-1 (7−37) itself is a 31 amino acid incretin peptide hormone of the growth hormone-releasing factor (GRF) superfamily secreted mainly by the L-cells of the distal ileum in response to glucose ingestion. Upon release into the circulation GLP-1 binds and activates its cognate G protein-coupled receptor (GLP-1R) expressed primarily on pancreatic β-cells (Adelhorst et al., 1994). Importantly, this action stimulates a proportional release of insulin into the circulation, resulting in uptake of glucose into tissues. In insulin-resistant individuals, however, blood glucose is persistently elevated in a condition known as prediabetes, a prelude to type 2 diabetes mellitus (T2DM) which can result in potentially fatal multiple organ damage. This metabolic disorder progresses as 'burnout' of pancreatic β-cells inevitably occurs as a result of chronic overstimulation of the pancreas by GLP-1 to produce insulin. The glucose-dependent release of insulin mediated via GLP-1 is particularly appealing as a safe point of therapeutic intervention in the treatment of T2DM since there is no risk of a potentially fatal hypoglycaemic episode, a distinct possibility with direct insulin administration (Müller et al., 2019). Native GLP-1 evolved to exert its pharmacological effects rapidly and to be inactivated just as rapidly, acting as a fleeting messenger and regulator of blood glucose. The sequence of native GLP-1 shows numerous proteolytic liabilities towards both digestive and circulatory proteases which together preclude oral absorption of the native unmodified peptide (Manandhar & Ahn, 2015) (Fig. 2.18).

The ubiquitous exopeptidase dipeptidyl peptidase-IV (DPP-IV) is abundant in the circulatory system and inactivates unprotected GLP-1 by cleavage of the N-terminal dipeptide His^7-Ala^8- through hydrolysis of the scissile Ala^8-Glu^9 amide bond, converting the native ligand from a potent

GLP-1 (7-37) HA^{8}-EGT^{11} FTSDV^{16} SSYLE^{21} GQAAK^{26} EFIAW^{31} LVK^{34}GR^{36} G-acid

Semaglutide H(Aib)^{8}-EGT^{11} FTSDV^{16} SSYLE^{21} GQAAK(-OEG-OEG-γE-C18-CO_2H)^{26} EFIAW^{31} LVR^{34}GR^{36} G-acid

Figure 2.18 Amino acid sequence of human GLP-1 (7–37) and a novel lipidated analog semaglutide.

agonist to a weak antagonist, providing a rapid independent stop to β-cell stimulation (Knudsen & Pridal, 1996). The combined hydrolytic actions of DPP-IV and neprilysin (NEP 24.11) both contribute to the very short circulatory half-life of just 2 ½ minutes of native GLP-1 in humans (Drucker, 2001). However, as DPP-IV has a high substrate specificity at positions proximal to the scissile amide bond, substitutions in this position can be expected to reduce or prevent degradation (Mentlein et al., 1993). In particular, proline and alanine are highly favoured at the P1 position and analogs bearing Pro^{8} or Ala^{8} are rapidly cleaved by DPP-IV. Other amino acids at this position have been shown to enhance the stability of these analogs towards DPP-IV, including Gly^{8}, Ser^{8}, Thr^{8}, Leu^{8} (Deacon et al., 1998; Ritzel et al., 1998), however, the nonnatural α-methyl residue Aib has proven most effective at maintaining both high receptor potency and DPP-IV resistance. To surmount the main causes of inactivation, semaglutide incorporates Aib in place of native Ala^{8}, conferring excellent resistance against DDP-IV. Although DPP-IV degradation is a major cause of GLP-1 functional loss, endoproteolysis by NEP 24.11 was also shown to contribute up to 50% towards the inactivation of native GLP-1 in vivo. Neprilysin shows a preference for cleavage on the amino-terminal side of hydrophobic residues and cleaves native GLP-1 at multiple sites accordingly. Specifically, NEP hydrolyzes peptide bonds Asp^{15}-Val^{16}, Ser^{18}-Tyr^{19}, Tyr^{19}-Leu^{20}, Glu^{27}-Phe^{28}, Phe^{28}-Ile^{29}, and Trp^{31}-Leu^{32} (Green et al., 2003; Plamboeck et al., 2005). Coupled with this information, NMR studies further revealed that GLP-1 consists of an N-terminal random coil segment (His^{7}-Thr^{13}), two helical stretches (Thr^{13}-Leu^{20} and Ala^{24}-Arg^{36}), and a less well-defined linker region (Glu^{21}-Gln^{23}). Adelhorst et al. performed an alanine scan on GLP-1, replacing each residue systematically with alanine to identify residues contributing to binding affinity and/or receptor activation (Adelhorst et al., 1994). From this work residues His^{7}, Gly^{10}, Asp^{15}, and Phe^{28} were revealed to be crucially required for receptor activation whilst His^{7}, Gly^{10}, Phe^{12}, Thr^{13}, Asp^{15}, Phe^{28}, Ile^{29}, Trp^{31} and Leu^{32} were shown to play an important role in receptor binding with Phe^{28}, Ile^{29}, Trp^{31}, and Leu^{32} making up an important hydrophobic surface crucial for receptor–ligand interaction. The dual roles of residues His^{7}, Gly^{10} and Phe^{28}, therefore, make these positions particularly critical to the function of GLP-1 (Hareter et al., 1997). Understanding both the potential proteolytic liabilities and structure–activity relationship of the native GLP-1 sequence allowed for the ideal placement of an octadecanoate lipid, appended through a short PEG linker to the ε-amine function of lysine[26]. Both the PEG linker and saturated long-chain lipid contribute significantly to the extended pharmacokinetic profile of semaglutide, binding with very high-affinity serum albumin which acts to shield the peptide backbone from proteolysis and significantly reduce excretion by kidney glomerular filtration. The site-specific effects on lipidation of the GLP-1 (7–37) scaffold of semaglutide were also investigated, revealing that receptor potency was substantially reduced in analogs where the lipid was located too near the N-terminus. These findings are in keeping with the alanine scan performed by Adelhorst et al. A systematic evaluation of both lipid length and class ranging from C_{12} to C_{20} saturated fatty acids and di-acids appended through a variety of linkages to lysine[26] revealed that C_{18} di-acid (1,18-octadecandioic acid) together with a chiral -γGlu-$(OEG)_2$- linker gave optimal albumin binding affinity, water solubility and GLP-1R potency. The meticulous efforts invested in engineering semaglutide demonstrate that the combination of carefully chosen physicochemical properties and appropriate formulation can allow for successful oral delivery of large peptides. Whilst such engineering successes may not always be possible with other peptide candidates, there is a renewed interest in exploring such strategies and the approval of Rybelsus has proven inspirational to the community.

2.6 Conclusions

Peptide therapeutics have proven vital in treating a wide range of physiological disorders and diseases; however, the majority of these treatments must still be given intravenously or by injection. Investment into new strategies for oral peptide therapeutic delivery is still booming even after 20 years of core research; however, single approaches alone have met with very little success. For larger peptides (> 1000 Da) only the combined-strategies of semi-stabilized peptide, very high potency, very high therapeutic index and smart formulation have been successful, as in the case of the oral GLP-1 agonist semaglutide Rybelsus. Whilst smaller, ideally cyclic peptides can show useful oral bioavailability, it is usually only those candidates which have been specifically engineered or which have evolved to be proteolytically resistant which have shown adequate oral bioavailability and safety profiles. Ideally, a peptide designed for oral delivery should exhibit a number of important physicochemical and pharmacological properties including high gastrointestinal and serum protease stability, good stability at low and neutral pH, significant lipophilicity, strong serum albumin association, excellent aqueous solubility and high target receptor potency. The combination of these properties coupled with a suitable formulation strategy should allow for consideration of oral delivery. As the manufacturing costs of non-natural amino acid derivatives continues to fall, the strategic incorporation of protease-resistant residues in positions greatest at risk of proteolysis becomes an ever more attractive method to stabilize peptide candidates, and exploiting synthetic methods such as lipidation and PEGylation should help extend the pharmacokinetic profile, leading to useful therapeutics. However, the reimbursable costs of the final drug product must always be an early consideration in the design of peptides intended for oral administration, a factor which excludes the oral option in therapeutic sectors where there are already effective cheaper alternatives. For life-threatening diseases, there may be scope for new best-in-class peptide treatments for which the reimbursable costs cover the significant development and manufacturing costs of peptide drugs adequately; however, this is frequently not the case at present, and small molecules remain dominant first-line treatments in many disease areas. To minimize the regulatory risk of clinical failure, most pharmaceutical organizations choose to focus on the repurposing of clinically validated peptide drug assets which are already approved for parenteral administration rather than on the development of new chemical entities (NCEs) designed specifically for oral delivery. Although this has led to the emergence of several useful oral peptide drugs over the past several decades, the number of approved peptides suitable for this recycling exercise is obviously limited in number. Only through the development and exploitation of new discovery technologies will the number of new peptide entities be suitable for oral absorption expansion (Muttenthaler et al., 2021).

References

Adelhorst, K., Hedegaard, B. B., Knudsen, L. B., & Kirk, O. (1994). Structure-activity studies of glucagon-like peptide-1. *Journal of Biological Chemistry*, *269*(9), 6275–6278. Available from https://doi.org/10.1016/s0021-9258(17)37366-0.

Adessi, C., & Soto, C. (2002). Converting a peptide into a drug: Strategies to improve stability and bioavailability. *Current Medicinal Chemistry*, *9*(9), 963–978. Available from https://doi.org/10.2174/0929867024606731.

Ali, A. M., Atmaj, J., Van Oosterwijk, N., Groves, M. R., & Dömling, A. (2019). Stapled peptides inhibitors: A new window for target drug discovery. *Computational and Structural Biotechnology Journal*, *17*, 263–281. Available from https://doi.org/10.1016/j.csbj.2019.01.012.

Aravinda, S., Shamala, N., & Balaram, P. (2008). Aib residues in peptaibiotics and synthetic sequences: Analysis of nonhelical conformations. *Chemistry and Biodiversity*, *5*(7), 1238–1262. Available from https://doi.org/10.1002/cbdv.200890112.

Augustijns, P. F., Bradshaw, T. P., Gan, L. S. L., Hendren, R. W., & Thakker, D. R. (1993). Evidence for a polarized efflux system in caco-2 cells capable of modulating cyclosporine a transport. *Biochemical and Biophysical Research Communications*, *197*(2), 360–365. Available from https://doi.org/10.1006/bbrc.1993.2487.

Bech, E. M., Pedersen, S. L., & Jensen, K. J. (2018). Chemical strategies for half-life extension of biopharmaceuticals: Lipidation and its alternatives. *ACS Medicinal Chemistry Letters*, *9*(7), 577–580. Available from https://doi.org/10.1021/acsmedchemlett.8b00226.

Bird, G. H., Madani, N., Perry, A. F., Princiotto, A. M., Supko, J. G., He, X., Gavathiotis, E., Sodroski, J. G., & Walensky, L. D. (2010). Hydrocarbon double-stapling remedies the proteolytic instability of a lengthy peptide therapeutic. *Proceedings of the National Academy of Sciences of the United States of America*, *107*(32), 14093–14098. Available from https://doi.org/10.1073/pnas.1002713107.

Biron, E., Chatterjee, J., Ovadia, O., Langenegger, D., Brueggen, J., Hoyer, D., Schmid, H. A., Jelinek, R., Gilon, C., Hoffman, A., & Kessler, H. (2008). Improving oral bioavailability of peptides by multiple N-methylation: Somatostatin analogues. *Angewandte Chemie - International Edition*, *47*(14), 2595–2599. Available from https://doi.org/10.1002/anie.200705797.

Bockus, A. T., McEwen, C. M., & Scott Lokey, R. (2013). Form and function in cyclic peptide natural products: A pharmacokinetic perspective. *Current Topics in Medicinal Chemistry*, *13*(6), 821–836. Available from https://doi.org/10.2174/15680266113130070005.

Brayden, D. J., Hill, T. A., Fairlie, D. P., Maher, S., & Mrsny, R. J. (2020). Systemic delivery of peptides by the oral route: Formulation and medicinal chemistry approaches. *Advanced Drug Delivery Reviews*, *157*, 2–36. Available from https://doi.org/10.1016/j.addr.2020.05.007.

D'souza, A. A., & Shegokar, R. (2016). Polyethylene glycol (PEG): A versatile polymer for pharmaceutical applications. *Expert Opinion on Drug Delivery*, *13*(9), 1257–1275. Available from https://doi.org/10.1080/17425247.2016.1182485.

Deacon, C. F., Knudsen, L. B., Madsen, K., Wiberg, F. C., Jacobsen, O., & Holst, J. J. (1998). Dipeptidyl peptidase IV resistant analogues of glucagon-like peptide-1 which have extended metabolic stability and improved biological activity. *Diabetologia*, *41*(3), 271–278. Available from https://doi.org/10.1007/s001250050903.

Di, L. (2015). Strategic approaches to optimizing peptide ADME properties. *AAPS Journal*, *17*(1), 134–143. Available from https://doi.org/10.1208/s12248-014-9687-3.

Drucker, D. J. (2001). Development of glucagon-like peptide-1-based pharmaceuticals as therapeutic agents for the treatment of diabetes. *Current Pharmaceutical Design*, 7(14), 1399–1412. Available from https://doi.org/10.2174/1381612013397401.

Drucker, D. J. (2020). Advances in oral peptide therapeutics. *Nature Reviews. Drug Discovery*, *19*(4), 277–289. Available from https://doi.org/10.1038/s41573-019-0053-0.

Fischer, N. H., Nielsen, D. S., Palmer, D., Meldal, M., & Diness, F. (2021). C-Terminal lactamization of peptides. *Chemical Communications*, *57*(7), 895–898. Available from https://doi.org/10.1039/d0cc06018f.

Formaggio, F., Moretto, V., Crisma, M., Toniolo, C., Kaptein, B., & Broxterman, Q. B. (2004). New tools for the control of peptide conformation: The helicogenic C α -methyl, C α -cyclohexylglycine*. *The Journal of Peptide Research, 63*(2), 161–170. Available from https://doi.org/10.1111/j.1399-3011.2003.00123.x.

Fu, A. Z., Qiu, Y., & Radican, L. (2009). Impact of fear of insulin or fear of injection on treatment outcomes of patients with diabetes. *Current Medical Research and Opinion, 25* (6), 1413–1420. Available from https://doi.org/10.1185/03007990902905724.

Gentilucci, L., de Marco, R., & Cerisoli, L. (2010). Chemical modifications designed to improve peptide stability: Incorporation of non-natural amino acids, pseudo-peptide bonds, and cyclization. *Current Pharmaceutical Design, 16*(28), 3185–3203. Available from https://doi.org/10.2174/138161210793292555.

Góngora-Benítez, M., Tulla-Puche, J., & Albericio, F. (2014). Multifaceted roles of disulfide bonds. peptides as therapeutics. *Chemical Reviews, 114*(2), 901–926. Available from https://doi.org/10.1021/cr400031z.

Green, B. D., Gault, V. A., Mooney, M. H., Irwin, N., Bailey, C. J., Harriott, P., Greer, B., Flatt, P. R., & O'Harte, F. P. M. (2003). Novel dipeptidyl peptidase IV resistant analogues of glucagon-like peptide-1(7-36)amide have preserved biological activities in vitro conferring improved glucose-lowering action in vivo. *Journal of Molecular Endocrinology, 31*(3), 529–540. Available from https://doi.org/10.1677/jme.0.0310529.

Guryanov, I., Bondesan, A., Visentini, D., Orlandin, A., Biondi, B., Toniolo, C., Formaggio, F., Ricci, A., Zanon, J., & Cabri, W. (2016). Innovative chemical synthesis and conformational hints on the lipopeptide liraglutide. *Journal of Peptide Science, 22* (7), 471–479. Available from https://doi.org/10.1002/psc.2890.

Hans, V. (1990). Basic pharmacology of desmopressin. *Drug Investigation, 2*, 2–8. Available from https://doi.org/10.1007/bf03258235.

Hareter, A., Hoffmann, E., Bode, H. P., Göke, B., & Göke, R. (1997). The positive charge of the imidazole side chain of histidine7 is crucial for GLP-1 action. *Endocrine Journal, 44*(5), 701–705. Available from https://doi.org/10.1507/endocrj.44.701.

Hu, Y., Hou, Y., Wang, H., & Lu, H. (2018). Polysarcosine as an alternative to PEG for therapeutic protein conjugation. *Bioconjugate Chemistry, 29*(7), 2232–2238. Available from https://doi.org/10.1021/acs.bioconjchem.8b00237.

Huber, R., & Bode, W. (1978). Structural basis of the activation and action of trypsin. *Accounts of Chemical Research, 11*(3), 114–122. Available from https://doi.org/10.1021/ar50123a006.

Hupe-Sodmann, K., Göke, R., Göke, B., Thole, H. H., Zimmermann, B., Voigt, K., & McGregor, G. P. (1997). Endoproteolysis of glucagon-like peptide (GLP)-1 (7-36) amide by ectopeptidases in RINm5F cells. *Peptides, 18*(5), 625–632. Available from https://doi.org/10.1016/S0196-9781(97)00123-Xo.

Ibrahim, Y. H. E. Y., Regdon, G., Hamedelniel, E. I., & Sovány, T. (2020). Review of recently used techniques and materials to improve the efficiency of orally administered proteins/peptides. *DARU, Journal of Pharmaceutical Sciences, 28*(1), 403–416. Available from https://doi.org/10.1007/s40199-019-00316-w.

Islam, R., Sviridov, D. O., Drake, S. K., Tunyi, J., Abdoulaeva, G., Freeman, L. A., Pastor, R. W., & Remaley, A. T. (2020). Incorporation of α-methylated amino acids into Apolipoprotein A-I mimetic peptides improves their helicity and cholesterol efflux potential. *Biochemical and Biophysical Research Communications, 526*(2), 349–354. Available from https://doi.org/10.1016/j.bbrc.2020.03.070.

Jensen, L., Helleberg, H., Roffel, A., van Lier, J. J., Bjørnsdottir, I., Pedersen, P. J., Rowe, E., Derving Karsbøl, J., & Pedersen, M. L. (2017). Absorption, metabolism and excretion of the GLP-1 analogue semaglutide in humans and nonclinical species. *European Journal of Pharmaceutical Sciences, 104*, 31–41. Available from https://doi.org/10.1016/j.ejps.2017.03.020.

Jing, X., & Jin, K. (2020). A gold mine for drug discovery: Strategies to develop cyclic peptides into therapies. *Medicinal Research Reviews*, *40*(2), 753–810. Available from https://doi.org/10.1002/med.21639.

Johannsson, G., Feldt-Rasmussen, U., Håkonsson, I. H., Biering, H., Rodien, P., Tahara, S., Toogood, A., & Rasmussen, M. H. (2018). Safety and convenience of once-weekly somapacitan in adult GH deficiency: A 26-week randomized, controlled trial. *European Journal of Endocrinology*, *178*(5), 491–499. Available from https://doi.org/10.1530/EJE-17-1073.

Kalra, S., & Sahay, R. (2020). A review on semaglutide: An oral glucagon-like peptide 1 receptor agonist in management of type 2 diabetes mellitus. *Diabetes Therapy*, *11*(9), 1965–1982. Available from https://doi.org/10.1007/s13300-020-00894-y.

Kalra, S., Kalra, B., & Agrawal, N. (2010). Oral insulin. *Diabetology and Metabolic Syndrome*, *2*(1). Available from https://doi.org/10.1186/1758-5996-2-66.

Knudsen, L. B., & Lau, J. (2019). The discovery and development of liraglutide and semaglutide. *Frontiers in Endocrinology*, 10. Available from https://doi.org/10.3389/fendo.2019.00155.

Knudsen, L. B., & Pridal, L. (1996). Glucagon-like peptide-1-(9-36) amide is a major metabolite of glucagon-like peptide-1-(7-36) amide after in vivo administration to dogs, and it acts as an antagonist on the pancreatic receptor. *European Journal of Pharmacology*, *318*(2–3), 429–435. Available from https://doi.org/10.1016/S0014-2999(96)00795-9.

Knudsen, L. B., Nielsen, P. F., Huusfeldt, P. O., Johansen, N. L., Madsen, K., Pedersen, F. Z., Thøgersen, H., Wilken, M., & Agersø, H. (2000). Potent derivatives of glucagon-like peptide-1 with pharmacokinetic properties suitable for once daily administration. *Journal of Medicinal Chemistry*, *43*(9), 1664–1669. Available from https://doi.org/10.1021/jm9909645.

Kolata, G. (1983). FDA speeds approval of cyclosporin. *Science (New York, NY)*. Available from https://doi.org/10.1126/science.221.4617.1273-a, 1273–1273.

Kong, X. D., Moriya, J., Carle, V., Pojer, F., Abriata, L. A., Deyle, K., & Heinis, C. (2020). De novo development of proteolytically resistant therapeutic peptides for oral administration. *Nature Biomedical Engineering*, *4*(5), 560–571. Available from https://doi.org/10.1038/s41551-020-0556-3.

Koshimizu, T. a, Nakamura, K., Egashira, N., Hiroyama, M., Nonoguchi, H., & Tanoue, A. (2012). Vasopressin V1a and V1b receptors: From molecules to physiological systems. *Physiological Reviews*, *92*(4), 1813–1864. Available from https://doi.org/10.1152/physrev.00035.2011.

Lambert, J. N., Mitchell, J. P., & Roberts, K. D. (2001). The synthesis of cyclic peptides. *Journal of the Chemical Society. Perkin Transactions*, *1*(5), 471–484. Available from https://doi.org/10.1039/b001942i.

Li, H., Aneja, R., & Chaiken, I. (2013). Click chemistry in peptide-based drug design. *Molecules (Basel, Switzerland)*, *18*(8), 9797–9817. Available from https://doi.org/10.3390/molecules18089797.

London, N., Movshovitz-Attias, D., & Schueler-Furman, O. (2010). The structural basis of peptide-protein binding strategies. *Structure (London, England: 1993)*, *18*(2), 188–199. Available from https://doi.org/10.1016/j.str.2009.11.012.

Lundquist, P., & Artursson, P. (2016). Oral absorption of peptides and nanoparticles across the human intestine: Opportunities, limitations and studies in human tissues. *Advanced Drug Delivery Reviews*, *106*, 256–276. Available from https://doi.org/10.1016/j.addr.2016.07.007.

Madsen, K., Knudsen, L. B., Agersoe, H., Nielsen, P. F., Thøgersen, H., Wilken, M., & Johansen, N. L. (2007). Structure-activity and protraction relationship of long-acting glucagon-like peptide-1 derivatives: Importance of fatty acid length, polarity, and bulkiness. *Journal of Medicinal Chemistry*, *50*(24), 6126–6132. Available from https://doi.org/10.1021/jm070861j.

Manandhar, B., & Ahn, J. M. (2015). Glucagon-like peptide-1 (GLP-1) analogs: Recent advances, new possibilities, and therapeutic implications. *Journal of Medicinal Chemistry*, *58*(3), 1020–1037. Available from https://doi.org/10.1021/jm500810s.

Mandsberg, N. K., Christfort, J. F., Kamguyan, K., Boisen, A., & Srivastava, S. K. (2020). Orally ingestible medical devices for gut engineering. *Advanced Drug Delivery Reviews*, *165–166*, 142–154. Available from https://doi.org/10.1016/j.addr.2020.05.004.

Matthews, D. M., & Adibi, S. A. (1976). Peptide absorption. *Gastroenterology*, *71*(1), 151–161. Available from https://doi.org/10.1016/S0016-5085(76)80117-5.

McKay, T. J., Phelan, A. W., & Plummer, T. H. (1979). Comparative studies on human carboxypeptidases B and N. *Archives of Biochemistry and Biophysics*, *197*(2), 487–492. Available from https://doi.org/10.1016/0003-9861(79)90271-6.

Medini, K., Harris, P. W. R., Menorca, A., Hards, K., Cook, G. M., & Brimble, M. A. (2016). Synthesis and activity of a diselenide bond mimetic of the antimicrobial protein caenopore-5. *Chemical Science*, 7(3), 2005–2010. Available from https://doi.org/10.1039/c5sc04187b.

Menacho-Melgar, R., Decker, J. S., Hennigan, J. N., & Lynch, M. D. (2019). A review of lipidation in the development of advanced protein and peptide therapeutics. *Journal of Controlled Release*, *295*, 1–12. Available from https://doi.org/10.1016/j.jconrel.2018.12.032.

Mentlein, R. (1999). Dipeptidyl-peptidase IV (CD26)-role in the inactivation of regulatory peptides. *Regulatory Peptides*, *85*(1), 9–24. Available from https://doi.org/10.1016/S0167-0115(99)00089-0.

Mentlein, R., Gallwitz, B., & Schmidt, W. E. (1993). Dipeptidyl-peptidase IV hydrolyses gastric inhibitory polypeptide, glucagon-like peptide-1(7–36)amide, peptide histidine methionine and is responsible for their degradation in human serum. *European Journal of Biochemistry*, *214*(3), 829–835. Available from https://doi.org/10.1111/j.1432-1033.1993.tb17986.x.

Metanis, N., Keinan, E., & Dawson, P. E. (2006). Synthetic seleno-glutaredoxin 3 analogues are highly reducing oxidoreductases with enhanced catalytic efficiency. *Journal of the American Chemical Society*, *128*(51), 16684–16691. Available from https://doi.org/10.1021/ja0661414.

Mitragotri, S., Burke, P. A., & Langer, R. (2014). Overcoming the challenges in administering biopharmaceuticals: Formulation and delivery strategies. *Nature Reviews. Drug Discovery*, *13*(9), 655–672. Available from https://doi.org/10.1038/nrd4363.

Moroz, E., Matoori, S., & Leroux, J. C. (2016). Oral delivery of macromolecular drugs: Where we are after almost 100 years of attempts. *Advanced Drug Delivery Reviews*, *101*, 108–121. Available from https://doi.org/10.1016/j.addr.2016.01.010.

Müller, T. D., Finan, B., Bloom, S. R., D'Alessio, D., Drucker, D. J., Flatt, P. R., Fritsche, A., Gribble, F., Grill, H. J., Habener, J. F., Holst, J. J., Langhans, W., Meier, J. J., Nauck, M. A., Perez-Tilve, D., Pocai, A., Reimann, F., Sandoval, D. A., Schwartz, T. W., & Tschöp, M. H. (2019). Glucagon-like peptide 1 (GLP-1). Molecular. *Metabolism: Clinical and Experimental*, *30*, 72–130. Available from https://doi.org/10.1016/j.molmet.2019.09.010.

Muttenthaler, M., King, G. F., Adams, D. J., & Alewood, P. F. (2021). Trends in peptide drug discovery. *Nature Reviews. Drug Discovery*. Available from https://doi.org/10.1038/s41573-020-00135-8.

Narayanan, A. S., & Anwar, R. A. (1969). The specificity of purified porcine pancreatic elastase. *The Biochemical Journal*, *114*(1), 11–17. Available from https://doi.org/10.1042/bj1140011.

Neurath, H., Walsh, K. A., & Winter, W. P. (1967). Evolution of structure and function of proteases. *Science (New York, NY)*, *158*(3809), 1638–1644. Available from https://doi.org/10.1126/science.158.3809.1638.

New, R. (2021). Oral delivery of biologics via the intestine. *Pharmaceutics*, *13*(1), 1−21. Available from https://doi.org/10.3390/pharmaceutics13010018.
Nielsen, D. S., Shepherd, N. E., Xu, W., Lucke, A. J., Stoermer, M. J., & Fairlie, D. P. (2017). Orally absorbed cyclic peptides. *Chemical Reviews*, *117*(12), 8094−8128. Available from https://doi.org/10.1021/acs.chemrev.6b00838.
Nir, Y., Paz, A., Sabo, E., & Potasman, I. (2003). Fear of injections in young adults: Prevalence and associations. *American Journal of Tropical Medicine and Hygiene*, *68*(3), 341−344. Available from https://doi.org/10.4269/ajtmh.2003.68.341.
Oswald, S. (2019). Organic anion transporting polypeptide (OATP) transporter expression, localization and function in the human intestine. *Pharmacology and Therapeutics*, *195*, 39−53. Available from https://doi.org/10.1016/j.pharmthera.2018.10.007.
Otero-Ramirez, M. E., Passioura, T., & Suga, H. (2018). Structural features and binding modes of thioether-cyclized peptide ligands. *Biomedicines*, *6*(4). Available from https://doi.org/10.3390/biomedicines6040116.
Patil, N. A., Tailhades, J., Hughes, R. A., Separovic, F., Wade, J. D., & Hossain, M. A. (2015). Cellular disulfide bond formation in bioactive peptides and proteins. *International Journal of Molecular Sciences*, *16*(1), 1791−1805. Available from https://doi.org/10.3390/ijms16011791.
Plamboeck, A., Holst, J. J., Carr, R. D., & Deacon, C. F. (2005). Neutral endopeptidase 24.11 and dipeptidyl peptidase IV are both mediators of the degradation of glucagon-like peptide 1 in the anaesthetised pig. *Diabetologia*, *48*(9), 1882−1890. Available from https://doi.org/10.1007/s00125-005-1847-7.
Pratley, R., Amod, A., Hoff, S. T., Kadowaki, T., Lingvay, I., Nauck, M., Pedersen, K. B., Saugstrup, T., & Meier, J. J. (2019). Oral semaglutide vs subcutaneous liraglutide and placebo in type 2 diabetes (PIONEER 4): a randomised, double-blind, phase 3a trial. *The Lancet*, *394*(10192), 39−50. Available from https://doi.org/10.1016/S0140-6736(19)31271-1.
Rabkin, R., & Dahl, David, C. (1993). Renal uptake and disposal of proteins and peptides. In K. L. Audus, & T. J. Raub (Eds.), *Biological barriers to protein delivery: Pharmaceutical biotechnology* (Vol. 4). Springer. Available from https://doi.org/10.1007/978-1-4615-2898-2_12.
Ramírez-Andersen, H. S., Behrens, C., Buchardt, J., Fels, J. J., Folkesson, C. G., Jianhe, C., Nørskov-Lauritsen, L., Nielsen, P. F., Reslow, M., Rischel, C., Su, J., Thygesen, P., Wiberg, C., Zhao, X., Wenjuan, X., & Johansen, N. L. (2018). Long-acting human growth hormone analogue by noncovalent albumin binding. *Bioconjugate Chemistry*, *29*(9), 3129−3143. Available from https://doi.org/10.1021/acs.bioconjchem.8b00463.
Rembratt, A., Graugaard-Jensen, C., Senderovitz, T., Norgaard, J. P., & Djurhuus, J. C. (2004). Pharmacokinetics and pharmacodynamics of desmopressin administered orally vs intravenously at daytime vs night-time in healthy men 55-70 years. *European Journal of Clinical Pharmacology*, *60*(6), 397−402. Available from https://doi.org/10.1007/s00228-004-0781-9.
Ritzel, U., Leonhardt, U., Ottleben, M., Rühmann, A., Eckart, K., Spiess, J., & Ramadori, G. (1998). A synthetic glucagon-like peptide-1 analog with improved plasma stability. *Journal of Endocrinology*, *159*(1), 93−102. Available from https://doi.org/10.1677/joe.0.1590093.
Santoprete, A., Capitò, E., Carrington, P. E., Pocai, A., Finotto, M., Langella, A., Ingallinella, P., Zytko, K., Bufali, S., Cianetti, S., Veneziano, M., Bonelli, F., Zhu, L., Monteagudo, E., Marsh, D. J., Sinharoy, R., Bianchi, E., & Pessi, A. (2011). DPP-IV-resistant, long-acting oxyntomodulin derivatives. *Journal of Peptide Science*, *17*(4), 270−280. Available from https://doi.org/10.1002/psc.1328.
Shen, H., Smith, D. E., & Brosius, F. C. (2001). Developmental expression of PEPT1 and PEPT2 in rat small intestine, colon, and kidney. *Pediatric Research*, *49*(6), 789−795. Available from https://doi.org/10.1203/00006450-200106000-00013.

Sridharan, K., Al Banna, R., Qader, A. M., & Husain, A. (2020). Evaluation of inter-patient variability in the pharmacodynamic indices of warfarin. *Expert Review of Cardiovascular Therapy*, *18*(11), 835–840. Available from https://doi.org/10.1080/14779072.2020.1814144.

Stähelin, H. F. (1996). The history of cyclosporin a (SandimmuneR) revisited: Another point of view. *Experientia*, *52*(1), 5–13. Available from https://doi.org/10.1007/BF01922409.

Stone, T. A., Cole, G. B., Nguyen, H. Q., Sharpe, S., & Deber, C. M. (2018). Influence of hydrocarbon-stapling on membrane interactions of synthetic antimicrobial peptides. *Bioorganic and Medicinal Chemistry*, *26*(6), 1189–1196. Available from https://doi.org/10.1016/j.bmc.2017.10.020.

Tamai, I. (2012). Oral drug delivery utilizing intestinal OATP transporters. *Advanced Drug Delivery Reviews*, *64*(6), 508–514. Available from https://doi.org/10.1016/j.addr.2011.07.007.

Tang, J. (2013). *Pepsin A. Handbook of proteolytic enzymes* (Vol. 1, pp. 27–35). Elsevier Ltd. Available from https://doi.org/10.1016/B978-0-12-382219-2.00003-X.

Taylor, J. W. (2002). The synthesis and study of side-chain lactam-bridged peptides. *Biopolymers - Peptide Science Section*, *66*(1), 49–75. Available from https://doi.org/10.1002/bip.10203.

Thombare, V. J., & Hutton, C. A. (2018). Bridged bicyclic peptides: Structure and function. *Peptide Science*, *110*(3). Available from https://doi.org/10.1002/pep2.24057.

Vasco, A. V., Brode, M., Méndez, Y., Valdés, O., Rivera, D. G., & Wessjohann, L. A. (2020). Synthesis of lactam-bridged and lipidated cyclo-peptides as promising anti-phytopathogenic agents. *Molecules (Basel, Switzerland)*, *25*(4). Available from https://doi.org/10.3390/molecules25040811.

Vickers, C., Hales, P., Kaushik, V., Dick, L., Gavin, J., Tang, J., Godbout, K., Parsons, T., Baronas, E., Hsieh, F., Acton, S., Patane, M., Nichols, A., & Tummino, P. (2002). Hydrolysis of biological peptides by human angiotensin-converting enzyme-related carboxypeptidase. *Journal of Biological Chemistry*, *277*(17), 14838–14843. Available from https://doi.org/10.1074/jbc.M200581200.

Watson, M. E., Scott, D., Jamieson, C., Layfield, R., & Mason, A. M. (2021). Design, synthesis and evaluation of E2-25K derived stapled peptides. *Peptide Science*, *113*(1). Available from https://doi.org/10.1002/pep2.24158.

White, A. M., de Veer, S. J., Wu, G., Harvey, P. J., Yap, K., King, G. J., Swedberg, J. E., Wang, C. K., Law, R. H. P., Durek, T., & Craik, D. J. (2020). Application and structural analysis of triazole-bridged disulfide mimetics in cyclic peptides. *Angewandte Chemie - International Edition*, *59*(28), 11273–11277. Available from https://doi.org/10.1002/anie.202003435.

Worthington, J. J., Reimann, F., & Gribble, F. M. (2018). Enteroendocrine cells-sensory sentinels of the intestinal environment and orchestrators of mucosal immunity. *Mucosal Immunology*, *11*(1), 3–20. Available from https://doi.org/10.1038/mi.2017.73.

Wu, H., & Huang, J. (2018). Optimization of protein and peptide drugs based on the mechanisms of kidney clearance. *Protein and Peptide Letters*, *25*(6), 514–521. Available from https://doi.org/10.2174/0929866525666180530122835.

Wu, Q., & Kuca, K. (2019). Metabolic pathway of cyclosporine A and its correlation with nephrotoxicity. *Current Drug Metabolism*, *20*(2), 84–90. Available from https://doi.org/10.2174/1389200219666181031113505.

Yang, N. J., & Hinner, M. J. (2015). Getting across the cell membrane: An overview for small molecules, peptides, and proteins. *Methods in Molecular Biology*, *1266*, 29–53. Available from https://doi.org/10.1007/978-1-4939-2272-7_3.

Yuen, T. Y., Brown, C. J., Xue, Y., Tan, Y. S., Ferrer Gago, F. J., Lee, X. E., Neo, J. Y., Thean, D., Kaan, H. Y. K., Partridge, A. W., Verma, C. S., Lane, D. P., &

Johannes, C. W. (2019). Stereoisomerism of stapled peptide inhibitors of the p53-Mdm2 interaction: An assessment of synthetic strategies and activity profiles. *Chemical Science*, *10*(26), 6457–6466. Available from https://doi.org/10.1039/c9sc01456j.

Zhang, R. Y., Thapa, P., Espiritu, M. J., Menon, V., & Bingham, J. P. (2018). From nature to creation: Going around in circles, the art of peptide cyclization. *Bioorganic and Medicinal Chemistry*, *26*(6), 1135–1150. Available from https://doi.org/10.1016/j.bmc.2017.11.017.

Zhao, R., Shi, P., Chen, J., Sun, S., Chen, J., Cui, J., Wu, F., Fang, G., Tian, C., Shi, J., Bierer, D., Liu, L., & Li, Y. M. (2020). Chemical synthesis and biological activity of peptides incorporating an ether bridge as a surrogate for a disulfide bond. *Chemical Science*, *11*(30), 7927–7932. Available from https://doi.org/10.1039/d0sc02374d.

Zorzi, A., Linciano, S., & Angelini, A. (2019). Non-covalent albumin-binding ligands for extending the circulating half-life of small biotherapeutics. *MedChemComm*, *10*(7), 1068–1081. Available from https://doi.org/10.1039/c9md00018f.

Zupančič, O., & Bernkop-Schnürch, A. (2017). Lipophilic peptide character – What oral barriers fear the most. *Journal of Controlled Release*, *255*, 242–257. Available from https://doi.org/10.1016/j.jconrel.2017.04.038.

CHAPTER THREE

A biopharmaceutics perspective on oral peptide developability and drug delivery

Ahil N. Ganesh, Sudhakar Garad and Manuel V. Sanchez-Felix
Novartis Institutes for BioMedical Research, Cambridge, MA, United States

3.1 Introduction

Peptides are the most prevalent endogenous signaling molecules; more than 7000 are known to exist (Fosgerau & Hoffmann, 2015). Natural peptides include hormones, neurotransmitters, growth factors, and antibacterial agents (Lin, 2009). Therapeutic peptides lie in the white space between conventional small molecule compounds and large biologics. Historically, therapeutic peptides have been defined by the US Food and Drug Administration as having an arbitrary cutoff of 50 amino acids (Carton & Strohl, 2013). Recently, Otvos et al. have stated that the limit should be expanded and that a more accurate definition of a therapeutic polypeptide is a polyamide or biopolymer (i.e., ester, thioester, or otherwise modified backbone) that can be synthesized via contemporary chemical peptide synthesis (Otvos & Wade, 2014). Most therapeutic peptides act as agonists offering high potency, high specificity, and low toxicity due to high binding affinity to their targets (Lin, 2009; Otvos & Wade, 2014). These advantages make peptides good candidates for inducing modulation of novel biologic targets where the binding sites are challenging to engage with conventional small molecules (Yang et al., 2020).

While many promising targets to hinder disease progression with therapeutic peptides exist, peptides fall outside the typical chemical space for oral bioavailability. Therapeutic peptides violate most of Lipinski's rule-of-5, which describes properties that are favorable for oral absorption of small molecules (Lipinski et al., 2001). As therapeutic peptides tend to have a large molecular size (>1000 Da) and a significant capacity for hydrogen

Oral Delivery of Therapeutic Peptides and Proteins.
DOI: https://doi.org/10.1016/B978-0-12-821061-1.00009-5

bonding, this limits absorption due to low permeability. Furthermore, the limitations require innovation in medicinal chemistry and formulation science to achieve the required oral bioavailability to progress a peptide candidate into a successful oral drug product.

The physicochemical properties of therapeutic peptides have made parenteral formulations the preferred choice for this modality. A review of approved and marketed therapeutic peptide products revealed that ~62% are administered as an injectable formulation (Ganesh et al., 2020). While oral delivery was the next most common route, it represented only 13% of products on the market. There are almost 2000 known active peptide programs in the development pipeline, from research to registration phases, highlighting the substantial interest in this modality (analysis of PharmaCircle database). Trends in routes of administration of development pipeline candidates (research to registration phases) reflect those of marketed therapies, for those projects where the route is known. Injectables account for 75% of the compounds in development, while oral is again the second most prevalent route of administration, accounting for ~9% of compounds. Injectable products are clearly still the preferred administration route for therapeutic peptides, mainly due to the fact that peptides have limited oral bioavailability. Upon employing technologies that can achieve consistent systemic bioavailability for oral peptide delivery, these distributions may shift. Innovation in both the areas of peptide design and formulation technologies will enable the full potential of oral peptide products to be realized.

Herein, we discuss oral peptide delivery from a biopharmaceutics perspective to help understand how peptides and formulations can be designed to induce flux across the intestinal barrier to achieve systemic exposure. Considerable efforts have gone into understanding developability of small molecule compounds for oral delivery. We aim to extend these considerations to peptide candidates to help guide the selection of chemical matter and formulations.

3.2 Lessons learned from small molecule developability

Deep understanding of the physicochemical and biopharmaceutics attributes early during the discovery phase is important to enable the successful development of therapeutic candidates. Developability assessments of small molecule new chemical entities have provided significant knowledge on how to best design and select compounds with properties that will enable oral absorption. The assessments have also guided the selection of appropriate formulation technologies to achieve the required exposure and bioavailability. Developability screens have been an important complement to efficacy and toxicity screens to select developable compounds. While these principles have been applied with great success to low molecular weight compounds, these same principles are very valuable to the selection of developable oral peptide candidates.

Several factors contribute to developability and appropriate studies can guide candidate selection through an iterative process. Key factors that are typically used to assess small molecule compound developability include solubility, permeability, and chemical stability (metabolic and solution). These factors are similarly translatable to understanding the development of orally bioavailable peptides. Keeping these biopharmaceutic considerations in mind, a multidisciplinary approach combining medicinal chemistry and formulation sciences can overcome the barriers precluding oral peptide bioavailability. Excellent reviews covering the developability of small molecules and appropriate screens are provided in refs (Butler & Dressman, 2010; Garad & Jain, 2015; Rosenberger et al., 2019).

3.2.1 Solubility

The oral bioavailability of a compound is significantly impacted by its solubility as only the soluble fraction is free to permeate across the intestinal barrier (see Section 3.5.1 for further discussion). Typically, the solubility of a peptide is not an issue due to peptidic linkages, however, depending on the chemistry of the amino acid side chains, peptides mainly fall into biopharmaceutics classification system (BCS) III (high solubility, low permeability) or IV (low solubility, low permeability). Solubility of a therapeutic peptide in biorelevant simulated gastrointestinal fluids is key to understanding how a compound may perform in vivo. As therapeutic peptides have low intrinsic permeability, solubility is the key parameter that can be modulated to improve flux. Due to the properties of peptides, efforts to characterize the tendency of peptides to self-assemble and aggregate under biologically relevant conditions are important (Zapadka et al., 2017). The formation of these higher-order structures can limit the utility of conventional shake flask, filtration, and ultracentrifugation-based methods for measuring solubility. Consequently, μ-flux methods are gaining

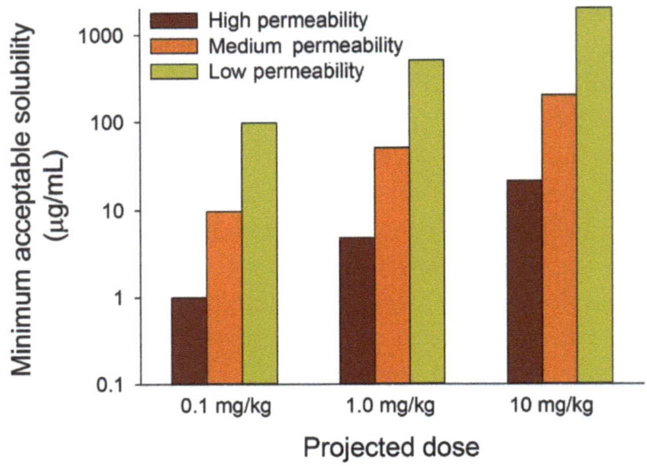

Figure 3.1 Minimum acceptable solubility for a therapeutic compound is dependent on its projected dose and permeability. For peptides, bars representing medium and low permeability compounds are of interest. *Image reproduced from Williams, H. D., et al. (2013). Strategies to address low drug solubility in discovery and development.* Pharmacological Reviews, *65(1), 315–499 with permission from American Society for Pharmacology and Experimental Therapeutics.*

traction to indicate how the specific activity of a compound influences its ability to cross a membrane (see Section 6.2.2 for further discussion). While solubility is a key parameter, if peptides with high solubility cannot be chemically designed, then formulation strategies will need to be used to increase the soluble fraction. Based on the authors' experience and published guidelines, a solubility threshold of 0.1 mg/mL can be used to classify compounds as highly or poorly soluble (Garad & Jain, 2015; Rosenberger et al., 2019). However, these threshold values will be largely based on individual company heuristics. Williams et al. highlighted that the minimum acceptable solubility for a new drug candidate depends on its permeability, a compound with low permeability must have a higher solubility than a highly permeable compound (Fig. 3.1) (Williams et al., 2013). This highlights the need to use very different solubility criteria for peptides with low or medium permeability.

3.2.2 Permeability

Permeability is the main limitation for the oral bioavailability of peptides. Most peptides have low or limited permeability (BCS III and IV) falling outside of the rule-of-5 in terms of molecular weight. Lipinski's rule-of-5 dictates that a compound with a molecular weight <500 will have medium-to-high passive permeability; however, peptides are typically >1000 Da and have little to no passive permeability resulting in oral bioavailability <3% (Tyagi et al., 2018) (Fig. 3.2). Lau and Dunn noted that the length (i.e., size) of therapeutic peptides in development has increased over the past four decades (Lau & Dunn, 2018). The average length of peptides entering clinical development in the 1980s versus 2000s was 9 and 19 amino acids, respectively. In the 1980s, 86% of peptides entering clinical development were <10 amino acids long while only 38% of peptides were shorter than 10 amino acids from 2000 to 2009. For peptides, it is also important to understand how conformational flexibility changes under different conditions and can be exploited to influence/improve permeability. Some peptides, such as the cyclic peptide cyclosporine, have been shown to adopt conformations that expose their hydrophobic groups that facilitate permeation across membranes (discussed further in the following sections).

3.2.3 Stability

Metabolism, both presystemic and systemic, is another key limitation of translating peptides for oral delivery. Presystemic metabolism is discussed

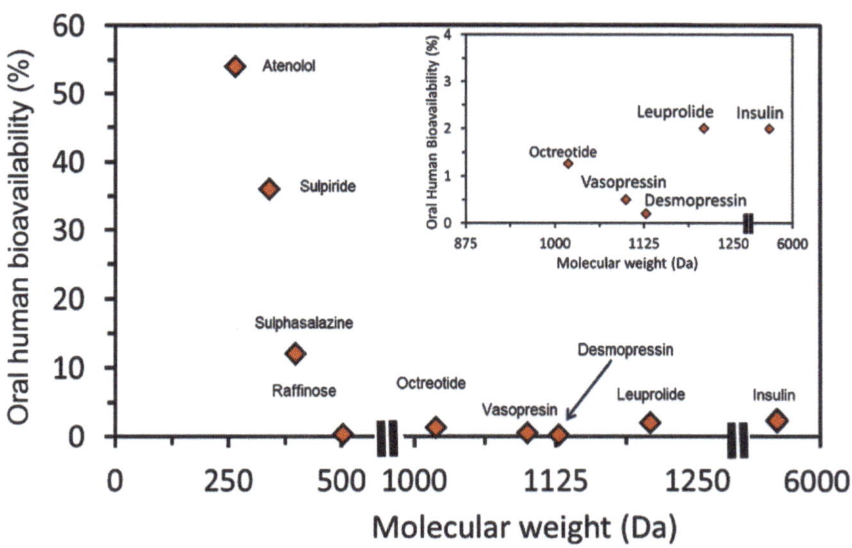

Figure 3.2 Oral bioavailability of small molecules and peptides as a function of molecular weight. Peptides consistently have oral bioavailability <3%. *Image reproduced from Tyagi, P., et al. (2018). Oral peptide delivery: Translational challenges due to physiological effects.* Journal of Controlled Release, 287, *167–176 with permission from Elsevier.*

in more detail in Section 3.1. The gastrointestinal tract (GIT) and plasma contain many proteases that degrade peptides. Understanding which protease(s) will contribute the most toward the degradation of the therapeutic peptide candidate can allow for mitigation of these effects either through chemical or formulation design. Studies have shown that the N-terminus of a peptide can be predictive of its plasma half-life; amino acids M, S, A, T, V, or G at the N-terminus are preferred over P, L, D, K, or R (Di, 2015). While formulation development may protect peptides from degradation in the GIT, stability against systemic degradation must be chemically designed into the peptide itself.

In addition to metabolism, the chemical stability of a therapeutic peptide should also be characterized in both the solid state and in solution. The characterization will be critical to formulate and manufacture the peptide into a drug product. A 20% degradation threshold can be used to indicate acceptable stability under various conditions; this is in line with FDA guidance that requires bioanalytical method accuracy to be within 20% of the nominal concentrations (FDA, F.a.D.A., 2018).

3.2.4 Ionization

Ionization state has a significant impact on both solubility and permeability, which will dictate oral bioavailability. Due to the diversity of side chains, peptides have many ionizable groups that must be considered and balanced to achieve the required properties. As the pH changes over the GIT, many of the amino acid side chains have pKa values that dictate the ionization state. The acidic side chain of aspartic acid and glutamic acid have pKa values of ~3.8 and ~4.2, respectively. The basic side chain of histidine has a pKa value ~6. In addition to the C- and N-terminus, these side chains confer ionic or zwitterionic nature to peptides under physiological conditions. The isoelectric point of peptides is an important property as solubility will be the lowest at this pH.

While ionization can increase solubility, it greatly reduces passive permeability across cell membranes. Interactions between molecules that possess multiple charges and endogenous ions in the local environment are complex processes to study, yet are poorly understood. The presence of charges can affect both chemical stability of the peptide and also physical stability with respect to aggregation. The combination of in silico ionization profile prediction and experimental determination can be used to select peptides with ionization profiles that will be most beneficial for absorption.

3.2.5 Inter- and intramolecular interactions

Due to the diversity of amino acid side chains, many opportunities exist for inter- and intramolecular interactions with peptides. The chemical diversity of peptides can be both a hindrance and an advantage; these groups can lead to many interactions due to charges, dipoles, van de Waal's forces, and hydrogen bonding (Wang et al., 2016). Intramolecular interactions can be important to maintain the peptide in an active conformation for target engagement. Peptides have many groups that can participate in hydrogen bonding as both donors and acceptors, often exceeding the limits set by the rule-of-5. Additionally, aromatic side chains can participate in π-π stacking interactions. Self-assembly of peptides into micelles or fibrillar structures is not uncommon and has even been observed with endogenous peptides (Sluzky et al., 1991; Thirumalai et al., 2003; Zapadka et al., 2017). The therapeutic octapeptide lanreotide has been shown to form nanotubes through interactions between its aromatic and aliphatic side chains (Valery et al., 2003). It is important to recognize where these intermolecular interactions may be a liability and design peptides to mitigate their effects. For example, cyclic peptides can have a propensity for planar stacking; however, tuning amino acid stereochemistry can reduce this likelihood (Ghadiri et al., 1993).

Intramolecular interactions can be used to shield polar groups to facilitate increased permeability (Lewis et al., 2015; Vorherr, 2015). The cyclic peptide cyclosporine has been shown to adopt different conformations depending on its environment (el Tayar et al., 1993). Intramolecular hydrogen bonding causes the peptide to adopt a conformation where its *N*-methyl groups are directed outwards in apolar environments; this facilitates the permeability of cyclosporine through cell membranes (Wang et al., 2018). However, when cyclosporine is bound to its protein target cyclophilin in aqueous environments, the internal hydrogen-bond interactions are broken and many of the polar side chains of cyclosporine are outwardly exposed. Careful design of a peptide can leverage these inter- and intramolecular interactions to obtain the required properties for solubility, stability, and permeability.

3.3 Barriers to oral peptide delivery

Bioavailability of orally administered therapeutic peptides is limited by many physiological barriers. The barriers can be biochemical in nature, leading to degradation of the therapeutic peptide cellular barriers that limit absorption, and/or even physicochemical barriers that limit the ability of the compound to remain in a soluble form for absorption. While some form of these barriers exists even for parenterally administered therapeutic peptides, there are key aspects that must be overcome to enable oral peptide delivery through either peptide or formulation design.

3.3.1 Biochemical barriers

As peptides do not have tertiary structures, biochemical factors such as pH have a reduced impact on their chemical stability relative to therapeutic proteins. However, pH is still an important factor in many degradation pathways such as oxidation and deamidation (Table 3.1) (Manning et al., 2010). Basit and colleagues studied the stability of 17 therapeutic peptides in gastric environments in vitro and found these peptides to be stable in

Table 3.1 Key degradation mechanisms of peptides during oral delivery. Cleavage site of proteases indicated by "/."

Degradation mechanism		**Amino acid liabilities**
Proteolytic cleavage	*Endopeptidases*	
	Trypsin	R/X and K/X
	α-Chymotrypsin	Y/X, W/X, F/X
	Pancreatic elastase	A/X,
	Exopeptidases	
	Carboxypeptidase A (C-terminus)	X/Y, except if Y = D, E, K, R, or P
	Carboxypeptidase B (C-terminus)	X/R and X/K
	Dipeptylpeptidase IV (N-terminus)	XP/Y or XA/Y
Deimination	Peptidyl arginine deiminases	R conversion to citrulline
Deamidation	pH	N and Q side chains
Oxidation	pH	C, H, M, W, and Y
Disulfide exchange with endogenous thiols	Exchange with endogenous thiols	C

both simulated gastric fluid (SGF) and simulated intestinal fluid (SIF) underlining the minimal impact of pH on peptide stability (Wang et al., 2015). Dimerization can be problematic in peptides where acidic environments can lead to disulfide exchange resulting in bridges between non-unique peptide molecules (Avanti et al., 2012).

Enzymatic degradation is the most signification biochemical barrier that needs to be overcome to enable oral peptide delivery. Due to the lack of tertiary structure, exposure of the amide bonds in peptides makes them more sensitive to degradation by peptidases (Wang et al., 2015). Exposure to endogenous enzymes of the GIT can lead to significant peptide degradation (Fjellestadpaulsen et al., 1995). These enzymes are localized to specific segments of the GI tract; pepsin and cathepsins are present in the stomach while chymotrypsin, carboxypeptidases, and pancreatic elastase are secreted by the pancreas into the intestinal lumen (Vlieghe et al., 2010). The brush border membrane of intestinal epithelial cells contains dipeptidyl peptidase IV, angiotensin-converting enzyme, and aminopeptidase A. These enzymes play important roles in the digestion of dietary intake and maintaining gastrointestinal pathophysiology, but must be mitigated to enable efficient delivery of peptides orally (Antalis et al., 2007). These enzymes have very specific cleavage sites, which can be liabilities if present in therapeutic peptides.

Basit et al. further investigated the stability of peptides in both simulated GI fluids with enzymes and fluids sourced from humans or pigs (Wang et al., 2015). Biorelevant concentrations of pepsin and pancreatin were added to SGF and SIF, respectively. Compared to stability in enzyme-free fluids, where minimal degradation was observed, significant degradation was observed for most peptides in fluids that contained biorelevant enzyme concentrations. Of the 17 peptides studies, 4 were completely degraded in human gastric fluid and 12 were completed degraded in human intestinal fluid; 2 out of 17 and 11 out of 17 were completely degraded in SGF and SIF, respectively. This study, among others, highlights the importance of understanding the susceptibility of a therapeutic peptide candidate to degradation by enzymes of the gastrointestinal tract.

The mucosal barrier lining the GIT provides a complex protective barrier against pathogens and acts as a lubricant during peristaltic movements to protect the epithelial layer (Cone, 2009; Johansson & Hansson, 2016). Mucosal layer thickness varies across the GI tract and is composed of numerous endogenous and exogenous components including mucins, glycoproteins, lipids, antibodies, and microflora (Atuma et al., 2001). Penetration through this mucosal layer is necessary before absorption. Factors such as size, charge, and lipophilicity of the compounds will dictate the ease in permeation (Khanvilkar et al., 2001). Strong interactions with the mucosal layer will result in mucoadhesive properties without achieving mucopenetration. While many studies have investigated the transport of small molecule therapeutics through mucosal layers (Khanvilkar et al., 2001; Larhed et al., 1998; Lock et al., 2018), only a few studies have investigated the diffusion of therapeutic peptides specifically (Larhed et al., 1997). There has been an ongoing debate regarding whether mucoadhesive or mucopenetrant drug delivery encapsulation strategies offer the greatest benefit for oral drug absorption. A recent review by Brayden et al. highlights literature focusing on mucopenetrant strategies as well as the limitations of previous mucoadhesive particle strategies due to the turnover of the intestinal mucosal layer (Brayden et al., 2020).

3.3.2 Cellular barrier

Permeability through a cellular barrier is required for systemic absorption after oral delivery. Transit across this barrier can be transcellular or paracellular (Fig. 3.3) (Brown et al., 2020). For transcellular absorption, compounds must permeate across the enterocytes of the intestinal epithelium (Artursson & Karlsson, 1991; Artursson et al., 2001). This permeation process is often overly simplified, as compounds must first enter the enterocytes, diffuse through the cell, and then exit into the basolateral compartment. Typically, paracellular transport is usually impeded via the presence of intercellular junction proteins (e.g., tight junctions, adherens junctions, and desmosomes) (Groschwitz & Hogan, 2009). Estimates for pore size of intestinal epithelium suggest a range of 5−15 Å, precluding transport of most molecules paracellularly (Linnankoski et al., 2010; Schwartz et al., 1995). Additionally, many receptors are luminally expressed to facilitate the transport of nutrients such as amino acids, lipids, and carbohydrates, and there are a few transporters that may facilitate the absorption of some therapeutic peptides (e.g., PEPT1 and PEPT2) (Giacomini et al., 2010; Leibach & Ganapathy, 1996; Terada & Inui, 2004). The exact route of absorption through this cellular barrier depends not only on the chemical properties of the peptide, but also on its physical state (i.e., aggregates) in the intestinal environment.

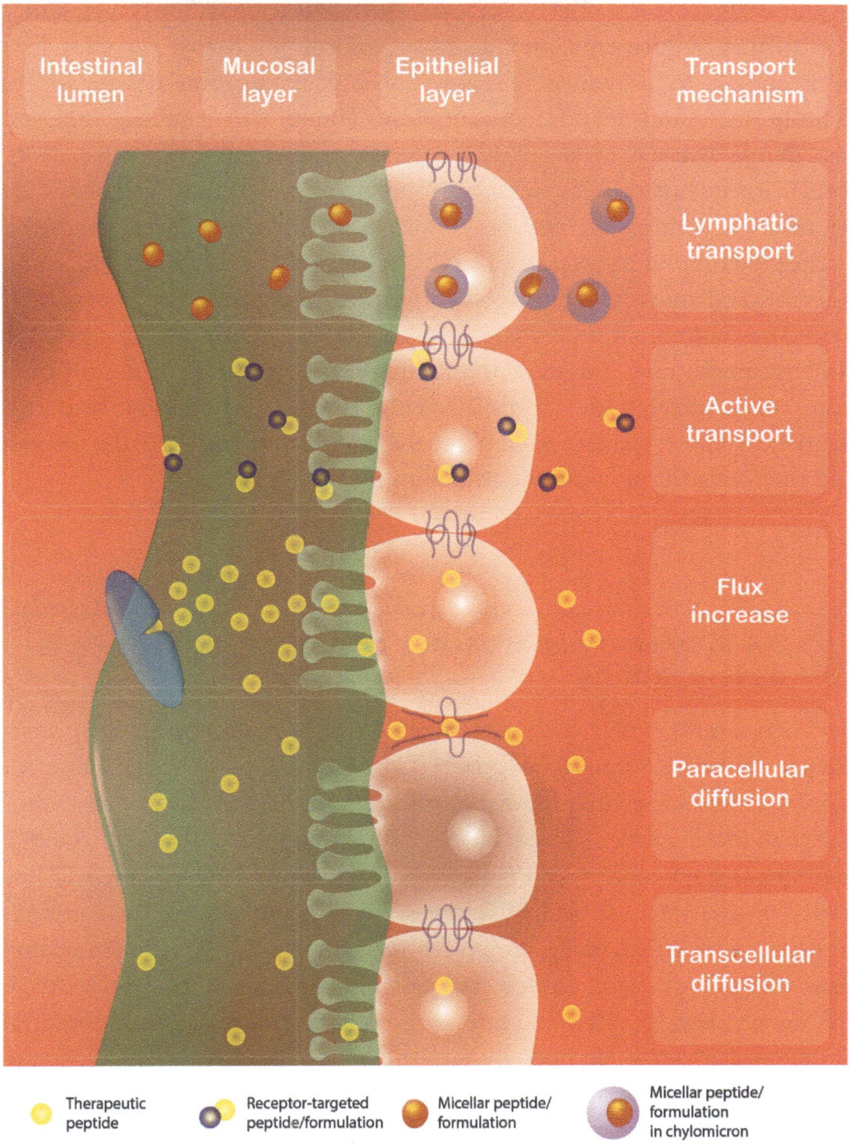

Figure 3.3 Therapeutic peptides must permeate across the intestinal epithelial barrier to achieve oral exposure. Many mechanisms have been proposed to overcome limitations due to peptide permeability.

3.3.3 Physicochemical properties of therapeutic peptides

To achieve systemic exposure, the formulation must maintain the therapeutic peptide in an appropriate physical state (i.e., fully dissolved) across all the different conditions before absorption, as aggregation/precipitation will further reduce flux and decrease absorption. Many variables have been shown to influence self-assembly and aggregation of therapeutic peptides including peptide concentration, amino acid sequence, pH, and net charge (Marek et al., 2012; Zapadka et al., 2016; Zapadka et al., 2017). Excipients such as surfactants, buffering agents, and osmolytes can be used to control aggregation. However, it is important to note that excipient-compound interactions will be dictated by physiological environments and must be maintained to prevent aggregation or precipitation. The GI environment itself changes in composition during transit; pH, bile content, microflora, enzymes, and fluid volume all change as the formulation transits through the GI tract (Kararli, 1995). It is important to understand how these changes will affect the interaction between the peptide and the formulation components (Kinnunen & Mrsny, 2014).

3.4 Chemistry approaches to oral peptide developability

As described previously, the chemistry of peptides renders them unfavorable for oral absorption and many chemical liabilities for degradation and bioconversion exist. Indeed, many endogenous peptides have very short half-lives (on the order of minutes) such as glucagon-like peptide-1 (GLP-1) and somatostatin (Gilbert & Pratley, 2020; Rai et al., 2015). Short half-lives limit the ability to use natural forms of these peptides even for parenteral administration. Additionally, amino acid side chain groups confer many attributes that provide significant resistance to the permeability of peptides across cell membranes, let alone the intestinal epithelial barrier. Natural evolution of peptides has provided inspiration on how the chemical design of peptides might facilitate oral bioavailability. The gold standard for oral peptide bioavailability is the naturally occurring cyclic peptide cyclosporine. While the exact contributing factors and mechanism resulting in the oral bioavailability of cyclosporine are poorly understood, its features have provided medicinal chemists with inspiration for the design of orally bioavailable cyclic peptides. This section highlights key medicinal chemistry strategies that have been used to improve peptide biochemical stability and permeability while maintaining the structure-activity relationship (SAR) that is required for potency and selectivity. Many examples exist in the literature of systematically applying these strategies to model peptides; however, limited systematic studies with peptides intended to be used as therapeutics are available in the literature. Examples of therapeutic peptides that are chemically modified analogs of endogenous peptides are listed in Table 3.2.

3.4.1 Amino acid sequence

Selection of the amino acid sequence is the first step toward improving the developability of a therapeutic peptide. Many different strategies are used to identify the minimal essential sequence that is needed to mimic the effects of an endogenous peptide. Methods such as truncation, deletion, and combinatorial deletion can be used to reduce the number of liabilities by reducing the number of amino acids. Alanine-scanning

Table 3.2 Summary of chemical modifications of endogenous peptides to produce therapeutic peptides with improved properties.

Therapeutic peptide	Endogenous peptide	Chemical modification
Semaglutide	Glucagon-like peptide 1	• 16-carbon fatty acid conjugated to Lys^{26} (γ-glutamic acid spacer) • Arginine replaces Lys^{34}
Liraglutide	Glucagon-like peptide 1	• 18-Carbon fatty acid conjugated to Lys^{26} (γ-glutamic acid and two 8-amino-3, 6-dioxaoctanoic acid) • α-Aminobutyric acid replaces Ala^{8}
Tesamorelin	Growth hormone-releasing factor	• *Trans*-3-hexenoic acid conjugated to N-terminal tyrosine
Buserelin	Luteinizing hormone-releasing hormone	• D-Serine(*t*-butyl) replaces Gly^{6} • Ethylamide replaces glycinamide at position 10
Octreotide	Somatostatin	• Truncated octapeptide • C-terminal threonine and N-terminal D-phenylalanine • D-Tryptophan replaces L-tryptophan in β-turn
Lanreotide	Somatostatin	• Truncated octapeptide • C-terminal threonine and N-terminal 2-napthyl-D-alanine • Valine, D-tryptophan, and tyrosine replace threonine, L-tryptophan, and phenylalanine in β-turn, respectively
Desmopressin	Vasopressin	• Deamination of Cys^{1} • D-Arg replaces Arg^{8}

methodologies are used to identify which functional groups are essential for target engagement (Fosgerau & Hoffmann, 2015). As described earlier, peptidases often have very specific cleavage sites that depend on amino acid sequence and selection of the peptide sequence should avoid these known liabilities (Werle & Bernkop-Schnurch, 2006).

3.4.2 Peptide cyclization

Cyclization of therapeutic peptides draws inspiration from nature to recapitulate the evolution of naturally occurring stable cyclic peptides (Bock et al., 2013). Examples such as cyclosporine and somatostatin have inspired the field of cyclic therapeutic peptide synthesis. Many different strategies can be employed to convert a linear peptide into a cyclic peptide, either through N-terminus-C-terminus cyclization, side-chain-to-side-chain cyclization, or side-chain-to-terminus cyclization (Table 3.3) (Nielsen et al., 2017; Vorherr, 2015).

Cyclization produces a number of benefits toward improving oral bioavailability of peptides. Importantly, cyclization confers stability against pH- and protease-mediated degradation due to the locked conformation of the backbone. Steric hindrance due to the more rigid backbone can protect the cyclic peptide's active site from endopeptidases, even if the cyclic peptide has the same amino acid sequence as a linear peptide with a known protease liability. Cyclization that involves the terminal amino acids can also protect peptides from exopeptidases. Rigidity of the backbone can be further increased through bicyclic arrangements (Tugyi et al., 2005). Many examples of conversion of linear peptides to cyclic analogs have shown maintenance of potency with improved stability against degradation (Nielsen et al., 2017).

In some cases, cyclization can also aid in conformational changes that facilitate permeability. As described in the previous section, the cyclic peptide cyclosporine has a conformation where its *N*-methylated amino acids are outwardly exposed in organic solvents (el Tayar et al., 1993; Yudin, 2015); this conformational difference is hypothesized to facilitate its membrane permeability, though the exact mechanism is still poorly understood (Kwon & Kodadek, 2007). However, some studies have shown that flexibility is not a requirement for cyclic peptides. Vorherr et al. studied a highly permeable cyclic hexapeptide that adopted the same conformation in both polar and nonpolar environments (Lewis et al., 2015). In this case, cyclization reduced the overall solvent-accessible polar surface area sufficiently enough to permit the permeability of this peptide.

Table 3.3 Examples of different chemistries that can be used to cyclize peptides. R represents the side chain of a single amino acid and R′ represents multiple amino acids (Remesic et al., 2016).

Cyclization type	Example chemical structure
N-terminus to C-terminus cyclization	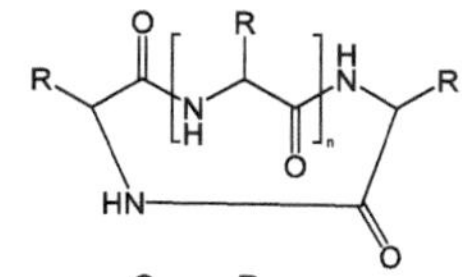
Side-chain to N-terminus cyclization	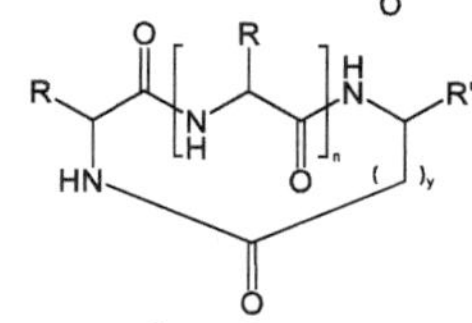
Side-chain to C-terminus cyclization	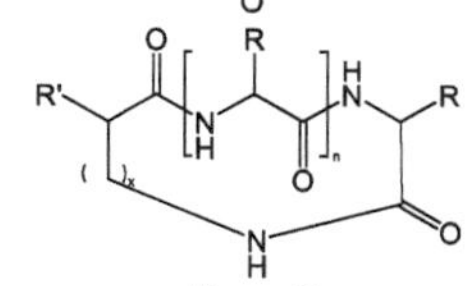
Disulfide side-chain to side-chain cyclization	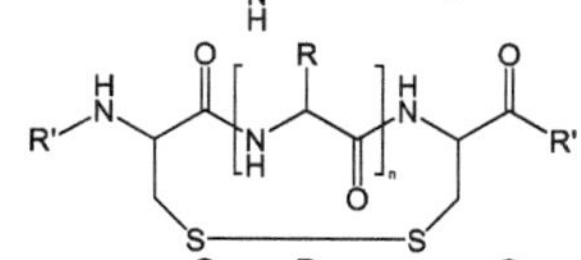
Thioether side-chain to side-chain cyclization	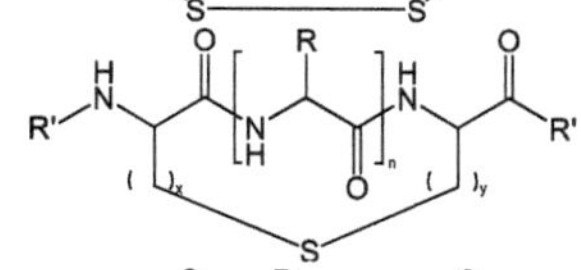
Lactam side-chain to side-chain cyclization	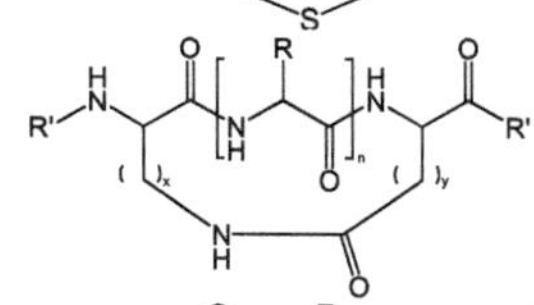
Olefin side-chain to side-chain cyclization	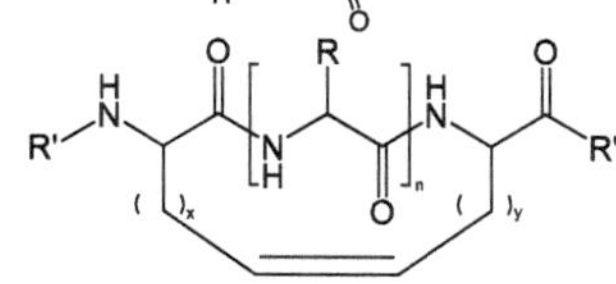

Source: Adapted from Remesic, M., et al. (2016). Cyclic opioid peptides. Current Medicinal Chemistry, 23(13), 1288–1303.

3.4.3 Hydrophobic modifications

Hydrophobic modifications can improve both stability and permeability of therapeutic peptides. A number of different approaches can be used to introduce hydrophobic modification including *N*-methylation, hydrocarbon stapling, and fatty acid conjugation.

N-methylation is the most common hydrophobic modification for therapeutic peptides. While there are many *N*-methylated natural peptides, inspiration for using this modification specifically for oral peptides is again drawn from cyclosporine, which has seven *N*-methyl groups. *N*-methylation has several benefits: it can reduce the number of hydrogen bonding sites and increase overall lipophilicity aiding in permeability. Vorherr et al. demonstrated how hydrophobic shielding due to *N*-methylation could enhance the permeability of a model hexapeptide (Tyr-Leu-D-Leu-Leu-Leu-D-Pro) (Vorherr et al., 2018). Again, they found that an appropriate selection of an *N*-methylation pattern that balances hydrophobic shielding and solubility was required to achieve oral bioavailability.

Another chemistry approach to improve peptide stability is hydrocarbon stapling, where a side-chain-to-side-chain or side-chain-to-terminus olefin bridge provides conformational rigidity to the peptide. This strategy is particularly useful for long peptides to achieve similar benefits to the cyclization of short peptides. Enfuvirtide is a 36 amino acid peptide (marketed as Fuzeon, Genentech) that lacks oral bioavailability and has poor in vivo stability. Bird et al. synthesized analogs of enfuvirtide that were stabilized by hydrocarbon stapling; these modifications increased plasma and GI stability (Bird et al., 2010). Hydrocarbon stapling was also used by Walensky et al. to maintain α-helix conformation of a linear peptide that improved stability against proteases and cell permeability (Walensky et al., 2004).

Albumin-binding strategies have been employed for therapeutics ranging from small molecules to proteins to achieve half-life extension (Kontermann, 2011; Kratz, 2008; Sleep et al., 2013; Zorzi et al., 2019). Conjugation to fatty acids promotes affinity for and binding to albumin, which enables therapeutic peptides to hitch-hike with the protein, extending their circulation time (Abdallah et al., 2020). Semaglutide and liraglutide are fatty acid conjugates designed to improve resistance against peptidases and increase half-life through albumin binding (Knudsen & Lau, 2019). While these strategies have been mostly explored in the context of parenterally administered therapeutics (Bech et al., 2018; Pollaro & Heinis, 2010), an oral version of semaglutide is now marketed. Employing a similar strategy would be of interest if consistent and higher bioavailability of these compounds can be achieved. It is important to keep in mind that modification of peptides with fatty acids will both increase their size, further limiting passive permeability, and decrease their solubility.

3.4.4 Nonnatural amino acids

Nonnatural amino acid variants have been successfully employed to improve the stability of therapeutic peptides. The use of D-amino acids increases stability against natural enzymes which only recognize L-amino acids (Garton et al., 2018; Miller et al., 1995; Rabideau & Pentelute, 2015). D-scanning at the early stages of sequence identification can be used to determine where stereochemistry might be changed without impacting target binding (Di, 2015). Even replacing a single L-amino acid with a D-amino acid can lead to significant improvements to stability. Desmopressin is a D-arginine analog of vasopressin and has a plasma half-life of 158 min versus 24 min for vasopressin (Sharman & Low, 2008). Mirror-image phage display can be used to identify active D-amino acid peptides. In this method, a phage display library of L-amino acid peptides is screened against a target protein that has been synthesized in D-amino acid form. The resulting peptide hits can then be synthesized in D-amino acid form and will have the same binding properties for endogenous targets (which are in L-amino acid form) (Schumacher et al., 1996; Sun et al., 2012).

Using amino acids with modified side chains can also impart improved stability. Buserelin is an analog of luteinizing hormone-releasing hormone (LH-RH) with a *t*-butyl-modified D-serine replacing glycine-6 and ethylamide replacing glycinamide in position 10 (Presant et al., 1985). These modifications confer both a longer duration of effect and greater pharmacodynamic effect compared to the natural LH-RH. Additionally, as liabilities are known when specific amino acids occur at the N- or C-terminus, N-acetylation or C-amidation modifications can be used to protect these peptides. Tesamorelin is an analog of growth hormone-releasing factor that has only a *trans*-3-hexenoyl modification at the N-terminal tyrosine; this single modification significantly increases plasma half-life (Ferdinandi et al., 2007).

3.4.5 Examples of chemically modified peptide analogs

i Somatostatin analogs

Due to the very short plasma half-life of somatostatin, several different analogs have been studied to turn this endogenous peptide into a therapeutic peptide (Fig. 3.4) (Anthony & Freda, 2009). Octreotide is one such analog that is used clinically and has a number of chemical modifications that enable its development (Katz & Erstad, 1989; Wolin, 2012). While somatostatin has 14 amino acids, octreotide is an octapeptide containing a

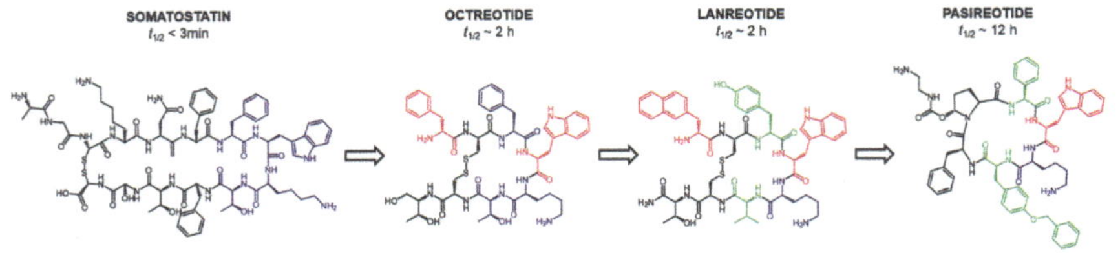

Figure 3.4 Somatostatin and therapeutic peptide analogs. Amino acids in blue indicate β-turn pharmacophores of somatostatin, amino acids in pink indicate D-isomers, and amino acids in green indicate substitutions. Plasma half-life is shown for each peptide (Golor et al. 2012). *Adapted from Golor, G., et al. (2012). A first-in-man study to evaluate the safety, tolerability, and pharmacokinetics of pasireotide (SOM230), a multireceptor-targeted somatostatin analog, in healthy volunteers.* Drug Design Development and Therapy, 6, *71–79 and Rai, U., et al. (2015). Therapeutic uses of somatostatin and its analogues: Current view and potential applications.* Pharmacology & Therapeutics, 152, *98–110.*

D-tryptophan, which is known to reduce cleavage by endopeptidases in endogenous somatostatin L-tryptophan. Octreotide also has a substituted phenylalanine for D-phenylalanine at the N-terminus and a reduced threonine to threoninol at the C-terminus. Together these modifications significantly improve the stability of octreotide conferring it with a much longer plasma half-life than somatostatin, 2 h versus <3 min, respectively. However, an oral octreotide formulation has only recently been approved using permeation enhancing technology by Chiasma (Chiasma, 2020).

ii GLP-1 analogs

GLP analogs are being intensely studied to produce therapeutic peptides where differentiation is required (Fig. 3.5) (Manandhar & Ahn, 2015; Yu et al., 2018). Endogenous GLP-1 itself is a potent regulator of postprandial hyperglycemia; however, it has a very short in vivo half-life on the order of minutes following degradation by dipeptidyl peptidase 4 (DPP4, cleavage site between Ala^{8} and Glu^{9}). The neutral endopeptidase (NEP) also cleaves GLP-1 at eight different sites (Deacon, 2004). The many clinically approved GLP-1 analogs include modifications that retain potency by preserving key amino acids (His^{7}, Gly^{10}, Phe^{12}, Thr^{13}, Asp^{15}, Phe^{28}, and Ile^{29}). Substitution to the other amino acids reduces degradation by DPP4 and NEP, as seen in exenatide. Fatty acid conjugation in liraglutide and semaglutide promotes albumin binding and extended plasma half-lives of these compounds. Similar to somatostatin analogs, all of these peptides were first approved as parenteral formulations. Only recently has an oral formulation of semaglutide been approved (Rybelsus, NovoNordisk)

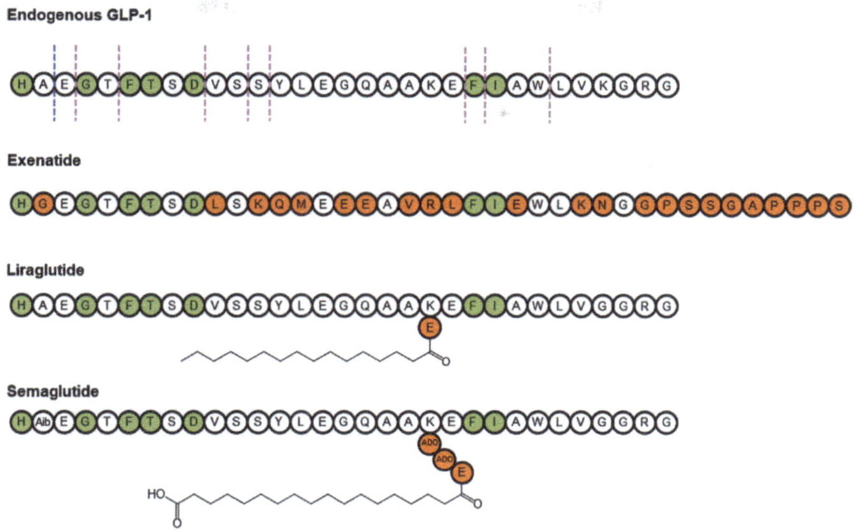

Figure 3.5 Amino acid sequences of endogenous GLP-1 and key analogs. Key amino acids for binding and potency are indicated in green, and modified amino acids are indicated in orange. Cleavage sites of endogenous GLP-1 by DPP4 and NEP are indicated by blue and purple dashed lines, respectively.

using permeation enhancement technology from Emisphere Technologies (discussed in subsequent sections).

3.5 Formulation approaches for peptide developability

3.5.1 Modulation of flux

Passive absorption of therapeutic agents in the GIT is a diffusion-controlled process governed by Fick's first law for diffusional flux. For time-independent linear concentration profiles, Fick's first law is expressed as follows:

$$J = P\Delta C$$

where J is the flux across a membrane, P is the permeability coefficient, and ΔC is the concentration gradient (i.e., concentration difference between donor and receiver compartments). Hence, flux across the intestinal epithelial barrier is dependent on the permeability of the compound and the concentration difference across the barrier. As discussed above, Lipinski's rule-of-5 provides key insights into properties favorable for the permeability of small molecules. Peptides are beyond-rule-of-5 compounds and show intrinsically low-to-moderate permeability with the chemical modifications described previously, thus modulation of concentration gradients provides the simplest approach to increase flux, and ultimately oral absorption.

Exploiting Fick's law to promote transport across the intestinal epithelium was originally proposed by Robinson et al. in 1987 (Gruber et al., 1987); this involves increasing the soluble peptide fraction in the luminal compartment so concentration gradients can drive the transport across the membrane by transcellular and paracellular routes (Fig. 3.3). It is important to note that the relationship only exists for fully dissolved peptides, so maintenance of the therapeutic in a free state is of utmost importance to increase flux, which can be challenging for peptides that fall in BCS class IV.

Peptides belong to BCS class III or IV, depending on their solubility. The ideal case for an oral therapeutic is a class I compound which has high solubility and high permeability. A significant fraction, if not all, of the administered dose is readily absorbed for class I compounds (Fig. 3.6A). However, as peptides fall under class III or IV, other strategies must be employed to increase flux and absorption. For therapeutic peptides that fall in class III (i.e., high solubility, but low permeability), to achieve the same absorbed dose as a class I compound ($F_{a,1}$), the peptide dose must be increased (D_2) so that concentration gradients can drive flux (Fig. 3.6B). Note, this is only true if the peptide remains fully soluble at the higher dose. For peptides that fall in class IV (i.e., low solubility and low permeability), there is a change in the rate of absorption (i.e., slope) once the solubility of the compound has been reached (Fig. 3.6C). In this case, when a dose (D_3) is administered, the fraction absorbed $F_{a,2}$ can be increased to $F_{a,1}$ by increasing the free fraction of the peptide via formulation to increase the solubility of the peptide.

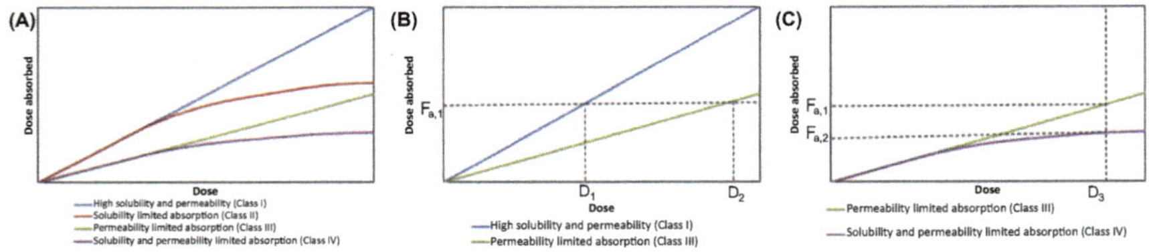

Figure 3.6 Absorption versus dose relationship for compounds with properties representative of BCS classes I–IV.

Liquid volumes in the GIT and the amount consumed to administer the oral dosage form are important variables that impact flux and thus oral bioavailability. Fluid volume and retention time vary along the GIT and are dependent on fed or fasted states (Schiller et al., 2005). Some studies have indicated that water can exist as discrete pockets along the GI tract. Mudie et al. measured intestinal volume following a 240 mL dose of water in fasted humans and found that while the water volume in the small intestines was on average 94 mL, it existed as 15 discrete pockets of 6 mL each (Mudie et al., 2014). The range in total volume in the small intestines was 15–264 mL; this wide range and existence of water pockets can introduce significant variability to flux when exploiting local concentrations. In addition to volumes, food can also greatly impact the solution state of the therapeutic peptide. Dietary components such as lipids can aid in solubilization of the compounds, increasing the overall soluble fraction and flux, or they can act as sinks, sequestering the compound in bile micelles, leading to a decrease in the overall free fraction and decreasing flux (Akiyama et al., 2020; Stappaerts et al., 2014).

All of the enabling oral peptide delivery strategies attempt to manage the inherent variability in flux due to liquid volume variations. In the case of the Emisphere Technology, strict guidelines are provided to the patients regarding when and how much water is consumed in a fasted state (i.e., no coffee or milk) to control variability. In the case of other technologies, enteric coatings are used to avoid variability in water consumption and increase flux by releasing the peptide in the duodenum where there is less liquid than in the stomach.

3.5.2 Enteric coatings

Enteric-coated capsules, tablets, or beads are used for a variety of oral formulations and often in combination with one or more additional oral strategy (ies) discussed in the following sections. Enteric coatings are often used to protect therapeutic compounds with limited stability in the gastric environment; though they are also used to protect the gastric mucosa from irritation caused by the therapeutic (Agyilirah, 1991; Maderuelo et al., 2019). In the context of therapeutic peptides, enteric coatings protect against degradation by enzymes found in the gastric milieu (see Table 3.4, e.g., of clinical stage oral peptide delivery strategies that use enteric coating). Enteric coatings also provide the added advantage of delivering the therapeutic to the small or large intestine where lower liquid volumes are present that lead to higher

Table 3.4 Examples of clinical stage drug delivery technologies for oral peptide delivery.

Drug Delivery Technology	Peptides	Mechanisms						Latest clinical phase	Dose	Bioavailability	Patient centricity comments
		Permeability enhancer	pH modulation	Enteric coating	Flux	Protease inhibitor	Device				
Eligen (Emisphere Technology)	Semaglutide (NovoNordisk)	Y	Y		Y			Marketed	3 mg, 7 mg, 14 mg once daily	• 1.22% (beagle) • 0.4%–1% (humans)	Requires dosing 30 min start of day, before any food/liquid intake with no more than 4 oz. of water only
POD (Oramed Pharma)	insulin	Y		Y	Y	Y		Phase 2b	8–24 mg daily		
Peptelligence (Enteris Biopharma)	Leuprolide, octreotide, difelikefalin	Y		Y	Y	Y		Phase 3	Leuprolide: 4 mg QD or BID, 10 mg BIDDifelikefalin: 0.25–1 mg QD		

Transient Permeability Enhance (Chiasma)	Octreotide	Y	Y		Approved	20 mg twice daily		
RaniPill (Rani Technologies)	Octreotide		Y	Y	Phase 1	100 μg	70% (<1−2 mg total)	Large solid unit that may not be suitable for patients with swallowing issues

Source: Reproduced from Ganesh, A.N., Heusser, C., Garad, S., Sanchez-Felix, M.V. (2020) Patient-centric design for peptide delivery: Trends in routes of administration and advancement in drug delivery technologies. Medicine in Drug Discovery with permission from Elsevier.

concentration of the drug, thus increasing flux and apparent permeability. Enteric coatings use polymers that remain insoluble in the acidic pH of the stomach and become soluble when the pH increases in the intestines. For example, Eudragits are copolymers of methylmethacrylate and ethylacrylate with methacrylic acid that have carboxylic acid groups that ionize at the pH range of 5–7 (Thakral et al., 2013). Next generation coatings can also take advantage of the gut microflora to achieve colon-specific delivery. The Phlora technology comprises two stimuli-responsive coatings: an outer pH-responsive polymer and an inner polysaccharide layer that are degraded by enzymes released by colon-specific bacteria (Varum et al., 2020a, 2020b). These types of coatings provide finer control over the location of drug release and reduce variability due to the requirement of multiple stimuli.

3.5.3 Protease inhibitors

Codelivery of therapeutic peptides with protease inhibitors can provide protection against degradation in the intestinal lumen, allowing for greater absorption by increasing flux. These types of formulations are typically designed when there is a specific peptidase liability identified. Some examples of the peptidase inhibitors include soybean trypsin inhibitor, aprotinin, and leupeptin (Liu et al., 2003; Tozaki et al., 1997; Yamamoto et al., 1994). The use of peptidase inhibitors must balance potency and safety to achieve the required increase in therapeutic peptide stability without any other adverse effects.

Peptidases, as with all enzymes, have optimal pH ranges which correspond to the local environment in which they are active. Coformulation with excipients that can modulate this local pH can reduce activity of these enzymes and prevent peptide degradation. For example, the activity of pancreatic enzymes is reduced in acidic pH environments; citric acid has been employed to reduce the local intestinal pH to prevent the degradation of calcitonin. This strategy has been shown to increase absorption in a beagle model (Lee et al., 1999). Furthermore, in a Phase 3 study, an oral calcitonin formulation using Peptelligence technology (Enteris Biopharma) with citric acid was superior to a nasal spray and placebo control (Binkley et al., 2012). Protease inhibitors ultimately improve oral bioavailability by increasing flux; reduction in peptide degradation improves the free fraction and concentration gradients can drive flux across the intestinal membrane.

3.5.4 Permeation enhancers

As described previously, peptides have inherently low permeability across cell membranes. Permeation enhancers aim to decrease the resistance provided by the intestinal epithelial barrier to improve flux. This can be achieved through chemical agents that transiently disrupt integrity of the cellular barrier. Chelators can be used to sequester calcium ions which are required for the tight junctions, thereby reducing the resistance to paracellular transport of compounds. Intermolecular interactions with complexing agents that modify the properties of the peptide can be used to promote transcellular permeation; for example, dipole–dipole interactions between semaglutide and *N*-[8-(2-hydroxybenzoyl)amino] caprylate are thought to contribute to the oral bioavailability of the peptide (Granhall et al., 2019). These strategies have been reviewed in depth elsewhere (Brayden et al., 2020).

The Peptelligence platform developed by Enteris Biopharma combines permeation enhancement with enteric coating and enzyme inhibition. It consists of a permeation enhancer (most commonly, an acyl carnitine; lauroyl-L-carnitine), an organic acid enzyme inhibitor and permeation enhancer (citric acid), and, as needed, antioxidants (to prevent methionine and tryptophan oxidation) in an enteric-coated tablet or capsule (Stern & Carl, 2013). Enteris has demonstrated the ability to increase the absorption of a number of therapeutics molecules with this technology (Fig. 3.7). There is a strong dependence on molecular weight with the technology, with large peptides (>2000 Da) still demonstrating <5% bioavailability, as observed in Fig. 3.2, highlighting all the other technologies. Fig. 3.7 also highlights that Enteris has a thorough mechanistic understanding of their formulation platform across a wide range of small molecules and peptides, which is critical for successful translation to the clinic. Furthermore, Enteris has demonstrated the ability to convert multiple simple prototype formulations into solid dosage forms that can be successfully developed and stable under ambient conditions.

3.5.5 Lipid-based formulations

Lipid-based formulations offer numerous advantageous for oral delivery of therapeutic peptides. They are able to overcome many solubility and aggregation limitations of peptides that fall in BCS class IV. Microemulsion generally consists of a preconcentrate that contains lipophilic, hydrophilic, and surfactant excipients that can be used to formulate compounds with similar

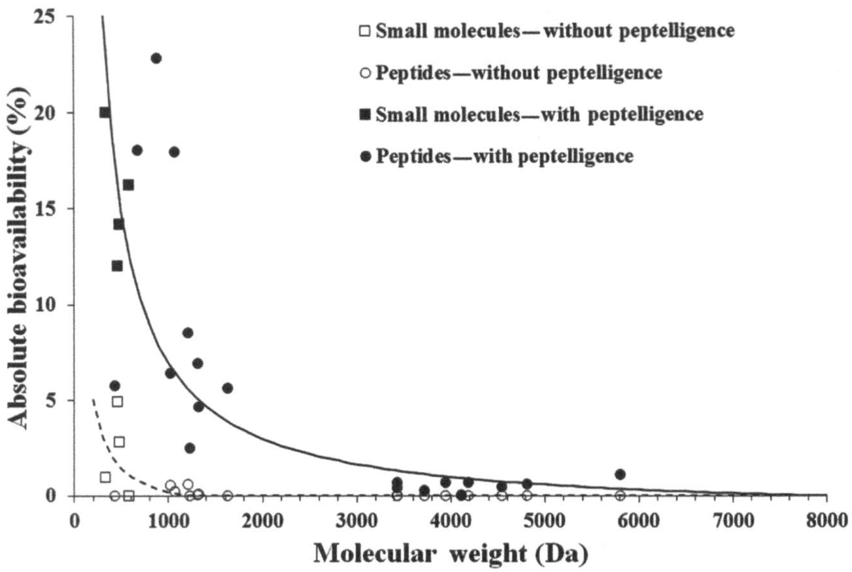

Figure 3.7 Peptelligence technology increases oral bioavailability for small molecules and peptides, yet greatly depends on molecular weight of the therapeutic. *Image kindly provided by Enteris Biopharma.*

properties. Some of these lipid formulations form micoremulsions in the GIT and are known to be taken up via a lymphatic absorption mechanism (Porter et al., 2007; Sun et al., 2011; Trevaskis et al., 2008). The composition of these microemulsions can have a significant impact on the extent of absorption via a lymphatic mechanism (Sun et al., 2011).

Oral lymphatic absorption is an endogenous mechanism for absorption of dietary fats and oils. The lymphatic transport mechanism requires the association of the therapeutic or formulation with chylomicrons, a specific type of lipoprotein, inside the enterocytes (Fig. 3.3 and Fig. 3.8). When chylomicrons are released from enterocytes this leads to absorption via the mesenteric lymphatic circulation to the systemic system versus typical hepatobiliary transport to the portal vein, which leads to transport through the liver before systemic circulation. This process bypasses the liver; however, the flow through lymphatic vessels is ~500-fold slower than vascular vessels. The literature contains a few examples of compounds and formulations that take advantage of intestinal lymphatics to achieve high bioavailability (Trevaskis et al., 2008). Lymphatic absorption of lipid formulations is a three-step mechanism: (1) bile association with the lipid formulation particles, (2) receptor-mediated internalization of the particles,

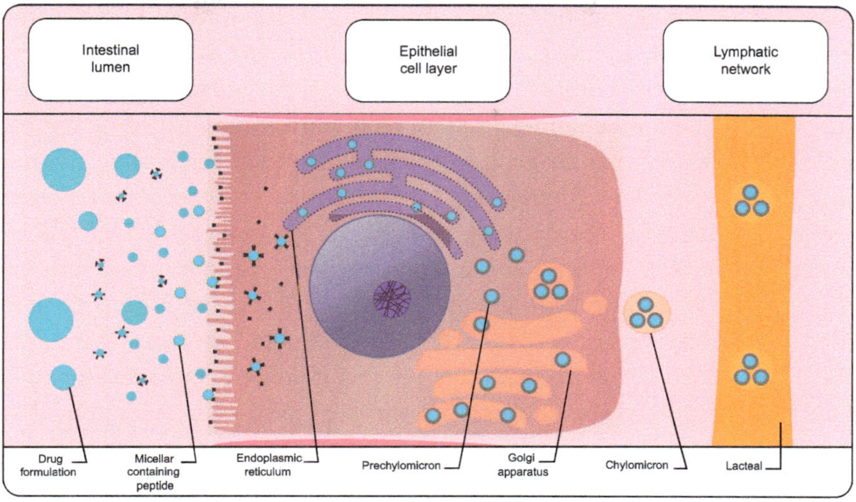

Figure 3.8 Steps involved in oral lymphatic absorption.

and (3) chylomicron association for transport out of cells into systemic lymphatic circulation. This adds additional complexity to the formulation design process and evaluation. Dietary fats and/or lipid-based formulations are required to stimulate chylomicron production to take advantage of a lymphatic absorption mechanism.

Some examples of marketed and clinical candidates that are known to be absorbed via lymphatic mechanisms using lipid-based formulations include testosterone undecanoate (Andriol) (Shackleford et al., 2003) and navitoclax (Choo et al., 2014; Yang et al., 2011). Exploiting the lymphatic mechanism enables high oral bioavailability for beyond rule-of-5 compounds. However, in the same way, components in the lipid formulation stimulate chylomicron production, and fats and oils in food elicit a similar effect (Bagchus et al., 2003; Wilson et al., 2010). This can lead to positive food effects that have been observed due to an increased production of bile and transfer of fats/oils/lipids to form chylomicrons, thus increasing absorption.

3.5.6 Novel transport-mediated uptake

While the intestinal epithelium expresses several transporters on the luminal surface, only a few strategies exploit specific transporters to achieve oral bioavailability. These processes specifically take advantage of natural transporters expressed on the surface of enterocytes for the transport of substrates including peptides, bile salts, and other nutrients (Estudante et al., 2013). One example of a potential transporter relevant for therapeutic peptides is the PEPT1 transporter which can influx di- and tripeptides from the intestinal lumen. Additionally, direct conjugation to dietary molecules such as vitamin B12 has been explored preclinically, but not progressed to the clinic (Petrus et al., 2009). Mrsny et al. recently reported a strategy utilizing a translocation domain of the cholera-derived cholix protein to enable apical-to-basal transport of a conjugated protein (Taverner et al., 2020). Enrollment for a Phase 2 trial has recently begun using the cholix platform for delivery of a biologic in ulcerative colitis (Transport, 2020). It is unclear whether this strategy is intended for only local delivery or systemic delivery, but this technology has the potential to be adapted for utilization with peptides.

Conjugation of bile acids to nanoparticles has achieved high oral bioavailability through lymphatic mechanisms described previously (Kim et al., 2018). Bae et al. used bile acid-modified liposomes to enhance bioavailability of exenatide ($\sim$20%) and other poorly permeable compounds by leveraging intestinal lymphatic absorption (Suzuki et al., 2019). The mechanism warrants further investigation in the context of therapeutic peptides as they fall outside the rule-of-5 and can be formulated with lipids. Currently, many targeted delivery strategies focus on transport into the cell, but lack mechanistic understanding of the more important transport out of the cell to achieve systemic exposure.

3.6 In silico and in vitro models

In silico and in vitro methods can be powerful tools to understand peptide and formulation properties that influence oral bioavailability. These methods can direct SAR changes to the molecule that enhance solubility, permeability, formulation selection, and optimization, including understanding of limitations (Hendriksen et al., 2003; Lewis et al., 2015).

3.6.1 In silico methods

In silico modeling tools can be leveraged during many stages of development to guide the selection of properties to achieve oral bioavailability. At the early stages of peptide design, in silico tools can be used to predict solubility, ionization profiles, lipophilicity, polar surface area, conformation, interactions with bilayer membranes, and aggregation propensity (Fuchs et al., 2018; Lewis et al., 2015; Wang et al., 2018; Zapadka et al., 2017). Due to the lack of tertiary structures in peptides, algorithms that predict aggregation perform better for peptides than proteins, though experimental validation is still important. Software that can predict the cleavage of peptide candidates by proteases can be used to screen and select sequences that could offer improved stability against enzymatic degradation (Radchenko et al., 2018; Song et al., 2019). While these methods have significantly improved over recent decades, the complexity of modeling large peptides can still be a limitation.

Physiologically based pharmacokinetic modeling is a powerful tool to understand absorption, distribution, metabolism, and excretion properties of new therapeutics. In combination with in vitro data (i.e., solubility, permeability, metabolic stability, etc.), software such as GastroPlus and Simcyp can be used to understand how formulation parameters influence oral bioavailability; these in silico models can aid in the iterative design of peptides and formulations without the need for large quantities of materials or costly in vivo studies (Chow et al., 2016; Daga et al., 2018). Parameter sensitivity analyses can identify factors that will most influence in vivo performance and highlight where resources should be focused to overcome limitations. While these methods have been successfully applied to conventional small molecules, input variables such as aggregation, regional stability (solution and protease metabolism), and transportation are not taken into account, limiting the generation of useful models for therapeutic peptides.

3.6.2 Characterization of peptide conformation and aggregation

Understanding of the conformational behavior of peptides is critical for the successful design of peptides that are permeable, stable, and soluble. Two applicable techniques include chromatography and nuclear magnetic resonance (NMR).

Many chromatographic methods can be used to understand the conformational properties of therapeutic peptides (Mant et al., 2007). Chromatographic methods can provide information about the overall characteristics of a peptide, including hydrophobic/hydrophilic nature, secondary structures, and ionization and aggregation state in different conditions. Recent advances in supercritical fluid chromatography have allowed for even deeper insight into peptide hydrophobicity and conformational flexibility (Goetz et al., 2014). Measurements of the solvent-accessible polar surface area by supercritical fluid chromatography can be used to understand how conformational changes might influence permeability, among other properties (Goetz et al., 2014; Vorherr et al., 2018). This can allow for the screening of multiple peptides with different modifications intended to promote permeability and identify those that will provide improvements. This method was used by Vorherr et al. to study several different *N*-methylated cyclic peptides and compare performance in in vitro permeability assays and in vivo bioavailability (Vorherr et al., 2018).

Similar to chromatographic methods, NMR can be used to characterize a range of properties pertaining to therapeutic peptides. Many studies have used NMR to investigate the conformational flexibility of cyclosporine in different solvent systems to understand how they influence permeability (el Tayar et al., 1993; Wang et al., 2018). Ideally, these studies compare peptide conformation in chloroform and water to identify differences between apolar and polar conditions; dimethylsulfoxide (DMSO) can be used as a substitute polar solvent for BCS class III peptides that have insufficient solubility in water. NMR measurement of interproton distances and coupling constants can be used to compare conformational differences between different peptide modifications (Lewis et al., 2015).

In addition to conformation, aggregation of peptides will greatly affect their in vivo performance. NMR methods can also be used to study aggregation states due to the broadening of peaks associated with slower tumbling of aggregate species (LaPlante et al., 2013). These experiments can identify which regions and interactions contribute to the aggregation of a given peptide (Fernandez et al., 2004; Wadhwani et al., 2008). Tayler dispersion analysis has been used to study peptide aggregation; by measuring changes to the diffusion coefficient and subvisible and suspended peptide aggregates that fall below the detection limits of some other techniques (Hogstedt et al., 2016; Lewis et al., 2015). This technique can be performed with small amounts of material and does not require any dilution that may result in deaggregation.

3.6.3 In vitro stability assessment

As biochemical degradation is a key limiting factor for oral peptide delivery, it is important to understand the solution and metabolic stability of peptides. Screening different peptide sequences, modifications, and formulations for their ability to reduce peptide degradation in vitro can provide insight on how these strategies may perform in vivo. Simulated intestinal fluids can be used to understand solution hydrolysis stability under the different buffer and pH conditions of the GI tract. Once solution stability has been confirmed, physiologically relevant concentrations of enzymes such as pepsin and pancreatin should be included to further evaluate presystemic degradation of the peptide. Basit et al. used in vitro stability experiments to understand how different components of aspirated and simulated GI fluids influence therapeutic peptide stability (Wang et al., 2015).

In addition to studying presystemic degradation, systemic metabolism of peptides should also be evaluated. Enzymes in the plasma and liver can metabolize many exogenous therapeutics limiting their systemic effects. In vitro assays studying the stability of peptides against these will enable the selection of candidates that are metabolically stable. Metabolism is highly species-dependent so screening against enzymes from both preclinical species and humans is crucial to understand the translation of a therapeutic from preclinical to clinical. Subcellular fractions of liver homogenates (S9 and microsome) can simulate simultaneous metabolism by many different enzymes. Metabolic stability against single enzymes can also be informative, particularly when specific metabolic liabilities are known. Using in vitro microsomal stability assays, Ma et al. were able to identify modifications that significantly decreased the metabolism of an α-aminoxy peptide (Ma et al., 2012). In addition to screening a series of modified cyclic peptides in liver microsome assays, Vorherr et al. investigated the metabolism of these peptides by individual CYP enzymes (Vorherr et al., 2019). By identifying metabolic hotspots, the authors were able to design analogs that had significantly increased oral bioavailability.

3.6.4 In vitro permeability and flux measurements

As permeation across the intestinal epithelial layer is a key limitation for peptides, in vitro assays to study this process are important in molecule and formulation design. Many in vitro assays exist for characterizing the permeability of therapeutic molecules; these assays have been developed to mimic key attributes of the intestinal epithelium while trying to reduce assay variability and build a calibration curve with known human data (Artursson & Karlsson, 1991; Artursson et al., 2001; Volpe, 2011). Immortalized human colorectal adenocarcinoma (Caco-2) or Madin-Darby canine kidney (MDCK) cell monolayers and the parallel artificial membrane permeability assay (PAMPA) are the most common in vitro methods used to assess the passive permeability of therapeutic compounds. While there is a good understanding of how the results of these assays can be correlated to human effective permeability, these assays still have many limitations. The PAMPA membrane contains no cellular components, while current cellular monolayers have differential expression of membrane transporters compared to the native intestinal epithelium. Additionally, current cellular monolayer assays cannot distinguish between paracellular and transcellular transport (Sun et al., 2008).

Flux is the major driving force that can be modulated to enable the oral delivery of peptides. As described in previous sections, flux is dependent on both the soluble fraction of the compound in the local environment and the permeability of the compound. In vitro flux assays can be used to study both of these aspects simultaneously. As therapeutic peptides can be in either BCS III or IV, understanding the solubility in biorelevant media will inform the requirements of the formulation. These assays can also incorporate additional processes that may occur in the GIT, such as lipolysis (Keemink & Bergstrom, 2018; Keemink et al., 2019). Common formulation excipients used to influence solubility and permeability may not always be tolerated by the cellular or artificial membranes used in these in vitro assays, leading to erroneous results. The use of in vitro tools can only be powerful if they are used in conjunction with each other and with a deep understanding of their limitations (Fig. 3.9, discussed further in the following).

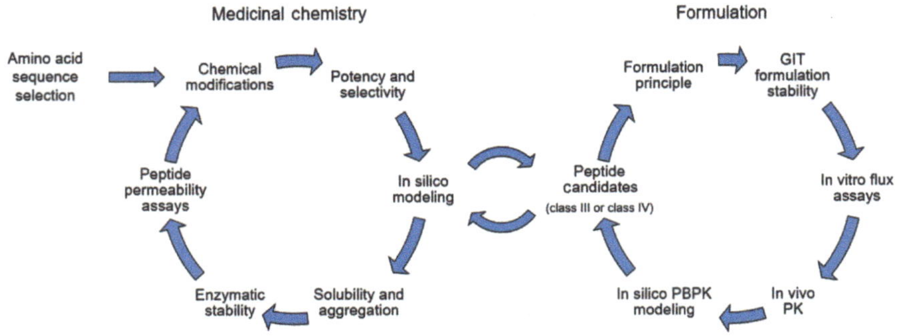

Figure 3.9 Flow scheme of therapeutic peptide design, formulation principle selection, and evaluation. Success criteria for each stage will be dictated by heuristic rules based on historical data.

3.7 Outlook

3.7.1 Reevaluating patient preferences

In therapeutic areas and indications where competition is very high, significant efforts are required to differentiate products and gain market access. This is often achieved by developing products to be more convenient for a patient, either through long-acting technologies to reduce the frequency of dosing or by converting a parenteral product into an oral product. As described previously, parenteral administration is still the most common route for peptide products on the market and in development. Many long-acting technologies exist for parenteral products that can increase patient preference due to the low dosing frequency (Ganesh et al., 2020).

The development of GLP-1 analogs is a prime example of such efforts (Yu et al., 2018). Design of analogs with extended half-lives allowed for a reduction in dosing frequency; Exenatide (Bayetta), first approved in 2005 requires twice-daily administration subcutaneously, but liraglutide (Victoza, 2010) and semglutide (Ozempic, 2017) require once daily and once-weekly dosing, respectively. Convenience of less frequent dosing plays an important role in patient adoption and market share; the market share value of Ozempic (once weekly) has been steadily increasing over the past year, while the Victoza (once daily) market share has been steadily decreasing over the same timeframe (NovoNordisk, 2020). Similar decreases to dosing frequency could be equally advantageous for oral peptide products. Additionally, strict restrictions on the quantity of water that needs to be consumed (to reduce concentration differences that cause flux variability) further reduce convenience and patient adherence; the current oral semaglutide (Rybelsus) label requires patients to take the dose at the start of the day, 30 min before any food, and with a specific quantity of only water (i.e., no coffee, milk, etc.) (Rybelsus, 2017). In addition, the GLP-1 analogs show on-target gastrointestinal adverse events, irrespective of administration route, that will further decrease the acceptance and adherence of an oral product (Bettge et al., 2017).

To be transformative, oral delivery technologies must be able to significantly increase bioavailability, reduce variability, decrease the frequency of dosing, and have no food effects to provide the greatest convenience for patients. While this set of requirements is a tall order to achieve, efforts should be guided by providing transformative medicines to patients. Oral delivery technologies capable of achieving weekly dosing, similar to current subcutaneous products while also achieving high bioavailability could satisfy these requirements. Some gastric-retentive technologies are already in development to provide weekly dosing (Altreuter et al., 2018). As highlighted, these technologies will still need to overcome the low and variable bioavailability of oral peptide delivery.

3.7.2 Inter- and intraspecies variability

Inter- and intraspecies variability remains a key challenge in drug development as products are progressed through preclinical studies to clinical trials. Translation across species is a problem for all delivery routes due to differences in anatomy and physiology leading to variability in bioavailability (Collins et al., 2020; McDonald et al., 2010). Many attributes of the GIT vary across preclinical species and humans, including factors such as gastric and intestinal volumes, retention times, pH, and bile content (Tyagi et al., 2018). For example, dogs have variable gastric pH in the fasted state, while pigs retain solid dosage forms in the stomach, and the use of enteric coatings is not feasible in rodent models (Davis et al., 2001; Papich & Martinez, 2015; Sagawa et al., 2009; Sanchez-Felix, 2019). These sources of variability can be challenging when efficacy and toxicity of peptides need to be evaluated in preclinical species. These factors can greatly influence the bioavailability of oral peptides, especially when flux is being exploited to increase absorption. Differences in metabolic enzymes and capacity further dictate the systemic fate of peptide therapeutics (Lin, 1995; Lin, 2009).

Intra-species variability is also an important consideration, especially in clinical settings where patient-to-patient variability of physiology and comorbidities can influence the performance of the formulation. Many studies have observed wide ranges in GI fluid volumes (Steingoetter et al., 2006; Yamashita et al., 2013). Mudie et al. found that the small intestine volumes in 12 healthy volunteers were 5−159 mL in fasted state and 15−264 mL after consumption of 250 mL of water. This variability of liquid volume will affect local concentrations and will greatly impact flux across the intestinal barrier. Intake of fats and oils from food also influences the variability of parameters that dictate flux (Boulby et al., 1997; Koziolek et al., 2014). Several studies have established the food effects on oral absorption, especially for class III and class IV compounds including peptides (Custodio et al., 2008; Fleisher et al., 1999; Riedmaier et al., 2020). As described previously, volume of water and food intake must be carefully controlled for the oral semaglutide formulation where dilution could diminish the intended effects of formulation components to increase flux.

3.7.3 Rationale design of a successful oral peptide product

Development of oral peptide therapeutics will require a collaborative effort between medicinal chemists and formulation scientists. Medicinal chemistry approaches have the greatest potential to address current limitations for oral peptide delivery, namely chemical stability and permeability. However, this is not always possible, and thus remaining liabilities will need to be overcome through formulation and delivery strategies. For example, following the work flow in Fig. 3.9, for a desired amino acid sequence, medicinal chemistry should attempt to produce a highly soluble, stable, and permeable peptide through iterative in vitro and in silico evaluation. If high permeability but not high solubility is able to be achieved, then a formulation principle that is able to increase the solubility of the peptide (e.g., lipid-based formulation) can be selected and further evaluated. Intrinsically, peptides can have either low or moderate permeability. Increasing permeability through medicinal chemistry efforts will be most impactful as even the best formulations currently only achieve $<5\%$ bioavailability for poorly permeable peptides with molecular weight >2000 Da, subsequently requiring the peptide to be very potent. On the other hand, with a moderately permeable peptide in combination with appropriate formulation strategies, achieving higher bioavailability may be possible and less potent peptides can be efficacious.

The assays outlined in Section 3.6 should be used to benchmark candidate peptides and formulations. Organizations will have their own heuristic rules for categorizing the assay results based on historical in vitro/in vivo correlation data and analysis of products in development or on the market. Incorporation of data into in silico models will provide insight on how individual factors and combinations of factors relating to SAR or the formulation are the most impactful and limiting in terms of achieving oral bioavailability. Together the in silico, in vitro, and in vivo assays should form a decision tree for evaluating both the therapeutic peptide and the formulation (e.g., Fig. 3.9). It is important to note that many peptides currently being evaluated clinically for oral delivery are based on products that have been previously approved for parenteral administration (e.g., octreotide and semaglutide). However, the translation is not straightforward as extensive clinical trials are required to not only show bioequivalence but also safety and efficacy studies need to be repeated because the new route of administration may lead to different distribution and metabolism. Brayden et al. have noted that all oral peptide products have required extensive Phase 3 studies to verify safety and efficacy (Brayden et al., 2020). A key challenge that remains to be fully addressed is the overall low and variable bioavailability of peptides delivered orally. Significant advances in oral peptide delivery have fostered the development of multiple oral peptides in the clinic and more recently onto the market. Critical to the continued advancement in this space is a fundamental understanding of the necessity to combine medicinal chemistry and formulation sciences as well as a clear understanding of the absorption mechanisms that will produce convenient oral products that the patient values.

Acknowledgments

The authors thank Carolyn Heusser for careful review of the manuscript. The authors thank Joshua Croteau for assistance with the preparation of this article and copyright permissions. The authors thank Olivia Sanchez-Felix and David Abraham for the generation of Figs. 3.3 and 3.8. The authors thank John Vrettos and Enteris Biopharma for the generation of Fig. 3.7.

References

Abdallah, M., Mullertz, O. O., Styles, I. K., Morsdorf, A., Quinn, J. F., Whitaker, M. R., & Trevaskis, N. L. (2020). Lymphatic targeting by albumin-hitchhiking: Applications and optimisation. *Journal of Controlled Release: Official Journal of the Controlled Release Society*, *327*, 117–128.

Agyilirah, G.A., Banker, G.S. (1991). Polymers for enteric coating applications, In Polymers for controlled drug delivery (Tarcha, P.J., ed.), CRC Press.

Akiyama, Y., Ito, S., Fujita, T., & Sugano, K. (2020). Prediction of negative food effect induced by bile micelle binding on oral absorption of hydrophilic cationic drugs. *European Journal of Pharmaceutical Sciences: Official Journal of the European Federation for Pharmaceutical Sciences*, *155*, 105543.

Altreuter, D. H., Kirtane, A. R., Grant, T., Kruger, C., Traverso, G., & Bellinger, A. M. (2018). Changing the pill: Developments toward the promise of an ultra-long-acting gastroretentive dosage form. *Expert Opinion on Drug Delivery*, *15*(12), 1189–1198.

Antalis, T. M., Shea-Donohue, T., Vogel, S. N., Sears, C., & Fasano, S. (2007). Mechanisms of disease: Protease functions in intestinal mucosal pathobiology. *Nature Clinical Practice. Gastroenterology & Hepatology*, *4*(7), 393–402.

Anthony, L., & Freda, P. U. (2009). From somatostatin to octreotide LAR: Evolution of a somatostatin analogue. *Current Medical Research and Opinion*, *25*(12), 2989–2999.

Artursson, P., & Karlsson, J. (1991). Correlation between oral-drug absorption in humans and apparent drug permeability coefficients in human intestinal epithelial (Caco-2) cells. *Biochemical and Biophysical Research Communications*, *175*(3), 880–885.

Artursson, P., Palm, K., & Luthman, K. (2001). Caco-2 monolayers in experimental and theoretical predictions of drug transport (Reprinted from Advanced Drug Delivery Reviews, 22, 67–84 (1996)). *Advanced Drug Delivery Reviews, 46*(1–3), 27–43.

Atuma, C., Strugala, V., Allen, A., & Holm, L. (2001). The adherent gastrointestinal mucus gel layer: Thickness and physical state in vivo. *American Journal of Physiology. Gastrointestinal and Liver Physiology, 280*(5), G922–G929.

Avanti, C., Permentier, H. P., van Dam, A., Poole, R., Jiskoot, W., Frijlink, H. W., & Hinrichs, W. L. J. (2012). A new strategy to stabilize oxytocin in aqueous solutions: II. Suppression of cysteine-mediated intermolecular reactions by a combination of divalent metal ions and citrate. *Molecular Pharmaceutics, 9*(3), 554–562.

Bagchus, W. M., Hust, R., Maris, F., Schnabel, P. G., & Houwing, N. S. (2003). Important effect of food on the bioavailability of oral testosterone undecanoate. *Pharmacotherapy, 23*(3), 319–325.

Bech, E. M., Pedersen, S. L., & Jensen, K. J. (2018). Chemical strategies for half-life extension of biopharmaceuticals: Lipidation and its alternatives. *ACS Medicinal Chemistry Letters, 9*(7), 577–580.

Bettge, K., Kahle, M., Abd El Aziz, M. S., Meier, J. J., & Nauck, M. A. (2017). Occurrence of nausea, vomiting and diarrhoea reported as adverse events in clinical trials studying glucagon-like peptide-1 receptor agonists: A systematic analysis of published clinical trials. *Diabetes, Obesity and Metabolism, 19*(3), 336–347.

Binkley, N., Bolognese, M., Sidorowicz-Bialynicka, A., Vally, T., Trout, R., Miller, C., Buben, C. E., Gilligan, J. P., & Krause, D. S. (2012). A phase 3 trial of the efficacy and safety of oral recombinant calcitonin: The oral calcitonin in postmenopausal osteoporosis (ORACAL) trial. *Journal of Bone and Mineral Research, 27*(8), 1821–1829.

Bird, G. H., Madani, N., Perry, A. F., Princiotto, A. M., Supko, J. G., He, X., Gavathiotis, E., Sodroski, J. G., & Walensky, L. D. (2010). Hydrocarbon double-stapling remedies the proteolytic instability of a lengthy peptide therapeutic. *Proceedings of the National Academy of Sciences of the United States of America, 107*(32), 14093–14098.

Bock, J. E., et al. (2013). Getting in shape: Controlling peptide bioactivity and bioavailability using conformational constraints. *ACS Chemical Biology, 8*(3), 488–499.

Boulby, P., et al. (1997). Use of echo planar imaging to demonstrate the effect of posture on the intragastric distribution and emptying of an oil/water meal. *Neurogastroenterology and Motility, 9*(1), 41–47.

Brayden, D. J., et al. (2020). Systemic delivery of peptides by the oral route: Formulation and medicinal chemistry approaches. *Advanced Drug Delivery Reviews.*

Brown, T. D., et al. (2020). Materials for oral delivery of proteins and peptides. *Nature Reviews Materials, 5*(2), 127–148.

Butler, J. M., & Dressman, J. B. (2010). The developability classification system: Application of biopharmaceutics concepts to formulation development. *Journal of Pharmaceutical Sciences, 99*(12), 4940–4954.

Carton, J.M. & Strohl, W.R. (2013) Chapter 4 – Protein therapeutics (introduction to biopharmaceuticals). In Introduction to biological and small molecule drug research and development (Ganellin, R. et al., eds.), pp. 127–159, Elsevier

Chiasma. (2020). Chiasma announces FDA approval of MYCAPSSA® (Octreotide) capsules, the first and only oral somatostatin analog.

Choo, E. F., et al. (2014). The role of lymphatic transport on the systemic bioavailability of the Bcl-2 protein family inhibitors navitoclax (ABT-263) and ABT-199. *Drug Metabolism and Disposition, 42*(2), 207–212.

Chow, E. C. Y., et al. (2016). Using physiologically based pharmacokinetic (PBPK), modeling to evaluate the impact of pharmaceutical excipients on oral drug absorption: Sensitivity analyses. *AAPS Journal, 18*(6), 1500–1511.

Collins, D. S., et al. (2020). Accelerating the development of novel technologies and tools for the subcutaneous delivery of biotherapeutics. *Journal of Controlled Release, 321*, 475–482.
Cone, R. A. (2009). Barrier properties of mucus. *Advanced Drug Delivery Reviews, 61*(2), 75–85.
Custodio, J. M., et al. (2008). Predicting drug disposition, absorption/elimination/transporter interplay and the role of food on drug absorption. *Advanced Drug Delivery Reviews, 60*(6), 717–733.
Daga, P. R., et al. (2018). Physiologically based pharmacokinetic modeling in lead optimization. 1. Evaluation and adaptation of GastroPlus to predict bioavailability of medchem series. *Molecular Pharmaceutics, 15*(3), 821–830.
Davis, S. S., et al. (2001). Gastrointestinal transit of dosage forms in the pig. *The Journal of Pharmacy and Pharmacology, 53*(1), 33–39.
Deacon, C. F. (2004). Circulation and degradation of GIP and GLP-1. *Hormone and Metabolic Research, 36*(11–12), 761–765.
Di, L. (2015). Strategic approaches to optimizing peptide ADME properties. *AAPS Journal, 17*(1), 134–143.
el Tayar, N., et al. (1993). Solvent-dependent conformation and hydrogen-bonding capacity of cyclosporin A: Evidence from partition coefficients and molecular dynamics simulations. *Journal of Medicinal Chemistry, 36*(24), 3757–3764.
Estudante, M., et al. (2013). Intestinal drug transporters: An overview. *Advanced Drug Delivery Reviews, 65*(10), 1340–1356.
(FDA), F.a.D.A. (2018) Bioanalytical method validation—guidance for industry.
Ferdinandi, E. S., et al. (2007). Non-clinical pharmacology and safety evaluation of TH9507, a human growth hormone-releasing factor analogue. *Basic & Clinical Pharmacology & Toxicology, 100*(1), 49–58.
Fernandez, C. O., et al. (2004). NMR of alpha-synuclein-polyamine complexes elucidates the mechanism and kinetics of induced aggregation. *Embo Journal, 23*(10), 2039–2046.
Fjellestadpaulsen, A., et al. (1995). Metabolism of vasopressin, oxytocin, and their analogs in the human gastrointestinal tract. *Peptides, 16*(6), 1141–1147.
Fleisher, D., et al. (1999). Drug, meal and formulation interactions influencing drug absorption after oral administration – Clinical implications. *Clinical Pharmacokinetics, 36*(3), 233–254.
Fosgerau, K., & Hoffmann, T. (2015). Peptide therapeutics: Current status and future directions. *Drug Discovery Today, 20*(1), 122–128.
Fuchs, J. A., et al. (2018). Lipophilicity prediction of peptides and peptide derivatives by consensus machine learning. *Medchemcomm, 9*(9), 1538–1546.
Ganesh, A.N., Heusser, C., Garad, S., Sanchez-Felix, M.V. (2020). Patient-centric design for peptide delivery: Trends in routes of administration and advancement in drug delivery technologies. *Medicine in Drug Discovery.*
Garad, S. & Jain, A. (2015). Developability assessment and risk management during drug discovery. In Discovering and Developing Molecules with Optimal Drug-Like Properties (Templeton, A.C. et al., eds.), pp. 3–47, Springer New York.
Garton, M., et al. (2018). Method to generate highly stable D-amino acid analogs of bioactive helical peptides using a mirror image of the entire PDB. *Proceedings of the National Academy of Sciences of the United States of America, 115*(7), 1505–1510.
Ghadiri, M. R., et al. (1993). Self-assembling organic nanotubes based on a cyclic peptide architecture. *Nature, 366*(6453), 324–327.
Giacomini, K. M., et al. (2010). Membrane transporters in drug development. *Nature Reviews. Drug Discovery, 9*(3), 215–236.
Gilbert, M. P., & Pratley, R. E. (2020). GLP-1 analogs and DPP-4 inhibitors in type 2 diabetes therapy: Review of head-to-head clinical trials. *Frontiers in Endocrinology, 11.*

Goetz, G. H., et al. (2014). EPSA: A novel supercritical fluid chromatography technique enabling the design of permeable cyclic peptides. *ACS Medicinal Chemistry Letters, 5* (10), 1167–1172.

Golor, G., et al. (2012). A first-in-man study to evaluate the safety, tolerability, and pharmacokinetics of pasireotide (SOM230), a multireceptor-targeted somatostatin analog, in healthy volunteers. *Drug Design Development and Therapy, 6*, 71–79.

Granhall, C., et al. (2019). Safety and pharmacokinetics of single and multiple ascending doses of the novel oral human GLP-1 analogue, oral semaglutide, in healthy subjects and subjects with type 2 diabetes. *Clinical Pharmacokinetics, 58*(6), 781–791.

Groschwitz, K. R., & Hogan, S. P. (2009). Intestinal barrier function: Molecular regulation and disease pathogenesis. *Journal of Allergy and Clinical Immunology, 124*(1), 3–20.

Gruber, P., et al. (1987). Some biological issues in oral, controlled drug delivery. *Advanced Drug Delivery Reviews, 1*(1), 1–18.

Hendriksen, B. A., et al. (2003). The composite solubility vs pH profile and its role in intestinal absorption prediction. *AAPS PharmSci, 5*(1).

Hogstedt, U. B., et al. (2016). Taylor Dispersion Analysis as a promising tool for assessment of peptide-peptide interactions. *European Journal of Pharmaceutical Sciences, 93*, 21–28.

Johansson, M. E., & Hansson, G. C. (2016). Immunological aspects of intestinal mucus and mucins. *Nature Reviews. Immunology, 16*(10), 639–649.

Kararli, T. T. (1995). Comparison of the gastrointestinal anatomy, physiology, and biochemistry of humans and commonly used laboratory-animals. *Biopharmaceutics & Drug Disposition, 16*(5), 351–380.

Katz, M. D., & Erstad, B. L. (1989). Octreotide, a new somatostatin analog. *Clinical Pharmacy, 8*(4), 255–273.

Keemink, J., & Bergstrom, C. A. S. (2018). Caco-2 cell conditions enabling studies of drug absorption from digestible lipid-based formulations. *Pharmaceutical Research, 35*, 4.

Keemink, J., et al. (2019). Lipolysis-permeation setup for simultaneous study of digestion and absorption in vitro. *Molecular Pharmaceutics, 16*(3), 921–930.

Khanvilkar, K., et al. (2001). Drug transfer through mucus. *Advanced Drug Delivery Reviews, 48*(2-3), 173–193.

Kim, K. S., et al. (2018). Oral nanoparticles exhibit specific high-efficiency intestinal uptake and lymphatic transport. *ACS Nano, 12*(9), 8893–8900.

Kinnunen, H. M., & Mrsny, R. J. (2014). Improving the outcomes of biopharmaceutical delivery via the subcutaneous route by understanding the chemical, physical and physiological properties of the subcutaneous injection site. *Journal of Controlled Release: Official Journal of the Controlled Release Society, 182*, 22–32.

Knudsen, L. B., & Lau, J. (2019). The discovery and development of liraglutide and semaglutide. *Frontiers in Endocrinology, 10*.

Kontermann, R. E. (2011). Strategies for extended serum half-life of protein therapeutics. *Current Opinion in Biotechnology, 22*(6), 868–876.

Koziolek, M., et al. (2014). Intragastric volume changes after intake of a high-caloric, high-fat standard breakfast in healthy human subjects investigated by MRI. *Molecular Pharmaceutics, 11*(5), 1632–1639.

Kratz, F. (2008). Albumin as a drug carrier: Design of prodrugs, drug conjugates and nanoparticles. *Journal of Controlled Release, 132*(3), 171–183.

Kwon, Y. U., & Kodadek, T. (2007). Quantitative comparison of the relative cell permeability of cyclic and linear peptides. *Chemistry & Biology, 14*(6), 671–677.

LaPlante, S. R., et al. (2013). Compound aggregation in drug discovery: Implementing a practical NMR assay for medicinal chemists. *Journal of Medicinal Chemistry, 56*(12), 5142–5150.

Larhed, A. W., et al. (1997). Diffusion of drugs in native and purified gastrointestinal mucus. *Journal of Pharmaceutical Sciences, 86*(6), 660–665.

Larhed, A. W., et al. (1998). The influence of intestinal mucus components on the diffusion of drugs. *Pharmaceutical Research, 15*(1), 66–71.

Lau, J. L., & Dunn, M. K. (2018). Therapeutic peptides: Historical perspectives, current development trends, and future directions. *Bioorganic & Medicinal Chemistry, 26*(10), 2700–2707.

Lee, Y. H., et al. (1999). Impact of regional intestinal pH modulation on absorption of peptide drugs: Oral absorption studies of salmon calcitonin in beagle dogs. *Pharmaceutical Research, 16*(8), 1233–1239.

Leibach, F. H., & Ganapathy, V. (1996). Peptide transporters in the intestine and the kidney. *Annual Review of Nutrition, 16*, 99–119.

Lewis, I., et al. (2015). A detailed investigation on conformation, permeability and PK properties of two related cyclohexapeptides. *International Journal of Peptide Research and Therapeutics, 21*(2), 205–221.

Lin, J. H. (1995). Species similarities and differences in pharmacokinetics. *Drug Metabolism and Disposition, 23*(10), 1008–1021.

Lin, J. H. (2009). Pharmacokinetics of biotech drugs: Peptides, proteins and monoclonal antibodies. *Current Drug Metabolism, 10*(7), 661–691.

Linnankoski, J., et al. (2010). Paracellular porosity and pore size of the human intestinal epithelium in tissue and cell culture models. *Journal of Pharmaceutical Sciences, 99*(4), 2166–2175.

Lipinski, C. A., et al. (2001). Experimental and computational approaches to estimate solubility and permeability in drug discovery and development settings (Reprinted from *Advanced Drug Delivery Reviews, 23*, 3–25 (1997)). *Advanced Drug Delivery Reviews, 46*(1–3), 3–26.

Liu, H., et al. (2003). Potential utility of various protease inhibitors for improving the intestinal absorption of insulin in rats. *Journal of Pharmacy and Pharmacology, 55*(11), 1523–1529.

Lock, J. Y., et al. (2018). Mucus models to evaluate the diffusion of drugs and particles. *Advanced Drug Delivery Reviews, 124*, 34–49.

Ma, B., et al. (2012). Effect of structural modification on the gastrointestinal stability and hepatic metabolism of alpha-aminoxy peptides. *Amino Acids, 43*(5), 2073–2085.

Maderuelo, C., et al. (2019). Enteric coating of oral solid dosage forms as a tool to improve drug bioavailability. *European Journal of Pharmaceutical Sciences, 138*.

Manandhar, B., & Ahn, J. M. (2015). Glucagon-like peptide-1 (GLP-1) analogs: Recent advances, new possibilities, and therapeutic implications. *Journal of Medicinal Chemistry, 58*(3), 1020–1037.

Manning, M. C., et al. (2010). Stability of protein pharmaceuticals: An update. *Pharmaceutical Research, 27*(4), 544–575.

Mant, C.T. et al. (2007) HPLC analysis and purification of peptides. In Peptide Characterization and Application Protocols (Fields, G.B., ed.), pp. 3–55, Humana Press

Marek, P. J., et al. (2012). Ionic strength effects on amyloid formation by amylin are a complicated interplay among debye screening, ion selectivity, and hofmeister effects. *Biochemistry, 51*(43), 8478–8490.

McDonald, T. A., et al. (2010). Subcutaneous administration of biotherapeutics: Current experience in animal models. *Current Opinion in Molecular Therapeutics, 12*(4), 461–470.

Miller, S. M., et al. (1995). Comparison of the proteolytic susceptibilities of homologous L-amino-acid, D-amino-acid, and *N*-substituted glycine peptide and peptoid oligomers. *Drug Development Research, 35*(1), 20–32.

Mudie, D. M., et al. (2014). Quantification of gastrointestinal liquid volumes and distribution following a 240 mL dose of water in the fasted state. *Molecular Pharmaceutics, 11* (9), 3039–3047.
Nielsen, D. S., et al. (2017). Orally absorbed cyclic peptides. *Chemical Reviews, 117*(12), 8094–8128.
NovoNordisk. (2020). Q2 2020 financial results.
Otvos, L., Jr., & Wade, J. D. (2014). Current challenges in peptide-based drug discovery. *Frontiers in Chemistry, 2*, 62.
Papich, M. G., & Martinez, M. N. (2015). Applying biopharmaceutical classification system (BCS) criteria to predict oral absorption of drugs in dogs: Challenges and pitfalls. *AAPS Journal, 17*(4), 948–964.
Petrus, A. K., et al. (2009). Traveling the vitamin B12 pathway: Oral delivery of protein and peptide drugs. *Angewandte Chemie (International (Ed.) in English), 48*(6), 1022–1028.
Pollaro, L., & Heinis, C. (2010). Strategies to prolong the plasma residence time of peptide drugs. *Medchemcomm, 1*(5), 319–324.
Porter, C. J. H., et al. (2007). Lipids and lipid-based formulations: Optimizing the oral delivery of lipophilic drugs. *Nature Reviews. Drug Discovery, 6*(3), 231–248.
Presant, C. A., et al. (1985). Buserelin as primary therapy in advanced prostatic-carcinoma. *Cancer, 56*(10), 2416–2419.
Rabideau, A. E., & Pentelute, B. L. (2015). A D-amino acid at the N-terminus of a protein abrogates its degradation by the N-end rule pathway. *ACS Central Science, 1*(8), 423–430.
Radchenko, T., et al. (2018). Software-aided approach to investigate peptide structure and metabolic susceptibility of amide bonds in peptide drugs based on high resolution mass spectrometry (vol 12, e0186461, 2017). *PLoS One, 13*(7).
Rai, U., et al. (2015). Therapeutic uses of somatostatin and its analogues: Current view and potential applications. *Pharmacology & Therapeutics, 152*, 98–110.
Remesic, M., et al. (2016). Cyclic opioid peptides. *Current Medicinal Chemistry, 23*(13), 1288–1303.
Riedmaier, A. E., et al. (2020). Use of physiologically based pharmacokinetic (PBPK) modeling for predicting drug-food interactions: An industry perspective. *AAPS Journal, 22*, 6.
Rosenberger, J., et al. (2019). Application of a refined developability classification system. *Journal of Pharmaceutical Sciences, 108*(3), 1090–1100.
Rybelsus. (2017). Sumaglutide tablets, for oral use. In Prescribing Information.
Sagawa, K., et al. (2009). Fed and fasted gastric pH and gastric residence time in conscious beagle dogs. *Journal of Pharmaceutical Sciences, 98*(7), 2494–2500.
Sanchez-Felix, M. V. (2019). Translation of oral delivery technologies: Challenges and understanding preclinical species. In *Controlled release society annual meeting*. Controlled Release Society.
Schiller, C., et al. (2005). Intestinal fluid volumes and transit of dosage forms as assessed by magnetic resonance imaging. *Alimentary Pharmacology & Therapeutics, 22*(10), 971–979.
Schumacher, T. N. M., et al. (1996). Identification of D-peptide ligands through mirror-image phage display. *Science (New York, NY), 271*(5257), 1854–1857.
Schwartz, R. M., et al. (1995). Paracellular intestinal transport of six-carbon sugars is negligible in the rat. *Gastroenterology, 109*(4), 1206–1213.
Shackleford, D. M., et al. (2003). Contribution of lymphatically transported testosterone undecanoate to the systemic exposure of testosterone after oral administration of two andriol formulations in conscious lymph duct-cannulated dogs. *Journal of Pharmacology and Experimental Therapeutics, 306*(3), 925–933.

Sharman, A., & Low, J. (2008). Vasopressin and its role in critical care. *Continuing Education in Anaesthesia Critical Care & Pain, 8*(4), 134–137.

Sleep, D., et al. (2013). Albumin as a versatile platform for drug half-life extension. *Biochimica Et Biophysica Acta-General Subjects, 1830*(12), 5526–5534.

Sluzky, V., et al. (1991). Kinetics of insulin aggregation in aqueous-solutions upon agitation in the presence of hydrophobic surfaces. *Proceedings of the National Academy of Sciences of the United States of America, 88*(21), 9377–9381.

Song, J. N., et al. (2019). iProt-Sub: a comprehensive package for accurately mapping and predicting protease-specific substrates and cleavage sites. *Briefings in Bioinformatics, 20* (2), 638–658.

Stappaerts, J., et al. (2014). Human and simulated intestinal fluids as solvent systems to explore food effects on intestinal solubility and permeability. *European Journal of Pharmaceutical Sciences, 63*, 178–186.

Steingoetter, A., et al. (2006). Effects of posture on the physiology of gastric emptying: A magnetic resonance imaging study. *Scandinavian Journal of Gastroenterology, 41*(10), 1155–1164.

Stern, W. M., Mehta, N., & Carl, S. (2013). Oral delivery of peptides by peptelligenceTM technology. *Drug Development and Delivery, 13*, 1–6.

Sun, H., et al. (2008). The Caco-2 cell monolayer: Usefulness and limitations. *Expert Opinion on Drug Metabolism & Toxicology, 4*(4), 395–411.

Sun, M., et al. (2011). Intestinal absorption and intestinal lymphatic transport of sirolimus from self-microemulsifying drug delivery systems assessed using the single-pass intestinal perfusion (SPIP) technique and a chylomicron flow blocking approach: Linear correlation with oral bioavailabilities in rats. *European Journal of Pharmaceutical Sciences, 43* (3), 132–140.

Sun, N., et al. (2012). Mirror image phage display—generating stable therapeutically and diagnostically active peptides with biotechnological means. *Journal of Biotechnology, 161* (2), 121–125.

Suzuki, K., et al. (2019). Long-term oral administration of Exendin-4 to control type 2 diabetes in a rat model. *Journal of Controlled Release, 294*, 259–267.

Taverner, A., et al. (2020). Cholix protein domain I functions as a carrier element for efficient apical to basal epithelial transcytosis. *Tissue Barriers, 8*(1), 1710429.

Terada, T., & Inui, K. (2004). Peptide transporters: Structure, function, regulation and application for drug delivery. *Current Drug Metabolism, 5*(1), 85–94.

Thakral, S., et al. (2013). Eudragit (R): A technology evaluation. *Expert Opinion on Drug Delivery, 10*(1), 131–149.

Thirumalai, D., et al. (2003). Emerging ideas on the molecular basis of protein and peptide aggregation. *Current Opinion in Structural Biology, 13*(2), 146–159.

Tozaki, H., et al. (1997). Degradation of insulin and calcitonin and their protection by various protease inhibitors in rat caecal contents: Implications in peptide delivery to the colon. *Journal of Pharmacy and Pharmacology, 49*(2), 164–168.

Transport, A. M. (2020). Study of the efficacy and safety of AMT-101 in subjects with ulcerative colitis (LOMBARD).

Trevaskis, N. L., et al. (2008). Lipid-based delivery systems and intestinal lymphatic drug transport: A mechanistic update. *Advanced Drug Delivery Reviews, 60*(6), 702–716.

Tugyi, R., et al. (2005). The effect of cyclization on the enzymatic degradation of herpes simplex virus glycoprotein D derived epitope peptide. *Journal of Peptide Science: an Official Publication of the European Peptide Society, 11*(10), 642–649.

Tyagi, P., et al. (2018). Oral peptide delivery: Translational challenges due to physiological effects. *Journal of Controlled Release, 287*, 167–176.

Valery, C., et al. (2003). Biomimetic organization: Octapeptide self-assembly into nanotubes of viral capsid-like dimension. *Proceedings of the National Academy of Sciences of the United States of America*, *100*(18), 10258–10262.

Varum, F., et al. (2020a). OPTICORE (TM), an innovative and accurate colonic targeting technology. *International Journal of Pharmaceutics*, *583*.

Varum, F., et al. (2020b). A dual pH and microbiota-triggered coating (Phloral (TM)) for fail-safe colonic drug release. *International Journal of Pharmaceutics*, *583*.

Vlieghe, P., et al. (2010). Synthetic therapeutic peptides: Science and market. *Drug Discovery Today*, *15*(1–2), 40–56.

Volpe, D. A. (2011). Drug-permeability and transporter assays in Caco-2 and MDCK cell lines. *Future Medicinal Chemistry*, *3*(16), 2063–2077.

Vorherr, T. (2015). Modifying peptides to enhance permeability. *Future Medicinal Chemistry*, 7(8), 1009–1021.

Vorherr, T., et al. (2018). Modulation of oral bioavailability and metabolism for closely related cyclic hexapeptides. *International Journal of Peptide Research and Therapeutics*, *24*(1), 35–48.

Vorherr, T. L. I., Wille, R., Lochmann, T., Berghausen, J., Huth, F., Schroer, K., Blanz, J., Aichholz, R., & Schaefer, M. (2019). On the importance of metabolic stability to achieve high oral exposures for cyclic peptides. *American Journal of Medicinal Chemistry*, *1*(1), 3–10.

Wadhwani, P., et al. (2008). Using a sterically restrictive amino acid as a F-19 NMR label to monitor and to control peptide aggregation in membranes. *Journal of the American Chemical Society*, *130*(49), 16515.

Walensky, L. D., et al. (2004). Activation of apoptosis in vivo by a hydrocarbon-stapled BH3 helix. *Science (New York, NY)*, *305*(5689), 1466–1470.

Wang, C. K., et al. (2018). Conformational flexibility is a determinant of permeability for cyclosporin. *Journal of Physical Chemistry B*, *122*(8), 2261–2276.

Wang, J., et al. (2015). Toward oral delivery of biopharmaceuticals: An assessment of the gastrointestinal stability of 17 peptide drugs. *Molecular Pharmaceutics*, *12*(3), 966–973.

Wang, J., et al. (2016). Peptide self-assembly: Thermodynamics and kinetics. *Chemical Society Reviews*, *45*(20), 5589–5604.

Werle, M., & Bernkop-Schnurch, A. (2006). Strategies to improve plasma half life time of peptide and protein drugs. *Amino Acids*, *30*(4), 351–367.

Williams, H. D., et al. (2013). Strategies to address low drug solubility in discovery and development. *Pharmacological Reviews*, *65*(1), 315–499.

Wilson, W. H., et al. (2010). Navitoclax, a targeted high-affinity inhibitor of BCL-2, in lymphoid malignancies: A phase 1 dose-escalation study of safety, pharmacokinetics, pharmacodynamics, and antitumour activity. *Lancet Oncology*, *11*(12), 1149–1159.

Wolin, E. M. (2012). The expanding role of somatostatin analogs in the management of neuroendocrine tumors. *Gastrointest Cancer Research*, *5*(5), 161–168.

Yamamoto, A., et al. (1994). Effects of various protease inhibitors on the intestinal-absorption and degradation of insulin in rats. *Pharmaceutical Research*, *11*(10), 1496–1500.

Yamashita, S., et al. (2013). Measurement of drug concentration in the stomach after intragastric administration of drug solution to healthy volunteers: Analysis of intragastric fluid dynamics and drug absorption. *Pharmaceutical Research*, *30*(4), 951–958.

Yang, J., et al. (2011). Abstract 1300: Evaluating the relative bioavailability of a new formulation of navitoclax (ABT-263) in both cancer patients and healthy volunteers. *Cancer Research*, *71*(8 Suppl.), 1300-1300.

Yang, W. Z., et al. (2020). The evolving druggability and developability space: Chemically modified new modalities and emerging small molecules. *AAPS Journal*, *22*(2).

Yu, M., et al. (2018). Battle of GLP-1 delivery technologies. *Advanced Drug Delivery Reviews*, *130*, 113–130.

Yudin, A. K. (2015). Macrocycles: Lessons from the distant past, recent developments, and future directions. *Chemical Science*, *6*(1), 30–49.

Zapadka, K. L., et al. (2016). A pH-induced switch in human glucagon-like peptide-1 aggregation kinetics. *Journal of the American Chemical Society*, *138*(50), 16259–16265.

Zapadka, K. L., et al. (2017). Factors affecting the physical stability (aggregation) of peptide therapeutics. *Interface Focus*, 7(6).

Zorzi, A., et al. (2019). Non-covalent albumin-binding ligands for extending the circulating half-life of small biotherapeutics. *Medchemcomm*, *10*(7), 1068–1081.

CHAPTER FOUR

Nanoparticle-assisted oral delivery of small and large peptides

Meenakshi Arora[1,2,3,4], Dianxiong Zou[1] and M.N.V. Ravi Kumar[1,2,3,4,5]
[1]College of Community Health Sciences, The University of Alabama, Tuscaloosa, AL, United States
[2]The Center for Convergent Bioscience and Medicine (CCBM), The University of Alabama, Tuscaloosa, AL, United States
[3]Department of Biological Sciences, The University of Alabama, Tuscaloosa, AL, United States
[4]Alabama Life Research Institute, The University of Alabama, Tuscaloosa, AL, United States
[5]Chemical and Biological Engineering, University of Alabama, Tuscaloosa, AL, United States

4.1 Introduction

The cost associated with bringing to market a new, FDA-approved prescription drug is estimated to be about $2.6 billion in 2013, a 135% increase from about $1.1 billion in 2003 (DiMasi et al., 2016). While there are numerous extrinsic and intrinsic factors contributing to such exorbitant price tag being the norm, perhaps the most salient is the high attrition rates inherent to drug development, where fewer than 0.1% of all bioactive compounds that have undergone preclinical testing reach phase I clinical trials, and less than 20% of these then proceed to receive regulatory approval (Hay et al., 2014; Takebe et al., 2018). Taking into account the substantial hurdles facing de novo drug development, as well as the economic and societal impacts of increasingly unaffordable treatment options, it is of little wonder that research related to the refinement or repurposing older drugs with well-understood pharmacological profiles, aided by nonconventional dosage forms, is quickly gaining appeal in the healthcare industry (Breckenridge & Jacob, 2019; Li et al., 2019; Polamreddy & Gattu, 2019; Pushpakom et al., 2019). One of the most exciting frontiers of such research is the implementation of colloidal nanoscale vehicles (e.g., nanoparticles (NPs)), which in principle, can address many of the issues that have traditionally limited a drug's therapeutic applications, such as low target tissue

Oral Delivery of Therapeutic Peptides and Proteins.
DOI: https://doi.org/10.1016/B978-0-12-821061-1.00007-1

bioavailability, high toxicity profiles, and poor release characteristics (Anselmo & Mitragotri, 2014; Davis & Illum, 1998). Much of the efforts behind enhancing drug specificity, safety and stability with NP-assisted delivery are also *à propos* to the problem of opening up new routes of administration for drugs that have heretofore been amenable to injections only. Throughout this chapter, we will focus on exploring recent advances made in the field of nanotechnology with respect to the oral delivery of peptides, which are arguably the most challenging class of compounds to administer by such route. The forthcoming examples will illustrate the promises as well as difficulties associated with the current state-of-art, and offer insights on designing the next-generation of nanoscale delivery vehicles that can be clinically feasible.

Oral delivery can be subcategorized into peroral, which utilizes the gastrointestinal (GI) tract, and transmucosal, which involves direct placement and absorption of drugs through the buccal as well as sublingual epithelia. Compared to peroral delivery, the transmucosal route is extremely fast acting. Accordingly, drugs with rapid-onset properties or those intended for localized effects have been successfully administered sublingually (Patel et al., 2011). Moreover, the buccal mucosa, which covers a significant portion of the oral cavity, offers relatively straightforward access to systemic circulation via the internal jugular vein, and consequently, drugs taken up by such route avoid not only hostile GI tract environments, but also extensive metabolism present within the enteric epithelium and liver. Despite these advantages, relatively few drugs utilize the transmucosal route; this is in large part due to the limited surface area, continuous salivary flux, and low permeability of the underlying buccal/sublingual membranes (Hao & Heng, 2003; Reddy et al., 2011). As such, the predominant route of oral drug delivery has historically been and continues to be through the GI tract. Thus, for the remainder of this chapter and for the sake of convenience, we will refer to the peroral route as simply the oral route.

The benefits associated with oral delivery are in general, unparalleled. These include medical conveniences such as increased patient adherence, ease of self-administration, negligible pain/discomfort, and reduced need for supervision. From an economic standpoint, orally dosed drugs also have little or no requirement for sterile manufacturing, and possess comparably longer shelf lives as well as more relaxed storage conditions thanks to the solid-phase dosage forms utilized in most cases (Homayun et al., 2019). These benefits combined to make oral delivery of drugs a highly attractive approach for managing a variety of diseases, particularly those that necessitate chronic medication. Fortunately, a wide spectrum of drugs can be given orally after changes are made to the excipient composition, such as with advanced tablet/capsule formulation and coating techniques (Kaur et al., 2019; Sastry et al., 2000). Notwithstanding recent advances in formulation technology, orally administered drugs, when juxtaposed against their parenterally-administered counterparts, are still far more susceptible to degradation and bio-elimination as a result of challenging pH and enzymatic conditions of the stomach and small instestines (Choonara et al., 2014). The abundant presence of 50–200 μm-thick mucosal layer serves to aggregate and destabilize large molecules, while the tightly-packed intestinal epithelium makes paracellular transport through the tight junctions impossible for all but the smallest molecules (<100–200 Daltons) (Antunes et al., 2013; Boegh & Nielsen, 2015; Choonara et al., 2014). Absorption through enterocytes (i.e., via transcellular transport) is also problematic for a large number of drugs as active Pgp-1 efflux and CYP450-mediated metabolism of xenobiotics are commonplace (Goldberg & Gomez-Orellana, 2003; Homayun et al., 2019). Finally, apart from those released into mesenteric lymphatic vessels, drugs that are able to enter systemic circulation do so through the hepatic portal vein, a route that further subjects orally administered drugs to first-pass metabolism and facilitates their renal/biliary elimination (Martinez & Amidon, 2002). Because of these challenges, oral drug delivery has been mostly confined to small and somewhat hydrophilic molecules, with the overall suitability of this route falling precipitously for molecules larger than 700 Daltons (Renukuntla et al., 2013).

Peptides are large molecules having less than 50 amino acids (about 5500 Daltons) that comprise the bulk of signaling ligands within the body, regulating such diverse processes as metabolism, blood pressure, angiogenesis, neurotransmission, and immunity (Owji et al., 2018). These versatile compounds can be generated using traditional synthesis techniques as well as by combinatorial chemistry along with high-throughput screening to acquire extremely high target specificity, and making them ideal therapeutic agents (Falciani et al., 2005; Lau & Dunn, 2018). While their enormous medical relevance is just beginning to be appreciated, peptides have been almost entirely precluded from the oral route, and for which injections remain the only viable choice. All of the aforementioned challenges facing a typical ingested drug are magnified in case of peptides (Choonara et al., 2014;

Goldberg & Gomez-Orellana, 2003). Because of their reliance on not just primary, but also secondary and tertiary structure for bioactivity, peptides are especially sensitive to the denaturing conditions of the GI tract, such as the dramatic change in pH between the stomach (pH 1–3) and the duodenum (pH 6–8). A variety of hydrolytic enzymes are also ubiquitously present, either secreted into the lumen as endopeptidases (e.g., trypsin, chymotrypsin, and elastase) and exopeptidases (e.g., carboxypeptidase A and aminopeptidase) or within the lysosomes of enterocytes (Altshuler et al., 2016; Antalis et al., 2007). Finally, aggregate-inducing mucous and a largely impermeable gut epithelial wall collectively result in very short half-lives for any orally administered, peptide-based drugs (Behzadi et al., 2017; Patricia & Dhamoon, 2019).

Still, there are multiple, actively-pursued approaches to overcome the obstacles present in the GI tract, and these are broadly categorized as using (1) prodrug/analogs, (2) absorption/permeation enhancers, (3) enzyme inhibitors, or (4) encapsulation within nano- or micro-scale particles (Brayden et al., 2020; Liu et al., 2018; Pauletti et al., 1997). Since peptide specificity and function are highly adapted to the existing structure, any modification to yield a more GI-permeable prodrug/analog can be potentially risky not only in terms of the loss of pharmacologic utility, but also the gain of off-target effects. Similar safety concerns can be found in the use of absorption/permeation enhancers and enzyme inhibitors (Choonara et al., 2014). While commonly deployed by researchers in oral drug delivery, the chronic use of these compounds is feared to disrupt the natural homeostasis of the gut, and contribute to the unwanted seepage of microbiota into systemic circulation (Maher & Brayden, 2012; Maher et al., 2019). Provided that the underlying materials used are safe, NPs have the versatility to become delivery vehicles for a variety of chemically diverse compounds, including peptides. This unique capability of NPs is becoming increasingly attractive to patients, physicians, healthcare system, and industry (Table 4.1), in terms of what each party is perceiving to be the most important factors that will maximize the potential of existing pharmaceutical portfolios.

NPs are a highly diversified groups of supramolecular drug carriers comprising synthetic polymers, peptides, carbohydrates, lipids, dendrimers, or inorganic substances such as silica. These carriers should ideally have compatibility with the transported drug, (e.g., favorable hydrogen bond or van der Waals interactions) (Homayun et al., 2019; Lee & Youn, 2016; Patra et al.,

Table 4.1 Priorities behind effective utilization of prescription drugs by various stakeholders in healthcare ecosystem.

Industrial perspective	Physician/pharmacist perspective
Improved utilization of costly (bio) pharmaceuticals (e.g., low-dose formulation, improved drug solubility/stability, controlled drug release, improved pharmacokinetic profile, targeted drug delivery) Drug product reformulation by using innovative health technology (e.g., expanded drug lifecycle, drug reintroduction) Maximizing the return of R&D investments	More effective and less toxic therapeutic interventions Patient-friendly drug product (e.g., self-administered drug product) Personalized therapy Simplified therapeutic procedures Providing targeted drug performance Accelerating the healing process Improved patient compliance/ adherence Improved medical/ pharmaceutical care
Health care system perspective	**Patient perspective**
Rational prescribing Overall reduction in health care costs (e.g., by increasing the drug efficacy, reducing the length of in-patient care stay, reducing personal health care costs, and the effective treatment of expensive major diseases) Improving the quality of health care services	Reducing the frequency of dosage Minimally invasive method of administration Improved therapeutic outcomes Reducing adverse drug effects Improved quality of life of patients

Source: Reprinted from Hafner A, Lovrić J, Lakoš GP, Pepić I. Nanotherapeutics in the EU: an overview on current state and future directions. Int J Nanomedicine. 2014 Feb 19;9:1005-23. doi: 10.2147/IJN.S55359. PMID: 24600222; PMCID: PMC3933707.

2018; Saini, Arora, et al., 2016; Xin et al., 2017). Physicochemical properties of NPs such as size, shape, charge, zeta-potential, and polydispersity index are all traditionally considered during synthesis, and serve as useful benchmarks to evaluate batch-to-batch variability (Bhattacharjee, 2016; Walkey & Chan, 2012). Furthermore, all NPs acquire a unique signature once they are in systemic circulation as a result of adsorbing a surface layer of plasma proteins to form what is called the protein corona, a structure that can subsequently dictate biological response as well as the physiological fate of NPs. The composition of protein coronas is dynamically influenced by the abovementioned physicochemical properties in addition to the presence of any surface modifications (Walkey & Chan, 2012). Therefore, careful tuning of these properties should be made by researchers, as subtle changes can have major impact on corona formation and NP behavior (Hoshyar et al., 2016; Nel et al., 2009; Tang et al., 2018).

Although carriers such as NPs protect the drug molecule against harsh GI environments, accomplishing actual transit across the intestinal epithelium intact is still quite difficult. To cross this highly viscous and mucin-laden barrier, NPs must adhere to the intestinal surface with greater affinity than they do to the overlying mucous, which is constantly being flushed down the GI tract by peristalsis, thereby facilitating the clearance of any particles trapped within (Ensign et al., 2012). Prolonged retention of the NPs increases the probability of NPs penetrating the mucus by diffusing through the gaps/leaks in the mucin mesh. Because of their small size (which is comparable to that of many viruses), NPs can be taken up either by phagocytosis among Peyer's patch-localized microfold cells, or by a combination of endocytic processes carried out by the general enterocyte population (Behzadi et al., 2017; Salatin & Yari Khosroushahi, 2017). Despite of various efforts in optimizing the surface morphology and physicochemical properties of NPs, the in vivo therapeutic efficiency remains relatively low for "passive" NP transport, (e.g., through simple diffusion and nonspecific, stochastic cellular uptake) (Vacha et al., 2011). However, studies have shown that conjugating the NP surface with a ligand specific for uniquely overexpressed receptor or antigen targets present on the surface of cells greatly enhances the internalization of NPs through the GI barrier. The exact mechanism behind such "active" transport process remains to be elucidated, but the presence of ligand-receptor interaction likely promotes the retention of NPs on the mucus-covered membrane, thereby conferring some protection from a speedy clearance by the moving GI environment (Karakoti et al., 2015; Mahmoudi et al., 2014; Verma & Stellacci, 2010).

The exploitation of surface-decorated ligands for receptor-mediated active NP transport is an area of active research not only for oral drug delivery but also targeted therapeutics in general. Properly chosen ligand attachments can serve to concentrate NPs to certain tissue types or aid in the absorption and transcytosis across otherwise impermeable physiological barriers, such that the ultimate consequence would be to lower systemic dosage needed to achieve the desired balance of biodistribution, cytotoxicity, and therapeutic efficiency (Ashfaq et al., 2017; Foroozandeh & Aziz, 2018; Sun et al., 2020; Wang & Kohane, 2017). For active transport, a plethora of surface receptors have been investigated to date (Ciacci et al., 1993; Herschman et al., 1982; Okumura & Takeda, 2017), but much more critical to the overall scheme's success involves the choice of an appropriate ligand (Attie & Raines, 1995; Chen et al., 2017).

Characteristics such as conformational and chemical stability, specificity, affinity towards the receptor, immunogenicity, as well as the ease and cost associated with NP conjugation must all be considered (Fig. 4.1). The pros and cons of various classes of targeting ligands, which are broadly divided into peptides, nonpeptides macromolecules, and small molecules, have been extensively discussed elsewhere in the literature (des Rieux et al., 2013; Yun et al., 2013).

Since their emergence in the 1970s, NPs have expanded both structurally and functionally into a diverse family comprising of, among others, polymeric NPs, micelles, liposomes, nanomaterials, polysaccharides, and nanocrystals (Cheng et al., 2015), some of which even possess triggering mechanisms that help to release the drug cargo into specific tissues, under pathophysiologically relevant contexts, or upon certain artificially introduced stimuli (Brown et al., 2019; Sun et al., 2020). Hybrid NPs utilizing a mixture of synthetic polymers, lipids, peptides, and polysaccharide, leveraging the advantages inherent to each material, are also quite common. These nanoscale carriers are often biodegradable or at least

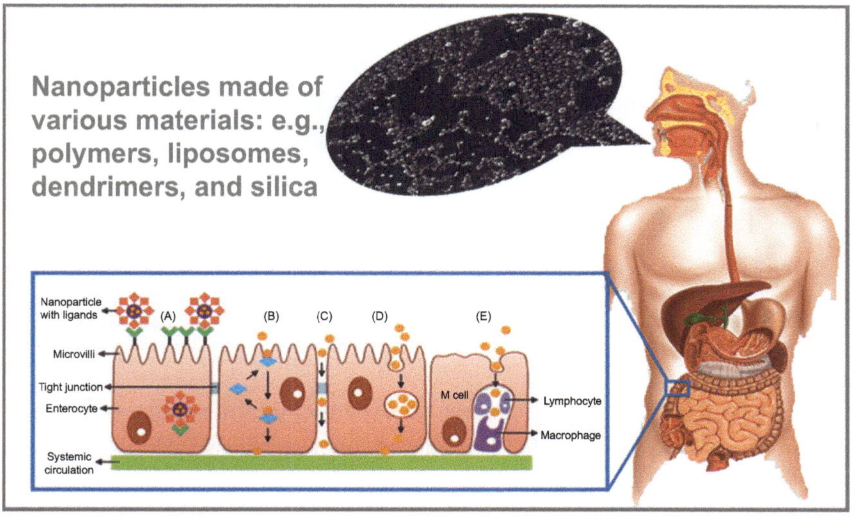

Figure 4.1 Schematic representation of the transport mechanisms of oral formulation: (a) receptor-mediated transport; (b) carrier-mediated transport; (c) paracellular transport; (d) transcellular transport; and (e) M cell-mediated transport (i.e., phagocytosis by M cells). *Adapted from Yun, Y., Cho, Y. W., Park, K. (2013). Nanoparticles for oral delivery: Targeted nanoparticles with peptidic ligands for oral protein delivery.* Advanced Drug Delivery Reviews, *65(6), 822–832. doi: 10.1016/j.addr.2012.10.007 2013, with copyright permission from Elsevier.*

biocompatible, but even more biosimilar versions, incorporating a colloidal system of natural ingredients such as plant, egg, or milk-based protein and lipids, are also being developed (McClements, 2018; Perry & McClements, 2020). Nonetheless, not all NPs discussed above are suited for oral delivery, and one of the key obstacles on making effective improvements and adaptations for this route has been the lack of a good, predictive model system. At the moment, testing of NPs in the lab remains mostly confined to animal models, in particular rodents, and concerns about their lack of correlation to actual human subjects have been well-documented (Hatton et al., 2015; Musther et al., 2014). As an alternative to in vivo models, Lundquist and Artursson has found that no in vitro system currently in use can accurately recapitulate various molecular and physiological constraints of the gut, but an Ussing chamber containing ex vivo sections of the human intestine may be a viable, albeit low throughput option (Lundquist & Artursson, 2016).

Clearly, the oral delivery of peptides by NPs can have medically groundbreaking consequences. But bearing in mind that an overwhelming majority of NPs have not reached clinical trials, it is reasonable to say that research in this field is still in its sophomoric stages, with much of the ongoing endeavors still focusing on a handful of simple, yet highly marketable model compounds. This chapter is by no means an exhaustive account of these endeavors, but we have selected some representative, recent examples of NP-based oral delivery with respect to four peptides of significant clinical value: insulin, cyclosporine, calcitonin, and exenatide.

4.2 Insulin

Insulin is one of the earliest therapeutic peptides that have been made commercially available. Developed by Eli Lilly in 1982, the drug has been primarily applied to treat type I, insulin-dependent diabetes (T1D), as well as later stages of type II insulin-independent diabetes (T2D) (Bliss, 1982; Vecchio et al., 2018). To date, insulin is given only in parenteral (primarily subcutaneous) dosage forms, but the slower release kinetics of insulin delivered by the oral route, combined with indirect access to blood, offers not only a less painful experience, but can also mitigate the possibility of anaphylaxis, hypoglycemia, and peripheral hyperinsulinemia brought about by frequent or incorrectly-dosed injections (Edgerton et al., 2019; Haastrup et al., 2018; Yu et al., 2017). While a large body of basic research has been accumulated on the development and validation of oral insulin dosage forms, in practice, few of these non-invasive alternatives have been tested at the clinical trials stage, and those that have are beset with a host of stability and immunogenicity issues (Gedawy et al., 2018; Kumar et al., 2018; Macedo et al., 2020; Sharma et al., 2015; Torosantucci et al., 2014; Wong et al., 2020).

A range of polymeric NPs, based on materials such as poly(lactic-co-glycolic acid) (PLGA), polycaprolactone (PCL), polylactic acid (PLA), chitosan, and dextran, were tried in oral insulin formulations, but with plenty of shortcomings associated with each delivery system, approval by the FDA still appeared improbable in the near term (Fonte et al., 2015; Luo et al., 2016; Sharma et al., 2015). To improve enteric uptake, one approach that researchers have taken is to incorporate mild permeation enhancers to the NPs. In vitro studies showed that heparin sulfate-appended PLGA NPs were capable of penetrating triple co-cultured intestinal models, and deliver the insulin cargo two to three folds better in comparison to unmodified NP and native insulin (Jaradat et al., 2020). The poly(ethylene glycol) (PEG)-capped PLGA NP is another promising option as PEG can enhance mucopenetration thanks to its amphiphilicity, while also having a well-established safety profile (Malathi et al., 2015).

Perhaps the most extensively studied permeation enhancer is the β-(1−4) linked polysaccharide chitosan, where key derivatives include, among others, N-trimethyl-chitosan, thiolated chitosan, and carboxyl-methyl chitosan (Ways et al., 2018). The ionic interaction between the cationic amino acid groups of chitosan and the anionic functional groups of sialic acid, a common component of mucins, results in a high degree of mucoadhesion, which improves retention time on the intestinal wall, and increases the probability of interdiffusion across the mucous mesh (Andreani et al., 2015; Builders et al., 2008; Mumuni et al., 2020). Furthermore, chitosan has been shown to induce transient and reversible openings of enteric tight junctions, thereby facilitating paracellular transport of the drug cargo (Werle et al., 2009; Yeh et al., 2011). For instance, in diabetic rats, N-trimethyl-chitosan chloride-coated PLGA NPs increased the pharmacological availability of oral insulin by twofold compared to unmodified NPs, while both mucoadhesion and tight junction opening were enhanced in HT29-MTX and Caco-2 cells, respectively (Sheng et al., 2015). Likewise, a PLGA NP comprising of an inner coating of chitosan and PEG, as well as an outer coating of bovine serum albumin was effective at inducing a 50% glycemia decrease in diabetic rats 8 h after oral dosing (Amaral et al., 2020).

Because of its self-gelation property, chitosan is able to not only serve as surface modifications for other NPs, but also form its own NPs. Formulation composed of methylated (amino benzyl) chitosan and insulin has been prepared; the resulting NP is subsequently used to fill preclinical capsules coated with Eudragit L100, which prevents gastric release of the capsule contents. Diabetic rats dosed with the chitosan-insulin-containing capsules had significantly lower levels of serum, liver, and kidney oxidative stress biomarkers as well as improved histopathological features when compared to the untreated control group (Mahjub et al., 2020). A multifunctional NP combining the polymeric trimethyl chitosan with fucoidan have also been devised to encapsulate insulin. Such NP appeared to possess enhanced hypoglycemic efficacy, as the effects of orally delivered insulin was compounded by the inhibition of endogenous α-glucosidase activity by the co-absorbed fucoidan polymers (Tsai et al., 2019). Lastly, dual encapsulation of bovine insulin and exendin-4, a glucagon receptor agonist useful for treating T2D, has been carried out in an NP vehicle containing poly (γ-glutamic acid) and chitosan. In diabetic rats, the resulting pH-responsive NP formulation appeared to both reduce blood glucose as a consequence of exogenous insulin as well as from exendin-4-stimulated secretions of endogenous insulin (Chuang et al., 2013). Thus, chitosan-based NPs have the versatility to encapsulate multiple peptides at the same time, whose relatively low achievable oral bioavailability may be overcome by the presence of pharmacologic synergism, as in the case of insulin and exendin-4.

Zein is the primary storage protein from *Zea mays* (maize). Its high biodegradability along with relative ease-of-use makes zein a good candidate material for the preparation of colloidal NPs (Labib, 2018; Pascoli et al., 2018). Carboxymethylated short-chain amylose zein with chitosan coating demonstrated high encapsulation efficiency ($>$89%) with insulin. The resulting orally deliverable NP also exhibited a 12-fold higher transepithelial permeability when compared to insulin alone. In mice, pharmacokinetics studies of plasma insulin revealed that these NPs have T_{max} (i.e., time to reason maximum drug concentration) at 3 h and lasting up to 8 h, while maximal blood glucose depression achieved was 38.8% of initial level compared to 29.4% of initial level for subcutaneous insulin. However, whereas the hypoglycemic effects of subcutaneous insulin dissipated 6–8 h post dosing, the zein NP insulin appeared to sustain its hypoglycemic effect beyond 8 h, consistent with the slower release kinetics typically seen in orally delivered drugs, and in stark contrast to the rapid drop in drug concentration after 2 h with subcutaneous injection (Ji et al., 2018, 2019). In another approach, zein NPs coated with a poly (anhydride)-thiamine conjugate, (which is another form of permeation enhancer), was used to encapsulate insulin. These NPs proclaimed high insulin loading capacity (at 80 μg/mg of polymer), and was effective in counteracting hyperglycemia-induced lipid metabolism in the nematode *C. elegans*, as well as causing an 60% decrease in glucose level after oral dosing in rats, comparable to the effects of subcutaneous insulin injection (Inchaurraga et al., 2020).

Liposomes are uni-, multi-lamellar vesicles of phospholipids and cholesterol that can accommodate both hydrophilic and hydrophobic drugs of various sizes, along with having the capacity to provide extra protection and permeability. While several milestones have been made in the use of liposomes for injections, both conventional (i.e., those constituting unmodified lipids) and novel (i.e., those containing partially modified or substituted constituents) liposomes have been investigated in the oral delivery of peptides, and important categories of liposomal NPs such as double liposomes, proliposomes, bilosomes, surface coated liposomes, and cell specific targeting liposomes have been extensively discussed in Wong et al. and He et al. reviews (Wong et al., 2018). One drawback of liposomes however, is their lack of GI stability. NPs made of outward-facing cholesteryl moiety and inward-facing Ca^{2+}-chelated phosphates can be modified with surface phospholipids and glycocholate to form a hydrophilic corona that is quite resistant to hydrolytic GI conditions. These NPs also encapsulated insulin efficiently (>95%) and exhibited heightened Caco-2 uptake (31-fold), and resulted in a hypoglycemic effect that was comparable in magnitude but greater in duration with respect to subcutaneous insulin in a T1D rat model (Ding et al., 2020). In another study, solid lipid NPs are prepared by fatty acid coacervation method to encapsulate insulin or insulin glargine analog in the lipid matrix through hydrophobic ion pairing. This technique also resulted in improved GI stability as well as enhanced permeation, with about 6% absolute insulin bioavailability achievable in the rat bloodstream (Muntoni et al., 2019). Beyond these rodent studies, lipid-based NPs have even proceeded to the clinical stage. In several ongoing phase 1/2 trials, Diasome Pharmaceuticals treated T2D patients with oral hepatic-directed vesicle-insulin (HDV-I). These formulations are lipid bio-NPs with a surface-appended, proprietary hepatocyte-targeting molecule (HTM). The HTM facilitates the capturing of HDV-I by hepatocytes following uptake from the gut through the hepatic portal vein. Dogs were administered four different doses: 0.05–0.4 IU/kg; all 4 doses statically lowered mean postprandial plasma glucose by up to 50% in comparison to placebo (Geho et al., 2014).

In addition to spherical NPs, nanocapsules containing an outer shell of bio-friendly material and an inner nanoscale core of liquid, drug-stabilizing material have also been proposed as oral insulin delivering candidates. Cationic polypeptide protamine nanocapsules layered with polysialic acid (PSA) was used to improve the pharmacokinetics of oral insulin, as PSA provided not just protection against proteolytic enzyme but its mucin properties were also quite helpful in mediating nanocapsule interaction with the intestinal wall. These nanocapsules exhibited good in vivo efficacy, such as a prolonged, 60% glucose reduction up to 8 h after administration to diabetic rats (Thwala et al., 2018) Similarly, polyarginine nanocapsules consisting of oleic acid, sodium deoxycholate and Span 80 were able to sustain the release of encapsulated insulin for more than 4 h, and successfully reduced glucose levels in normoglycemic rats (Niu et al., 2017).

NPs comprising of materials other than the aforementioned have also been investigated for their ability to deliver insulin orally. For instance, nanocomposites made of iron, as a metal-organic framework, together with sodium dodecyl sulfate were successful in crossing the Caco-2 monolayer, and yielded promising in vivo rodent results in terms of enhancing blood insulin levels over a 6 h period post dosing (Zhou et al., 2020). Large surface area, tunable porosity, and the ease of functionalization make silica NPs a good choice for protein and peptide delivery. Scientist have synthesized silica NPs coated with various permeation enhancers from the previously discussed chitosan and PEG to sodium alginate (Andreani et al., 2014) and succinylated β-lactoglobulin (Juere et al., 2020) for systematic in vitro analysis with Caco-2 and HepG cell lines. The supramolecular structure arising from self-assembling oligosaccharide cyclodextrin is yet another foundation to build viable drug carriers. NPs based on cyclodextrin, which has mucopenetrating properties, could be modified with additional excipients such as PEGylated phospholipids and dextran sulfate; when used to encapsulating insulin, these NPs led to relatively high bioavailability (5.5%) in comparison to subcutaneous administration (Presas et al., 2018).

Targeted delivery of insulin has also been undertaken with various oral NPs formulations utilizing surface ligands that direct the particles to certain cellular receptors/antigens. For instance, vitamin B_{12}-conjugated chitosan and sodium alginate NPs were used to encapsulate insulin. When tagged with fluorescent FITC for in vivo studies, these NPs were confirmed to engage in receptor-mediated cellular uptake. Measurement of plasma insulin and glucose levels showed about fourfold increase in peptide bioavailability, with sustained hypoglycemic effect up to 12 h in comparison with NP prepared without vitamin B_{12} functionalization (Verma et al., 2016). In addition, receptor-mediated oral delivery of insulin was explored using the immunoglobulin fragment crystallizable (Fc) region conjugated to NPs to target the GI tract-expressed, neonatal Fc receptor (FcRn). This targeted NPs showed about 10-fold increase in absorption across the intestine relative to nontargeted controls. A dose dependent (0.66 and 1.1 U/kg) glucose response was also observed in WT mice, but this difference was abolished in the FcRn knockout mice. In the same WT mice, approximately 33% of glucose reduction was observed that persisted over a 10 h period (Pridgen et al., 2013). Finally, dendritic oligopeptide-modified PLGA have been proposed as a good carrier vehicle owing to their enormous number of potential ligand-modifying sites. An example of such dendrimeric NP utilizes glutamic acid conjugation to endow the particles with the ability to exploit multiple endocytic pathways, such as those involving oligopeptide and amino acid transporters (Bai et al., 2020). In rats, a relative insulin bioavailability of 10.04% was attained with these NPs, which was 1.89-fold higher than the level measured in standard, PEG-functionalized NPs.

Our lab became interested in a slightly different approach to targeted, oral NP delivery. Rather than using ligands that can be outcompeted by unpredictable levels of endogenous analogs, we reasoned that a noncompetitive ligand could result in more consistent and robust NP uptake. Thus, we focused on a plant xanthanoid molecule: gambogic acid (GA), which was found to induce cancer apoptosis via binding to transferrin receptor 1 (TfR1), but in a manner independent of external concentrations of transferrin – the receptor's natural ligand. TfR1 is highly expressed by the GI epithelium; in Caco-2 cells as well as ex vivo intestinal tissues, conjugation of GA to PLGA NPs (PLGA-GA) resulted in noncompetitive binding of these particles to TfR1, even in presence of excess transferrin. Subsequent oral dosing of insulin-encapsulated PLGA-GA to rats resulted in a fivefold increase in plasma insulins levels, when compared to the ligand-free PLGA NPs (Ganugula, Arora, Guada, et al., 2017).

While a majority of NP-assisted oral delivery strategies are showing promise in preclinical studies, the ongoing translational effort has been at a standstill for various reasons that include but are not limited to: (1) the inability for delivery strategies to meet pathophysiological needs; (2) most, if not all studies are short-term; (3) probably wrong choice of models as most studies use streptozotocin (STZ)-induced T1D, (4) random vs fasting levels, (5) dose regimens, (6) poor insulin loading in many carrier materials, etc. The best results for NP-assisted oral insulin delivery could probably be obtained when tested early in T2D.

4.3 Cyclosporine

Cyclosporin A (CsA) is an 11 amino acid, cyclic peptide immunosuppressant that is highly effective at mitigating or preventing T cell-mediated allograft rejections (Laupacis et al., 1982). Since its widespread adoption in transplant medicine, CsA has been gradually repurposed to treat other inflammatory disorders such as psoriasis, rheumatoid arthritis, dry eye syndrome, and even traumatic brain/cardiac injury (Guada, Beloqui, Kumar, et al., 2016; Johns & Littlejohn, 1999; Lallemand et al., 2017; Oh et al., 2018). The drug's well-established mechanism of action involves binding to the intracellular protein cyclophilin, the resulting drug/protein complex inhibits calcineurin, a key regulatory phosphatase whose substrates include the NF-AT family of transcription factors. Without calcineurin-mediated dephosphorylation, NF-AT is unable to translocate to the nucleus, and the transcription of its target genes critical to proper T cell functioning, such as IL-2, IL-4 and IFN-γ, is drastically diminished (Handschumacher et al., 1984; Russell et al., 1992; Tsuda et al., 2012). The low therapeutic index as well as high intra- and inter-individual variability of CsA necessitated multiple generations of modification to the drug, with the most recent water-free microemulsion from Novartis called Neoral being one of the most popular among clinicians (Singh & Narsipur, 2011; Taber et al., 2005).

Despite its shortcomings, CsA is the drug of choice for most organ transplantations, since in comparison to other commonly used immunosuppresants such as prednisone, tacrolimus, mycophenolate, and rapamycin, CsA appears to exhibit relatively milder myelotoxicity and more rapid therapeutic action

(Laupacis et al., 1982; Lim et al., 1996). Although CsA is available in a variety of formulations, including IV injections, gels, ophthalmic emulsions, and topical ointments, the oral route is preferred due to more favorable patient compliance. The aforementioned Neoral as well as its related drug Sandimmune are both modified oral formulations of CsA, but both still suffer from bioavailability issues, while bitter taste, low solubility, and low permeability remain unresolved problems (Group, 2002; Patel & Wairkar, 2019). The therapeutic index of CsA is also low so that routine monitoring of the plasma drug concentration is almost always necessary, and frequent dosing adjustments to achieve the intended plasma concentration are the norm. In many cases, high concentrations of CsA in circulation do occur with regular doses, and prolonged overexposure to CsA can lead to common side-effects such as hypertension and renal injury, as well as less common side-effects such as jaundice, infection, and sepsis (Graham, 1994; Kim et al., 2018; Tedesco & Haragsim, 2012).

Although NP-based encapsulation of CsA is quite challenging due to the drug's high molecular weight, a few carriers have shown evidence of being quite capable of providing improved bioavailability and reduced toxicity of this essential drug (Patel & Wairkar, 2019). The flash nano precipitation (FNP) technology involves kinetic mixing processes that control the nucleation of drug compounds, together with the adsorption of a stabilizer to yield consistently sized and highly concentrated NPs (Pustulka et al., 2013). The nanomatrix containing CsA and mannitol was prepared using FNP, and dissolution studies showed 30–40-fold better drug release, and up to threefold in vivo bioavailability improvement in rats, when compared against amorphous CsA. These differences were attributed to the increased NP wettability and dispersibility acquired from mannitol incorporation (Suzuki et al., 2017).

Chitosan-modified nanocochleates, composed of multilayered, cylindrical particles of positively charged calcium ions and negatively charged phospholipid, have also been devised as a delivery vehicle for CsA. These NPs led to roughly threefold increase in plasma drug levels compared pure CsA, a difference attributed mainly to the rearrangement of tight junction proteins ZO-1 and Claudin-4 by chitosan (Liu et al., 2017). Yet another potential CsA vehicle can be found in bilosomes, which are bile salt-containing liposomes that possess useful pharmacologic properties such as resistance to physiological emulsifier-induced disruptions, inhibition of enzymatic activity in the particle vicinity, and enhancement GI absorption (Ahmad et al., 2017; Chauhan et al., 2015). However, bilosomes suffer from multiple disadvantages if stored in liquid form, as leakage of the drug load, sedimentation, and agglomeration can occur. To overcome these problems, Guan et al. (Guan et al., 2011, 2016) incorporated gelatin, a common pharmaceutical excipient, to the interior of bilosomes, and these freeze-dried bilosomes achieved 1.23-fold better oral bioavailability compared to Neoral in beagle dogs. Due to their hydrophobicity, bilosomes were anticipated to follow the lymphatic pathway of absorption, and this was indeed the case as evidence by a mesenteric lymphatic/jugular vein-cannulated rat model.

Guada, Lasa-Saracibar, et al. (2016), Guada, Lana, et al. (2016), and Guada, Beloqui, Alhouayek, et al. (2016) have prepared Precirol matrix lipid NPs (LNs) to encapsulate CsA, while obviating the need for any toxic alcohol additives by using low concentrations of mild surfactants such as Tween-80 in the excipient to solubilized the drug. These LNs were comparable to conventional dosage forms in terms of shelf life, and Caco-2 monolayer assays showed that LNs could permeate the epithelial membrane without compromising tight junction integrity. In vivo oral administration of LNs in mice indicated that CsA had similar plasma bioavailability, as well as comparable drug deposition profile in tissues such as kidney, liver, heart, lungs and lymph nodes. The pharmacodynamics of LNs compared favorably against Neoral, where significant decreases to lymphocyte count after 15 days of administration was not matched by the commercial drug.

To treat gut-associated disorders such as inflammatory bowel disease (IBD), NPs can also be used as a means to prevent unwanted systemic uptake of CsA in the upper GI tract. For instance, CsA-encapsulated PLGA NPs demonstrated marginally better efficacy compared to Neoral, with respect to ameliorating lower GI tract inflammation caused by the dextran sodium sulfate (DSS)-induced ulcerative colitis in mice (Melero et al., 2017). Interestingly, the aforementioned LN-deliverable CsA did not produce any significant effects under the same DSS model, suggesting that perhaps the type of nanomaterial used was able to influence drug localization to the upper or lower GI tract (Guada, Beloqui, Alhouayek, et al., 2016). NPs of CsA prepared by a combination of PLGA and the pH-responsive Eudragit FS30D copolymer were also employed in the murine DSS-induced ulcerative colitis. In vivo results revealed significant downregulation of neutrophil-originated myeloperoxidase activity (threefold), TNF-α (threefold), as well as IL-6 (fourfold) in the treated group vs the untreated disease control. These improvements were in part explained by the dual protection to CsA offered by and PLGA and Eudragit FS30D, the latter of which restricted the release of CsA in the stomach and gave sustained releases above pH 7.4 (Naeem et al., 2018).

Similar to our strategy concerning oral insulin, our lab has applied noncompetitive, transferrin receptor 1 (TfR1)-mediated delivery to CsA as well. Gambogic acid (GA)-conjugated PLGA NPs (PLGA-GA) resulted in oral CsA achieving earlier onset and more robust plasma bioavailability in comparison to unconjugated PLGA NPs loaded with CsA. PLGA-GA also exhibited altered tissue distribution compared to unconjugated systems, with elevated brain but reduced hepatic enrichment (Saini, Ganugula, et al., 2016). Since commercial PLGA contains only a single conjugatable functional group (i.e., the terminal carboxyl moiety) per polymer, limiting the maximum number of ligands per NP, we prepared a hybrid polymer (Ganugula, Arora, Saini, et al., 2017) utilizing branching chains of PLA and PEG. This new polymer system consisted of 5–12 periodically spaced carboxyl functional groups per polymer, which significantly increased potential ligand density and leading to more favorable ligand-receptor stoichiometry. The GA-conjugated NPs synthesized from these polymers (called P2Ns-GA) outclassed the original PLGA-GA in multiple model systems. We have recently used the P2Ns-GA to encapsulate CsA in order to treat systemic lupus erythematosus (SLE) in a lymphoproliferative murine model. The NP-treated group received half of the number of doses in comparison to Neoral-treated group, but demonstrated almost 18-fold higher CsA bioavailability in the lymph nodes. The high lymphatic enrichment was possibly caused by P2Ns-GA particles engaging with TfR1-expressing $CD3^+$ and $CD20^+$ lymphocytes in the gut-associated lymphatic tissue (GALT), such as lacteals or Peyer's patches. In terms of therapeutic efficacy, CsA-containing P2Ns-GA normalized pathological hallmarks such as plasma cytokines, anti-dsDNA IgG, and glomerulonephritis in the SLE mice, whereas Neoral was completely ineffective with respect to these parameters (Ganugula et al., 2020).

4.4 Calcitonin

Calcitonin is a 32 amino acid peptide secreted by parafollicular cells of the thyroid; its key physiological function involves lowering blood calcium and phosphate levels by preventing the resorption of bone by osteoclasts (Azria et al., 1995; Gozdzialska et al., 2016; Potts, 1992; Watson et al., 1981). Homologs of this important signaling protein are found not just in mammals but also nonmammal vertebrates such as reptiles, avians, and fish. Calcitonin of salmon origin (sCT), extracted from the organism's ultimobranchial glands, resembles human calcitonin but possesses approximately 50-fold higher potency (Felsenfeld & Levine, 2015). This protein has since been reproduced in large quantities either by recombinant DNA technology or by peptide synthesis (Cetin et al., 2008; Hughes et al., 1975). In the clinic, sCT is used for the treatment of a variety of bone-related disorders including post-menopausal osteoporosis, Paget's disease, bone metastases, phantom limb pain, spinal stenosis and hypercalcemia (Carroll & Matfin, 2010; Jain & Chatterjee, 2020; Munoz-Torres et al., 2004; Peng et al., 2015; Singer, 2020; Viana & Payne, 2015). In addition to calcium management, there is evidence that sCT can regulate the renal electrolyte balance, and administration of sCT has even been associated with positive outcomes in the treatment of bipolar and neurodegenerative conditions (Carney, 1997; Svenningsson et al., 2017; Vik & Yatham, 1998).

Although sCT has been on the market from more than three decades, it has garnered relatively few FDA-approved prescriptions. Synthetic (Miacalcin from Novartis and the discontinued Calcimar from Sanofi) as well as recombinant (Fortical from Upsher-Smith Laboratories) variants of the drug are available in either injectable or nasal spray formats. Besides being uncomfortable, subcutaneous or intramuscular sCT injections currently suffer from short serum half-life (about 1 h, based on FDA estimates), thereby necessitating frequent dosing for some patients (Novartis MIACALCIN, 2020a). While the intranasal option is noninvasive, in the case of sCT, this route suffers from poor bioavailability (3% vs 71% for subcutaneous injections according to FDA), and cannot be properly utilized by those with chronic rhinorrhea or congested nasal passage (Novartis MIACALCIN, 2020b).

NPs based on hydroxyapatite, a naturally occurring mineral of calcium phosphate, encapsulated sCT at 85% efficiency. Sublingual delivery of this formulation in rabbits resulted in a relative bioavailability of approximately 15% compared to subcutaneous sCT, but the otherwise short life of the injectable drug was increased by almost twofold with hydroxyapatite NPs (Kotak & Devarajan, 2020). In another NP embodiment, dual encapsulation of sCT and puerarin, a protease inhibitor, in chitosan-modified PLGA NPs were used improve the GI uptake of sCT. These NPs released about 65% of sCT and 100% puerarin in vitro. FITC-labeled sCT from NPs exhibited greater uptake in Caco-2 cells compared to nonencapsulated sCT. In the rat pharmacokinetic study, dual encapsulated NPs had 1.74-fold higher bioavailability than NPs encapsulating only sCT; the 500 IU/kg oral NP dose (with sCT and puerarin) also compared favorably with a 20 IU/kg intravenous sCT dose, attaining an absolute bioavailability of 12.5% and longer drug half-life (Liu et al., 2019).

The chitosan derivative, *N*-trimethyl-chitosan chloride (TMC) and oligoarginine (Arg8), another permeation enhancer, were used to modify liposomes containing sCT. The utilization of two cationic permeation enhancers was found to be synergistic, leading to a prolonged reduction in calcemia (40%) in the rat plasma, and increased the pharmacological efficacy of sCT more than 16.sixfold and 1.sixfold compared to nonliposome or TMC-only liposome oral delivery of sCT, respectively (Huang et al., 2014). Nanocapsules comprising principally of polyarginine (in particular Arg8) have been previously discussed in terms of insulin delivery, but they are also increasingly being utilized for other peptides, such as sCT, owing to their favorable cell penetration properties, high loading efficiency, stability in simulated intestinal fluids, and low cytotoxicity (Gonzalez-Paredes et al., 2017; Lollo et al., 2017; Niu et al., 2017).

4.5 Exenatide

Exenatide (marketed as Byetta by Eli Lilly, or Bydureon by AstraZeneca) is a 39 amino acid, synthetic peptide based on exendin-4, a hormone found in the lizard *Heloderma suspectum*. Exenatide belongs to a class of compounds called glucagon-like peptide 1 (GLP-1) receptor agonists. These drugs mimic the physiological effects of incretins, which include hormones such as GLP-1 or gastric inhibitory peptide (GIP) secreted by the gut to reduce appetite, augment insulin secretion from pancreatic β cells, and prevent glucagon secretion from pancreatic α cells (American Diabetes, 2020; Doggrell, 2018; Hjerpsted et al., 2018). In type II diabetes (T2D), GLP-1 receptor agonists such as exenatide, along with sulfonylurea, dipeptidyl peptidase-4 (DPP4) inhibitors, and sodium-glucose co-transporter-2 inhibitors, have been prescribed as a second line of treatment after the patient develops resistance to the standard metformin therapy (Doggrell, 2018).

Exenatide was the first GLP-1 receptor agonist to come into market, and its formulations typically consisted of twice daily subcutaneous injections given prior to food intake. After receiving FDA approval in 2005 however, the drug was beset by multiple black box warnings of thyroid tumors, medullary thyroid carcinoma and an endocrine disorder called multiple endocrine neoplasia syndrome type 2 (Drab, 2016; Filippatos et al., 2014; Goldenberg, 2012; Nauck & Friedrich, 2013; Romley et al., 2012; Tseng et al., 2015). These risks are not unique to exenatide; in fact, more recent GLP-1 receptor agonists, such as liraglutide (Victoza) and dulaglutide (Trulicity) also exhibit tumorigenic side-effects related to the thyroid. Nonetheless, GLP-1 receptor agonists are generally well-tolerated, with an overall lower incidence of hypoglycemia and stroke compared to other second line T2D therapeutics (Zheng et al., 2018). The only oral dosage form of GLP-1 receptor agonists currently available is semaglutide (Rybelsus from Novo Nordisk), which is a proprietary tablet formulation comprising of highly modified, DPP4-resistant GLP-1 and salcaprozate sodium (SNAC), an absorption enhancer. Key disadvantages of Rybelsus however, are low bioavailability (0.4%-1%), high dose, stringent prandial requirements, and higher incidence of diabetic retinopathy (3% vs 1.8% in the placebo group) (NovoNordisk RYBELSUS, 2020).

While oral delivery of exenatide remains the key focus of research, efforts have been made to use polymeric particles to extend the release kinetics of the parenteral drug. Lecithin NPs packaged into PLGA microspheres as well as dimpled PLGA microparticles both exhibited in vitro exenatide release up to 60 days, and sustained *in situ* (as single subcutaneous injection) release up to 30 days in rats (Dong et al., 2020; Zhu et al., 2019). Likewise, polyplex nanocarriers encapsulating exenatide was synthesized from poly(L-lysine)-poly(ethylene glycol)-poly(L-lysine). The exenatide from these subcutaneously delivered particles were found to be effective at mitigating ischemia/reperfussion-induced myocardial injury as well as pancreatic injury in rats. Furthermore, significant reductions in reactive oxygen species, serum IL-4 and IL-6 were observed in the rodent models (Ma et al., 2019; Zhang, Qian, et al., 2018).

Nanostructured lipid carriers (NLC) are partially crystalized forms of solid lipid NPs that could be used for oral peptide delivery. NLCs are prepared by mixing different chain lengths of solid and liquid phase lipids; the resulting matrix core is relatively unordered and can accommodate drugs both in larger quantities and more varied morphologies (Gaba et al., 2015). Interestingly, NLCs alone could stimulate endogenous GLP-1 secretion, both in vitro from enteroendocrinal L-cells and in vivo from the murine gut. However, NLCs encapsulating exenatide or liraglutide appeared to be trapped in the intestinal mucous layer, and were unable to cross the GI barrier (Shrestha et al., 2018). In another set of experiments, using NPs comprised of dextran and albumin, and further embedding these within gastro-resistant Eudragit L microparticles was effective in the oral delivery of exenatide. These NP formulations led to decreases in blood glucose as well as glycated hemoglobin (HbA_{1c}) that were comparable in magnitude to the effects of subcutaneous exenatide (Byetta) injections (Soudry-Kochavi et al., 2018).

As in previously discussed peptides, NPs based on zein or liposomes have also been applied to exenatide oral delivery. Hybrid zein NPs prepared by complexing exenatide with cholic acid and phosphatidylcholine greatly increased loading capacity, whereas in other embodiment, zein NPs containing exenatide disperse with hypromellose phthalate, an absorption enhancer and casein, a GI protectant, were also prepared. Both zein NPs led to increased insulin secretion (up to 60% relative to baseline), and promoted pancreatic β cell proliferation (up to 120% relative to baseline) after 7 weeks of administration (Bao et al., 2020). Exenatide-loaded cationic liposomes comprising of 1,2-dioleoyl-3-trimethylammonium-propane and 1,2-dioleoyl-sn-glycero-3-phosphocholine were also prepared. These NPs were later coated with anionic chondroitin sulfate-g-glycocholic acid, a modified form of bile acid, in order to improve GI stability. A single oral dose (200 μg/kg) of the liposomal exenatide attained a relative oral bioavailability of 19.5% compared to subcutaneous administration (20 μg/kg). Whereas the subcutaneous exenatide group showed a sudden drop in plasma drug concentration, as expected from the drug's short half-life, but the liposomal exenatide was able to sustain drug release for up to 5 days. The HbA_{1c} profile of the orally dosed group was also markedly better than the subcutaneous group over a period of 4 weeks (Suzuki et al., 2019).

Interestingly, Lamson et al. demonstrated that plain silica NPs with net negative surface charge and size ≤ 200 nm induced tight junction relaxation and conferred greater intestinal absorption without the use of additional permeation enhancers (Lamson et al., 2020). Similarly, silica NPs-crosslinked to a self-assembled core of L-arginine, Zn^{2+}, and exenatide is a simple but highly scalable delivery vehicle. The core excipients served to stabilize the peptide drug, whereas the cationic L-arginine also exhibited permeation enhancement properties. In mice, these NPs, (which orally delivered exenatide at 1 mg/kg dose,) were found to prevent plasma glucose elevation by about 16% after an oral glucose challenge test, (compared to about 77% prevention by 10 μg/kg subcutaneous exenatide) (Hristov et al., 2020). Finally, large pore dendritic silica NPs with a pore size of about 10 nm were prepared to encapsulate exenatide at a notably high loading capacity of 35%–40%. These NPs, after further coating with chitosan, effectively delivered the drug across single (i.e., Caco-2 monolayer) as well as triple co-cultured intestinal models (Abeer et al., 2019).

The conjugation of highly specific and relatively nontoxic peptides to the surface of NPs has been discussed in prior sections in terms of active, receptor-mediated delivery. Here, a cationic cell penetrating peptide—the low molecular weight protamine (LMWP) with sequence VSRRRRRRGGRRRRC has shown cell internalization proficiency at a level similar to that of HIV-1 TAT (trans-activator of transcript). LMWP-conjugated dextran PLGA NPs successfully overcame multiple intestinal barriers, and in mice, Zn^{2+}-stabilized exenatide was orally delivered by these NPs to achieve a relative bioavailability of 8.4%, and an 80% reduction in glycemia. Consistent with the slower release kinetics of oral formulations, the T_{max} of LMWP-conjugated NPs was delayed to 8 h post-administration, in comparison to 15 min for the subcutaneous exenatide (Song et al., 2019). In another study, LMWP was used to conjugate a similar polymeric NP system consisting of PEG and PLGA, which in turn encapsulated the exenatide/Zn^{2+} complex. In diabetic rats, these NPs also led to marked reduction in glycemia (57% at hours of treatment), and resulted in a relative bioavailability of 7.44% (Zhang, Shi, Song, Sun, et al., 2018).

Transferrin receptor 1 (TfR1) and neonatal Fc receptor (FcRN)-targeted delivery of exenatide have been investigated using the cognate ligands transferrin (Tf) and Fc, respectively. Either ligand could be conjugated to PEG-PLGA NPs using standard methods such as sulfhydryl-reactive crosslinker chemistry with Traut's reagent (2-iminothiolane) and maleimide-functionalized PEG-PLGA. The Tf-conjugated NPs were able to encapsulate Zn^{2+}-stabilized exenatide, and attain a relative bioavailability of 6.45% from a 100 μg/kg oral dose. Furthermore, the TfR1-directed delivery resulted in a 1.67-fold higher plasma exenatide concentration in comparison to the ligand-free PEG-PLGA NPs. In a subsequent pharmacodynamics study, diabetic mice were shown to experience drug-induced hypoglycemia (approximately 50% reduction in blood glucose) nearly to the same extent with oral Tf-conjugated NPs (100 μg/kg) as with a subcutaneous exenatide injection (10 μg/kg) (Zhang, Shi, Song, Duan, et al., 2018). Likewise, the same research group performed Fc conjugation to PEG-PLGA NPs using similar synthesis methods. In diabetic mice, a corresponding drop of approximately 30% in blood glucose level was observed in the Fc-conjugated NP group (100 μg/kg), whereas a slight increase in blood glucose level was observed in the ligand-free NP controls (100 μg/kg). In both studies, the authors used DiR: a near-infrared fluorescent dye, to image the NPs transiting through the GI tract. They found that both Tf and Fc greatly increased the GI retention time of the particles, which was consistent with the presence of high-affinity, receptor-mediated particle interactions in the gut (Shi et al., 2018).

4.6 Future perspectives

A common theme observed in the few successful or partially successful clinical trials for oral peptide delivery has been the deployment of permeation enhancers, which nonspecifically open up avenues of absorption for macromolecules to cross the intestinal barriers. However, using the ligand-receptor approach to improve absorption should be a better alternative to permeation enhancers, as the specificity of the former would result in a far less risky safety profile. Medicinal chemists are always keen on the invention of an analog or prodrug of proteins and peptides that exhibits better physicochemical behavior. Still, we must keep in mind that any covalent modification may lead to another dark tunnel with unexpected permeation, metabolism, efficacy, pharmacokinetics, pharmacodynamics, and clearance. Perhaps the most straightforward path through the current tunnel is to keep the peptide drugs in their native states, and instead work around the drug formulation, where changes to the carrier size (i.e., to micro- or nanoscale) and surface properties have been correlated with improved physiological uptake (des Rieux et al., 2006; Hoshyar et al., 2016; Mosquera et al., 2018; Vance et al., 2015).

Despite years of efforts in this field, we believe that the technological barriers to formulating a clinically-viable NP may not be as insurmountable as a pessimist would contend. As in the case of the GLP-1 analog Rybelsus, a peptide drug given at correspondingly high oral doses can nonetheless show efficacy even with low systemic bioavailability (i.e., about 1%). Since many of the NPs discussed in this chapter exhibit good loading capacity and can theoretically carry a concentrated drug cargo, we may have already developed a proficient-enough oral delivery system for insulin, calcitonin and other peptides that has simply not been tested in humans yet. Moving forward, there will always be a place for individual labs to come up with creative formulations that push the boundaries of nanotechnology, but the current emphasis of this medically-oriented field should be on translation.

While there are numerous, highly inventive NP prototypes being developed around the world, much of the current work is still heavily reliant on cell cultures and rodent models, with more expensive experimental options (e.g., those involving dogs, pigs, and monkeys) as well as human trials out of reach for the typical university lab. Yet another more pressing problem lies within the largely uncoordinated nature of basic research, where it is often difficult to compare data from different labs to determine how to optimize a particular NP system. Similar to the clinical trial consortia, which evolved out of a need to perform large scale studies with limited resources, a nanomedicine consortium would pool and coordinate resources from multiple institutions with complementary expertise so that more ambitious projects could be planned. Such was the rationale behind the formation of TRANS-INT (2012–2017), a €11 million EU consortium comprising of sixteen partners from academia and industry, which systematically investigated multiple candidates for NP-assisted oral peptide delivery. While choosing to employ only in vitro and preclinical models at the outset, TRANS-INT nevertheless identified up to 13 NP families with therapeutic potentials, leading to 25 publications and 3 patent applications (TRANS-INT, 2020). Future advancements in nanomedicine may very well depend on more consortia like TRANS-INT, which not only possess the means to advocate larger funding and to conduct more costly preclinical and clinical research, but also the mandate to openly share their discoveries with the wider scientific community.

Acknowledgments

This work is supported in part by grants R01DK125372, R01DK128597, R01AI155908 and R01EY028169 from the National Institute of Health.

References

Abeer, M. M., Meka, A. K., Pujara, N., Kumeria, T., Strounina, E., Nunes, R., Costa, A., Sarmento, B., Hasnain, S. Z., Ross, B. P., & Popat, A. (2019). Rationally designed dendritic silica nanoparticles for oral delivery of exenatide. *Pharmaceutics*, *11*(8).

Ahmad, J., Singhal, M., Amin, S., Rizwanullah, M., Akhter, S., Kamal, M. A., Haider, N., Midoux, P., & Pichon, C. (2017). Bile salt stabilized vesicles (bilosomes): A novel nano-pharmaceutical design for oral delivery of proteins and peptides. *Current Pharmaceutical Design*, *23*(11), 1575–1588.

Altshuler, A. E., Kistler, E. B., & Schmid-Schonbein, G. W. (2016). Autodigestion: Proteolytic degradation and multiple organ failure in shock. *Shock*, *45*(5), 483–489.

Amaral, M., Martins, A. S., Catarino, J., Faisca, P., Kumar, P., Pinto, J. F., Pinto, R., Correia, I., Ascensao, L., Afonso, R. A., Gaspar, M. M., Charmier, A. J., Figueiredo, I. V., & Reis, C. P. (2020). How Can biomolecules improve mucoadhesion of oral insulin? a comprehensive insight using ex-vivo, in silico, and in vivo models. *Biomolecules*, *10*(5).

American Diabetes, A. (2020). 9. Pharmacologic approaches to glycemic treatment: Standards of medical care in diabetes-2020. *Diabetes Care*, *43*(Suppl 1), S98–S110.

Andreani, T., Kiill, C. P., de Souza, A. L., Fangueiro, J. F., Fernandes, L., Doktorovova, S., Santos, D. L., Garcia, M. L., Gremiao, M. P., Souto, E. B., & Silva, A. M. (2014). Surface engineering of silica nanoparticles for oral insulin delivery: Characterization and cell toxicity studies. *Colloids and Surfaces. B, Biointerfaces*, *123*, 916–923.

Andreani, T., Miziara, L., Lorenzon, E. N., de Souza, A. L., Kiill, C. P., Fangueiro, J. F., Garcia, M. L., Gremiao, P. D., Silva, A. M., & Souto, E. B. (2015). Effect of mucoadhesive polymers on the in vitro performance of insulin-loaded silica nanoparticles: Interactions with mucin and biomembrane models. *European Journal of Pharmaceutics and Biopharmaceutics: Official Journal of Arbeitsgemeinschaft fur Pharmazeutische Verfahrenstechnik e.V*, *93*, 118–126.

Anselmo, A. C., & Mitragotri, S. (2014). An overview of clinical and commercial impact of drug delivery systems. *Journal of Controlled Release: Official Journal of the Controlled Release Society*, *190*, 15–28.

Antalis, T. M., Shea-Donohue, T., Vogel, S. N., Sears, C., & Fasano, A. (2007). Mechanisms of disease: Protease functions in intestinal mucosal pathobiology. *Nature Clinical Practice. Gastroenterology & Hepatology*, *4*(7), 393–402.

Antunes, F., Andrade, F., Ferreira, D., Morck Nielsen, H., & Sarmento, B. (2013). Models to predict intestinal absorption of therapeutic peptides and proteins. *Current Drug Metabolism*, *14*(1), 4–20.

Ashfaq, U. A., Riaz, M., Yasmeen, E., & Yousaf, M. Z. (2017). Recent advances in nanoparticle-based targeted drug-delivery systems against cancer and role of tumor microenvironment. *Critical Reviews in Therapeutic Drug Carrier Systems*, *34*(4), 317–353.

Attie, A. D., & Raines, R. T. (1995). Analysis of receptor-ligand interactions. *Journal of Chemical Education*, *72*(2), 119–124.

Azria, M., Copp, D., & Zanelli, J. (1995). 25 years of salmon calcitonin: From synthesis to therapeutic use. *Calcified Tissue International*, *57*(6), 405–408.

Bai, Y., Zhou, R., Wu, L., Zheng, Y., Liu, X., Wu, R., Li, X., & Huang, Y. (2020). Nanoparticles with surface features of dendritic oligopeptides as potential oral drug delivery systems. *Journal of Materials Chemistry B*, *8*(13), 2636–2649.

Bao, X., Qian, K., & Yao, P. (2020). Oral delivery of exenatide-loaded hybrid zein nanoparticles for stable blood glucose control and beta-cell repair of type 2 diabetes mice. *Journal of Nanobiotechnology*, *18*(1), 67.

Behzadi, S., Serpooshan, V., Tao, W., Hamaly, M. A., Alkawareek, M. Y., Dreaden, E. C., Brown, D., Alkilany, A. M., Farokhzad, O. C., & Mahmoudi, M. (2017). Cellular uptake of nanoparticles: Journey inside the cell. *Chemical Society Reviews, 46* (14), 4218–4244.

Bhattacharjee, S. (2016). DLS and zeta potential—what they are and what they are not? *Journal of Controlled Release: Official Journal of the Controlled Release Society, 235*, 337–351.

Bliss, M. (1982). Banting's, Best's, and Collip's accounts of the discovery of insulin. *Bulletin of the History of Medicine, 56*(4), 554–568.

Boegh, M., & Nielsen, H. M. (2015). Mucus as a barrier to drug delivery—understanding and mimicking the barrier properties. *Basic & Clinical Pharmacology & Toxicology, 116* (3), 179–186.

Brayden, D. J., Hill, T. A., Fairlie, D. P., Maher, S., & Mrsny, R. J. (2020). Systemic delivery of peptides by the oral route: Formulation and medicinal chemistry approaches. *Advanced Drug Delivery Reviews.*

Breckenridge, A., & Jacob, R. (2019). Overcoming the legal and regulatory barriers to drug repurposing. *Nature Reviews. Drug Discovery, 18*(1), 1–2.

Brown, T. D., Whitehead, K. A., & Mitragotri, S. (2019). Materials for oral delivery of proteins and peptides. *Nature Reviews Materials*, 1–22.

Builders, P. F., Kunle, O. O., Okpaku, L. C., Builders, M. I., Attama, A. A., & Adikwu, M. U. (2008). Preparation and evaluation of mucinated sodium alginate microparticles for oral delivery of insulin. *European Journal of Pharmaceutics and Biopharmaceutics: Official Journal of Arbeitsgemeinschaft fur Pharmazeutische Verfahrenstechnik e.V, 70*(3), 777–783.

Carney, S. L. (1997). Calcitonin and human renal calcium and electrolyte transport. *Mineral and Electrolyte Metabolism, 23*(1), 43–47.

Carroll, R., & Matfin, G. (2010). Endocrine and metabolic emergencies: Hypercalcaemia. *Therapeutic Advances in Endocrinology and Metabolism, 1*(5), 225–234.

Cetin, M., Youn, Y. S., Capan, Y., & Lee, K. C. (2008). Preparation and characterization of salmon calcitonin-biotin conjugates. *AAPS PharmSciTech, 9*(4), 1191–1197.

Chauhan, M. K., Sahoo, P. K., Rawat, A. S., Singh, A., Bamrara, A., & Sharma, D. (2015). Bilosomes: A novel approach to meet the challenges in oral immunization. *Recent Patents on Drug Delivery & Formulation, 9*(3), 201–212.

Chen, J., Almo, S. C., & Wu, Y. (2017). General principles of binding between cell surface receptors and multi-specific ligands: A computational study. *PLoS Computational Biology, 13*(10), e1005805.

Cheng, C. J., Tietjen, G. T., Saucier-Sawyer, J. K., & Saltzman, W. M. (2015). A holistic approach to targeting disease with polymeric nanoparticles. *Nature Reviews. Drug Discovery, 14*(4), 239–247.

Choonara, B. F., Choonara, Y. E., Kumar, P., Bijukumar, D., du Toit, L. C., & Pillay, V. (2014). A review of advanced oral drug delivery technologies facilitating the protection and absorption of protein and peptide molecules. *Biotechnology Advances, 32*(7), 1269–1282.

Chuang, E.-Y., Nguyen, G. T. H., Su, F.-Y., Lin, K.-J., Chen, C.-T., Mi, F.-L., Yen, T.-C., Juang, J.-H., & Sung, H.-W. (2013). Combination therapy via oral co-administration of insulin-and exendin-4-loaded nanoparticles to treat type 2 diabetic rats undergoing OGTT. *Biomaterials, 34*(32), 7994–8001.

Ciacci, C., Mahida, Y. R., Dignass, A., Koizumi, M., & Podolsky, D. K. (1993). Functional interleukin-2 receptors on intestinal epithelial cells. *The Journal of Clinical Investigation, 92*(1), 527–532.

Davis, S. S., & Illum, L. (1998). Drug delivery systems for challenging molecules. *International Journal of Pharmaceutics, 176*(1), 1–8.

des Rieux, A., Fievez, V., Garinot, M., Schneider, Y.-J., & Préat, V. (2006). Nanoparticles as potential oral delivery systems of proteins and vaccines: A mechanistic approach. *Journal of Controlled Release: Official Journal of the Controlled Release Society, 116*(1), 1–27.

des Rieux, A., Pourcelle, V., Cani, P. D., Marchand-Brynaert, J., & Preat, V. (2013). Targeted nanoparticles with novel non-peptidic ligands for oral delivery. *Advanced Drug Delivery Reviews, 65*(6), 833–844.

DiMasi, J. A., Grabowski, H. G., & Hansen, R. W. (2016). Innovation in the pharmaceutical industry: New estimates of R&D costs. *Journal of Health Economics, 47*, 20–33.

Ding, Y., Wang, Q., Liu, G., Feng, Y., & Zhou, W. (2020). Cholesterol moieties as building blocks for assembling nanoparticles to achieve effective oral delivery of insulin. *Biomaterials Science, 8*(14), 3979–3993.

Doggrell, S. A. (2018). Sgemaglutide in type 2 diabetes - is it the best glucagon-like peptide 1 receptor agonist (GLP-1R agonist)? *Expert Opinion on Drug Metabolism & Toxicology, 14*(3), 371–377.

Dong, N., Zhu, C., Jiang, J., Huang, D., Li, X., Quan, G., Liu, Y., Tan, W., Pan, X., & Wu, C. (2020). Development of composite PLGA microspheres containing exenatide-encapsulated lecithin nanoparticles for sustained drug release. *Asian Journal of Pharmaceutical Sciences., 15*(3), 347–355.

Drab, S. R. (2016). Glucagon-like peptide-1 receptor agonists for type 2 diabetes: A clinical update of safety and efficacy. *Current Diabetes Reviews, 12*(4), 403–413.

Edgerton, D. S., Scott, M., Farmer, B., Williams, P. E., Madsen, P., Kjeldsen, T., Brand, C. L., Fledelius, C., Nishimura, E., & Cherrington, A. D. (2019). Targeting insulin to the liver corrects defects in glucose metabolism caused by peripheral insulin delivery. *JCI Insight*, 5.

Ensign, L. M., Schneider, C., Suk, J. S., Cone, R., & Hanes, J. (2012). Mucus penetrating nanoparticles: Biophysical tool and method of drug and gene delivery. *Advanced Materials, 24*(28), 3887–3894.

Falciani, C., Lozzi, L., Pini, A., & Bracci, L. (2005). Bioactive peptides from libraries. *Chemistry & Biology, 12*(4), 417–426.

Felsenfeld, A. J., & Levine, B. S. (2015). Calcitonin, the forgotten hormone: Does it deserve to be forgotten? *Clininical Kidney Journal, 8*(2), 180–187.

Filippatos, T. D., Panagiotopoulou, T. V., & Elisaf, M. S. (2014). Adverse effects of glp-1 receptor agonists. *The Review of Diabetic Studies: RDS, 11*(3–4), 202–230.

Fonte, P., Araujo, F., Silva, C., Pereira, C., Reis, S., Santos, H. A., & Sarmento, B. (2015). Polymer-based nanoparticles for oral insulin delivery: Revisited approaches. *Biotechnology Advances, 33*(6 Pt 3), 1342–1354.

Foroozandeh, P., & Aziz, A. A. (2018). Insight into cellular uptake and intracellular trafficking of nanoparticles. *Nanoscale Research Letters, 13*(1), 339.

Gaba, B., Fazil, M., Ali, A., Baboota, S., Sahni, J. K., & Ali, J. (2015). Nanostructured lipid (NLCs) carriers as a bioavailability enhancement tool for oral administration. *Drug Delivery, 22*(6), 691–700.

Ganugula, R., Arora, M., Guada, M., Saini, P., & Kumar, M. N. R. (2017). Noncompetitive active transport exploiting intestinal transferrin receptors for oral delivery of proteins by tunable nanoplatform. *ACS Macro Letters, 6*(2), 161–164.

Ganugula, R., Arora, M., Saini, P., Guada, M., & Kumar, M. (2017). Next generation precision-polyesters enabling optimization of ligand-receptor stoichiometry for modular drug delivery. *Journal of the American Chemical Society, 139*(21), 7203–7216.

Ganugula, R., Arora, M., Zou, D., Agarwal, S. K., Mohan, C., & Kumar, M. R. (2020). A highly potent lymphatic system–targeting nanoparticle cyclosporine prevents glomerulonephritis in mouse model of lupus. *Science Advances, 6*(24), eabb3900.

Gedawy, A., Martinez, J., Al-Salami, H., & Dass, C. R. (2018). Oral insulin delivery: Existing barriers and current counter-strategies. *The Journal of Pharmacy and Pharmacology*, *70*(2), 197–213.

Geho, W. B., Rosenberg, L. N., Schwartz, S. L., Lau, J. R., & Gana, T. J. (2014). A single-blind, placebo-controlled, dose-ranging trial of oral hepatic-directed vesicle insulin add-on to oral antidiabetic treatment in patients with type 2 diabetes mellitus. *Journal of Diabetes Science and Technology (Online)*, *8*(3), 551–559.

Goldberg, M., & Gomez-Orellana, I. (2003). Challenges for the oral delivery of macromolecules. *Nature Reviews. Drug Discovery*, *2*(4), 289–295.

Goldenberg, M. M. (2012). Pharmaceutical approval update. *P T*, *37*(3), 137–139.

Gonzalez-Paredes, A., Torres, D., & Alonso, M. J. (2017). Polyarginine nanocapsules: A versatile nanocarrier with potential in transmucosal drug delivery. *International Journal of Pharmaceutics*, *529*(1–2), 474–485.

Gozdzialska, A., Jaskiewicz, J., Knapik-Czajka, M., Drag, J., Gawlik, M., Ciesla, M., Kulis, A., Zarzycki, D., & Lipik, E. (2016). Association of calcium and phosphate balance, vitamin d, pth, and calcitonin in patients with adolescent idiopathic scoliosis. *Spine*, *41*(8), 693–697.

Graham, R. M. (1994). Cyclosporine: Mechanisms of action and toxicity. *Cleveland Clinic Journal of Medicine*, *61*(4), 308–313.

Group, I. N. R. T. S. (2002). Cyclosporine microemulsion (neoral®) absorption profiling and sparse-sample predictors during the first 3 months after renal transplantation. *American Journal of Transplantation*, *2*(2), 148–156.

Guada, M., Beloqui, A., Alhouayek, M., Muccioli, G. G., Dios-Vieitez Mdel, C., Preat, V., & Blanco-Prieto, M. J. (2016). Cyclosporine a-loaded lipid nanoparticles in inflammatory bowel disease. *International Journal of Pharmaceutics*, *503*(1–2), 196–198.

Guada, M., Beloqui, A., Kumar, M. N., Preat, V., Dios-Vieitez Mdel, C., & Blanco-Prieto, M. J. (2016). Reformulating cyclosporine A (CsA): More than just a life cycle management strategy. *Journal of Controlled Release: Official Journal of the Controlled Release Society*, *225*, 269–282.

Guada, M., Lana, H., Gil, A. G., Dios-Vieitez Mdel, C., & Blanco-Prieto, M. J. (2016). Cyclosporine A lipid nanoparticles for oral administration: Pharmacodynamics and safety evaluation. *European Journal of Pharmaceutics and Biopharmaceutics: Official Journal of Arbeitsgemeinschaft fur Pharmazeutische Verfahrenstechnik e.V*, *101*, 112–118.

Guada, M., Lasa-Saracibar, B., Lana, H., Dios-Vieitez Mdel, C., & Blanco-Prieto, M. J. (2016). Lipid nanoparticles enhance the absorption of cyclosporine a through the gastrointestinal barrier: In vitro and in vivo studies. *International Journal of Pharmaceutics*, *500*(1–2), 154–161.

Guan, P., Lu, Y., Qi, J., Niu, M., Lian, R., Hu, F., & Wu, W. (2011). Enhanced oral bioavailability of cyclosporine A by liposomes containing a bile salt. *International Journal of Nanomedicine*, *6*, 965–974.

Guan, P., Lu, Y., Qi, J., & Wu, W. (2016). Readily restoring freeze-dried probilosomes as potential nanocarriers for enhancing oral delivery of cyclosporine A. *Colloids and Surfaces. B, Biointerfaces*, *144*, 143–151.

Haastrup, M. B., Henriksen, J. E., Mortz, C. G., & Bindslev-Jensen, C. (2018). Insulin allergy can be successfully managed by a systematic approach. *Clinical and Translational Allergy*, *8*, 35.

Handschumacher, R. E., Harding, M. W., Rice, J., Drugge, R. J., & Speicher, D. W. (1984). Cyclophilin: A specific cytosolic binding protein for cyclosporin A. *Science*, *226*(4674), 544–547.

Hao, J., & Heng, P. W. (2003). Buccal delivery systems. *Drug Development and Industrial Pharmacy*, *29*(8), 821–832.

Hatton, G. B., Yadav, V., Basit, A. W., & Merchant, H. A. (2015). Animal farm: Considerations in animal gastrointestinal physiology and relevance to drug delivery in humans. *Journal of Pharmaceutical Sciences*, *104*(9), 2747–2776.

Hay, M., Thomas, D. W., Craighead, J. L., Economides, C., & Rosenthal, J. (2014). Clinical development success rates for investigational drugs. *Nature Biotechnology*, *32*(1), 40–51.

Herschman, H. R., Simpson, D. L., & Cawley, D. B. (1982). Toxic ligand conjugates as tools in the study of receptor-ligand interactions. *Journal of Cellular Biochemistry*, *20*(2), 163–176.

Hjerpsted, J. B., Flint, A., Brooks, A., Axelsen, M. B., Kvist, T., & Blundell, J. (2018). Semaglutide improves postprandial glucose and lipid metabolism, and delays first-hour gastric emptying in subjects with obesity. *Diabetes, Obesity & Metabolism*, *20*(3), 610–619.

Homayun, B., Lin, X., & Choi, H. J. (2019). Challenges and recent progress in oral drug delivery systems for biopharmaceuticals. *Pharmaceutics*, *11*(3).

Hoshyar, N., Gray, S., Han, H., & Bao, G. (2016). The effect of nanoparticle size on in vivo pharmacokinetics and cellular interaction. *Nanomedicine*, *11*(6), 673–692.

Hristov, D., McCartney, F., Beirne, J., Mahon, E., Reid, S., Bhattacharjee, S., Penarier, G., Werner, U., Bazile, D., & Brayden, D. J. (2020). Silica-coated nanoparticles with a core of zinc, l-arginine, and a peptide designed for oral delivery. *ACS Applied Materials & Interfaces*, *12*(1), 1257–1269.

Huang, A., Su, Z., Li, S., Sun, M., Xiao, Y., Ping, Q., & Deng, Y. (2014). Oral absorption enhancement of salmon calcitonin by using both N-trimethyl chitosan chloride and oligoarginines-modified liposomes as the carriers. *Drug Delivery*, *21*(5), 388–396.

Hughes, J. L., Seyler, J. K., & Liu, R. C.-H. (1975) *Synthesis of salmon calcitonin*. Google Patents.

Inchaurraga, L., Martinez-Lopez, A. L., Martin-Arbella, N., & Irache, J. M. (2020). Zein-based nanoparticles for the oral delivery of insulin. *Drug Delivery and Translational Research*.

Jain, P. N., & Chatterjee, A. (2020). A randomized placebo-controlled trial evaluating the analgesic effect of salmon calcitonin in refractory bone metastasis pain. *Indian Journal Palliative Care*, *26*(1), 4–8.

Jaradat, A., Macedo, M. H., Sousa, F., Arkill, K., Alexander, C., Aylott, J., & Sarmento, B. (2020). Prediction of the enhanced insulin absorption across a triple co-cultured intestinal model using mucus penetrating PLGA nanoparticles. *International Journal of Pharmaceutics*, *585*, 119516.

Ji, N., Hong, Y., Gu, Z., Cheng, L., Li, Z., & Li, C. (2018). Preparation and characterization of insulin-loaded zein/carboxymethylated short-chain amylose complex nanoparticles. *Journal of Agricultural and Food Chemistry*, *66*(35), 9335–9343.

Ji, N., Hong, Y., Gu, Z., Cheng, L., Li, Z., & Li, C. (2019). Chitosan coating of zein-carboxymethylated short-chain amylose nanocomposites improves oral bioavailability of insulin in vitro and in vivo. *Journal of Controlled Release: Official Journal of the Controlled Release Society*, *313*, 1–13.

Johns, K. R., & Littlejohn, G. O. (1999). The safety and efficacy of cyclosporine (Neoral) in rheumatoid arthritis. *The Journal of Rheumatology*, *26*(10), 2110–2113.

Juere, E., Caillard, R., Marko, D., Del Favero, G., & Kleitz, F. (2020). Smart protein-based formulation of dendritic mesoporous silica nanoparticles: Toward oral delivery of insulin. *Chemistry*, *26*(23), 5195–5199.

Karakoti, A. S., Shukla, R., Shanker, R., & Singh, S. (2015). Surface functionalization of quantum dots for biological applications. *Advances in Colloid and Interface Science*, *215*, 28–45.

Kaur, G., Arora, M., & Ravi Kumar, M. N. V. (2019). Oral drug delivery technologies-a decade of developments. *The Journal of Pharmacology and Experimental Therapeutics, 370*(3), 529–543.

Kim, B. R., Yang, S., Doh, E. J., Choi, C. W., & Youn, S. W. (2018). Risk factors affecting adverse effects of cyclosporine a in a real-world psoriasis treatment. *Annals of Dermatology, 30*(2), 143–149.

Kotak, D. J., & Devarajan, P. V. (2020). Bone targeted delivery of salmon calcitonin hydroxyapatite nanoparticles for sublingual osteoporosis therapy (SLOT). *Nanomedicine: Nanotechnology, Biology, and Medicine, 24*, 102153.

Kumar, V., Choudhry, I., Namdev, A., Mishra, S., Soni, S., Hurkat, P., Jain, A., & Jain, D. (2018). Oral Insulin: Myth or Reality. *Current Diabetes Reviews, 14*(6), 497–508.

Labib, G. (2018). Overview on zein protein: A promising pharmaceutical excipient in drug delivery systems and tissue engineering. *Expert Opinion on Drug Delivery, 15*(1), 65–75.

Lallemand, F., Schmitt, M., Bourges, J. L., Gurny, R., Benita, S., & Garrigue, J. S. (2017). Cyclosporine A delivery to the eye: A comprehensive review of academic and industrial efforts. *European Journal of Pharmaceutics and Biopharmaceutics: Official Journal of Arbeitsgemeinschaft fur Pharmazeutische Verfahrenstechnik e.V, 117*, 14–28.

Lamson, N. G., Berger, A., Fein, K. C., & Whitehead, K. A. (2020). Anionic nanoparticles enable the oral delivery of proteins by enhancing intestinal permeability. *Nature Biomedical Engineering, 4*(1), 84–96.

Lau, J. L., & Dunn, M. K. (2018). Therapeutic peptides: Historical perspectives, current development trends, and future directions. *Biorganic & Medicinal Chemistry., 26*(10), 2700–2707.

Laupacis, A., Keown, P. A., Ulan, R. A., McKenzie, N., & Stiller, C. R. (1982). Cyclosporin A: A powerful immunosuppressant. *Canadian Medical Association Journal, 126*(9), 1041–1046.

Lee, E. S., & Youn, Y. S. (2016). Albumin-based potential drugs: Focus on half-life extension and nanoparticle preparation. *Journal of Pharmaceutical Investigation., 46*(4), 305–315.

Li, C., Wang, J., Wang, Y., Gao, H., Wei, G., Huang, Y., Yu, H., Gan, Y., Wang, Y., Mei, L., Chen, H., Hu, H., Zhang, Z., & Jin, Y. (2019). Recent progress in drug delivery. *Acta Pharmaceutica Sinica B, 9*(6), 1145–1162.

Lim, K. K., Su, W. P., Schroeter, A. L., Sabers, C. J., Abraham, R. T., & Pittelkow, M. R. (1996). Cyclosporine in the treatment of dermatologic disease: An update. *Mayo Clinic Proceedings. Mayo Clinic, 71*(12), 1182–1191.

Liu, C., Kou, Y., Zhang, X., Cheng, H., Chen, X., & Mao, S. (2018). Strategies and industrial perspectives to improve oral absorption of biological macromolecules. *Expert Opinion on Drug Delivery, 15*(3), 223–233.

Liu, L., Yang, H., Lou, Y., Wu, J. Y., Miao, J., Lu, X. Y., & Gao, J. Q. (2019). Enhancement of oral bioavailability of salmon calcitonin through chitosan-modified, dual drug-loaded nanoparticles. *International Journal of Pharmaceutics, 557*, 170–177.

Liu, M., Zhong, X., & Yang, Z. (2017). Chitosan functionalized nanocochleates for enhanced oral absorption of cyclosporine A. *Scientific Reports, 7*, 41322.

Lollo, G., Gonzalez-Paredes, A., Garcia-Fuentes, M., Calvo, P., Torres, D., & Alonso, M. J. (2017). Polyarginine nanocapsules as a potential oral peptide delivery carrier. *Journal of Pharmaceutical Sciences, 106*(2), 611–618.

Lundquist, P., & Artursson, P. (2016). Oral absorption of peptides and nanoparticles across the human intestine: Opportunities, limitations and studies in human tissues. *Advanced Drug Delivery Reviews, 106*(Pt B), 256–276.

Luo, Y. Y., Xiong, X. Y., Tian, Y., Li, Z. L., Gong, Y. C., & Li, Y. P. (2016). A review of biodegradable polymeric systems for oral insulin delivery. *Drug Delivery, 23*(6), 1882–1891.

Ma, Z., Qian, P., Shen, R., Hu, B., He, X., Gao, F., Shen, B., Zhang, N., Shan, Y., Shen, X., Gao, T., & Jin, L. (2019). Pharmacological signatures of the exenatide nanoparticles against hepatic ischemia/reperfusion-induced pancreatic injury. *Transplantation Proceedings*, *51*(3), 960–965.

Macedo, A., Filipe, P., Thome, N. G., Vieira, J., Oliveira, C., Teodosio, C., Ferreira, R., Roque, L., & Fonte, P. (2020). A brief overview of the oral delivery of insulin as an alternative to the parenteral delivery. *Current Molecular Medicine*, *20*(2), 134–143.

Maher, S., Brayden, D. J., Casettari, L., & Illum, L. (2019). Application of permeation enhancers in oral delivery of macromolecules: An update. *Pharmaceutics*, *11*(1), 41.

Maher, S., & Brayden, D. J. (2012). Overcoming poor permeability: Translating permeation enhancers for oral peptide delivery. *Drug Discovery Today: Technologies*, *9*(2), e113–e119.

Mahjub, R., Najafabadi, F. K., Dehkhodaei, N., Kheiripour, N., Ahmadabadi, A. N., Asl, S. S., Taheri-Azandariani, M., & Ranjbar, A. (2020). Eudragit L-100 capsules/aromatize and quaternerize chitosan for insulin nanoparticle oral delivery on toxic oxidative stress in rat liver and kidney. *Pharmaceutical nanotechnology*.

Mahmoudi, M., Meng, J., Xue, X., Liang, X. J., Rahman, M., Pfeiffer, C., Hartmann, R., Gil, P. R., Pelaz, B., Parak, W. J., Del Pino, P., Carregal-Romero, S., Kanaras, A. G., & Tamil Selvan, S. (2014). Interaction of stable colloidal nanoparticles with cellular membranes. *Biotechnology Advances*, *32*(4), 679–692.

Malathi, S., Nandhakumar, P., Pandiyan, V., Webster, T. J., & Balasubramanian, S. (2015). Novel PLGA-based nanoparticles for the oral delivery of insulin. *International Journal of Nanomedicine*, *10*, 2207–2218.

Martinez, M. N., & Amidon, G. L. (2002). A mechanistic approach to understanding the factors affecting drug absorption: A review of fundamentals. *Journal of Clinical Pharmacology*, *42*(6), 620–643.

McClements, D. J. (2018). Encapsulation, protection, and delivery of bioactive proteins and peptides using nanoparticle and microparticle systems: A review. *Advances in Colloid and Interface Science*, *253*, 1–22.

Melero, A., Draheim, C., Hansen, S., Giner, E., Carreras, J. J., Talens-Visconti, R., Garrigues, T. M., Peris, J. E., Recio, M. C., Giner, R., & Lehr, C. M. (2017). Targeted delivery of Cyclosporine A by polymeric nanocarriers improves the therapy of inflammatory bowel disease in a relevant mouse model. *European Journal of Pharmaceutics and Biopharmaceutics: Official Journal of Arbeitsgemeinschaft fur Pharmazeutische Verfahrenstechnik e. V*, *119*, 361–371.

Mosquera, J., Garcia, I., & Liz-Marzan, L. M. (2018). Cellular uptake of nanoparticles vs small molecules: A matter of size. *Accounts of Chemical Research*, *51*(9), 2305–2313.

Mumuni, M. A., Kenechukwu, F. C., Ofokansi, K. C., Attama, A. A., & Diaz, D. D. (2020). Insulin-loaded mucoadhesive nanoparticles based on mucin-chitosan complexes for oral delivery and diabetes treatment. *Carbohydrate Polymers*, *229*, 115506.

Munoz-Torres, M., Alonso, G., & Raya, M. P. (2004). Calcitonin therapy in osteoporosis. *Treatments in Endocrinology*, *3*(2), 117–132.

Muntoni, E., Marini, E., Ahmadi, N., Milla, P., Ghe, C., Bargoni, A., Capucchio, M. T., Biasibetti, E., & Battaglia, L. (2019). Lipid nanoparticles as vehicles for oral delivery of insulin and insulin analogs: Preliminary ex vivo and in vivo studies. *Acta Diabetologica*, *56*(12), 1283–1292.

Musther, H., Olivares-Morales, A., Hatley, O. J., Liu, B., & Hodjegan, A. R. (2014). Animal vs human oral drug bioavailability: Do they correlate? *European Journal of Pharmaceutical Sciences: Official Journal of the European Federation for Pharmaceutical Sciences*, *57*, 280–291.

Naeem, M., Bae, J., Oshi, M. A., Kim, M. S., Moon, H. R., Lee, B. L., Im, E., Jung, Y., & Yoo, J. W. (2018). Colon-targeted delivery of cyclosporine A using dual-functional Eudragit((R)) FS30D/PLGA nanoparticles ameliorates murine experimental colitis. *International Journal of Nanomedicine*, *13*, 1225–1240.

Nauck, M. A., & Friedrich, N. (2013). Do GLP-1-based therapies increase cancer risk? *Diabetes Care*, *36*(Suppl 2), S245–S252.

Nel, A. E., Madler, L., Velegol, D., Xia, T., Hoek, E. M., Somasundaran, P., Klaessig, F., Castranova, V., & Thompson, M. (2009). Understanding biophysicochemical interactions at the nano-bio interface. *Nature Materials*, *8*(7), 543–557.

Niu, Z., Tedesco, E., Benetti, F., Mabondzo, A., Montagner, I. M., Marigo, I., Gonzalez-Touceda, D., Tovar, S., Dieguez, C., Santander-Ortega, M. J., & Alonso, M. J. (2017). Rational design of polyarginine nanocapsules intended to help peptides overcoming intestinal barriers. *Journal of Controlled Release: Official Journal of the Controlled Release Society*, *263*, 4–17.

Novartis MIACALCIN® (calcitonin-salmon), Reference 3118265 https://www.accessdata.fda.gov/drugsatfda_docs/label/2012/017808s034lbl.pdf (accessed 09/01/2020a).

Novartis MIACALCIN® (calcitonin-salmon) nasal spray label, Reference 4151413 https://www.accessdata.fda.gov/drugsatfda_docs/label/2017/020313s036lbl.pdf (accessed 09/01/2020b).

NovoNordisk RYBELSUS (semaglutide) tablets, for oral use, Reference 4494169 https://www.accessdata.fda.gov/drugsatfda_docs/label/2019/213051s000lbl.pdf (accessed 09/01/2020).

Oh, E. H., Koh, W. S., Shin, J. M., Kim, J. E., Ko, J. Y., & Ro, Y. S. (2018). Clinical experience of cyclosporin treatment in patients with psoriasis and psoriatic arthritis. *The Journal of Dermatology*, *45*(3), 329–330.

Okumura, R., & Takeda, K. (2017). Roles of intestinal epithelial cells in the maintenance of gut homeostasis. *Experimental & Molecular Medicine*, *49*(5), e338.

Owji, H., Nezafat, N., Negahdaripour, M., Hajiebrahimi, A., & Ghasemi, Y. (2018). A comprehensive review of signal peptides: Structure, roles, and applications. *European Journal of Cell Biology*, *97*(6), 422–441.

Pascoli, M., de Lima, R., & Fraceto, L. F. (2018). Zein nanoparticles and strategies to improve colloidal stability: A mini-review. *Frontiers in Chemistry*, *6*, 6.

Patel, D., & Wairkar, S. (2019). Recent advances in cyclosporine drug delivery: Challenges and opportunities. *Drug Delivery and Translational Research*, *9*(6), 1067–1081.

Patel, V. F., Liu, F., & Brown, M. B. (2011). Advances in oral transmucosal drug delivery. *Journal of Controlled Release: Official Journal of the Controlled Release Society*, *153*(2), 106–116.

Patra, J. K., Das, G., Fraceto, L. F., Campos, E. V. R., Rodriguez-Torres, M. D. P., Acosta-Torres, L. S., Diaz-Torres, L. A., Grillo, R., Swamy, M. K., Sharma, S., Habtemariam, S., & Shin, H. S. (2018). Nano based drug delivery systems: Recent developments and future prospects. *Journal of Nanobiotechnology*, *16*(1), 71.

Patricia, J. J., & Dhamoon, A. S. (2019). *Physiology*, Digestion.

Pauletti, G. M., Gangwar, S., Siahaan, T. J., Aubé, J., & Borchardt, R. T. (1997). Improvement of oral peptide bioavailability: Peptidomimetics and prodrug strategies. *Advanced Drug Delivery Reviews*, *27*(2–3), 235–256.

Peng, K., Chen, L., Peng, J., Xing, F., & Xiang, Z. (2015). Effects of calcitonin on lumbar spinal stenosis: A systematic review and *meta*-analysis. *International Journal of Clinical and Experimental Medicine*, *8*(2), 2536–2544.

Perry, S. L., & McClements, D. J. (2020). Recent advances in encapsulation, protection, and oral delivery of bioactive proteins and peptides using colloidal systems. *Molecules (Basel, Switzerland)*, *25*(5).

Polamreddy, P., & Gattu, N. (2019). The drug repurposing landscape from 2012 to 2017: Evolution, challenges, and possible solutions. *Drug Discovery Today*, *24*(3), 789–795.
Potts, J. T., Jr. (1992). Chemistry of the calcitonins. *Bone and Mineral*, *16*(3), 169–173.
Presas, E., McCartney, F., Sultan, E., Hunger, C., Nellen, S., Alvarez, C. V., Werner, U., Bazile, D., Brayden, D. J., & O'Driscoll, C. M. (2018). Physicochemical, pharmacokinetic and pharmacodynamic analyses of amphiphilic cyclodextrin-based nanoparticles designed to enhance intestinal delivery of insulin. *Journal of Controlled Release*, *286*, 402–414.
Pridgen, E. M., Alexis, F., Kuo, T. T., Levy-Nissenbaum, E., Karnik, R., Blumberg, R. S., Langer, R., & Farokhzad, O. C. (2013). Transepithelial transport of Fc-targeted nanoparticles by the neonatal fc receptor for oral delivery. *Science Translational Medicine*, *5*(213), 213ra167.
Pushpakom, S., Iorio, F., Eyers, P. A., Escott, K. J., Hopper, S., Wells, A., Doig, A., Guilliams, T., Latimer, J., McNamee, C., Norris, A., Sanseau, P., Cavalla, D., & Pirmohamed, M. (2019). Drug repurposing: Progress, challenges and recommendations. *Nature Reviews. Drug Discovery*, *18*(1), 41–58.
Pustulka, K. M., Wohl, A. R., Lee, H. S., Michel, A. R., Han, J., Hoye, T. R., McCormick, A. V., Panyam, J., & Macosko, C. W. (2013). Flash nanoprecipitation: Particle structure and stability. *Molecular Pharmaceutics*, *10*(11), 4367–4377.
Reddy, P. C., Chaitanya, K., & Rao, Y. M. (2011). A review on bioadhesive buccal drug delivery systems: Current status of formulation and evaluation methods. *DARU. Sci.*, *19*(6), 385.
Renukuntla, J., Vadlapudi, A. D., Patel, A., Boddu, S. H., & Mitra, A. K. (2013). Approaches for enhancing oral bioavailability of peptides and proteins. *International Journal of Pharmaceutics*, *447*(1–2), 75–93.
Romley, J. A., Goldman, D. P., Solomon, M., McFadden, D., & Peters, A. L. (2012). Exenatide therapy and the risk of pancreatitis and pancreatic cancer in a privately insured population. *Diabetes Technology & Therapeutics*, *14*(10), 904–911.
Russell, G., Graveley, R., Seid, J., al-Humidan, A. K., & Skjodt, H. (1992). Mechanisms of action of cyclosporine and effects on connective tissues. *Seminars in Arthritis and Rheumatism*, *21*(6 Suppl 3), 16–22.
Saini, P., Arora, M., & Kumar, M. R. (2016). Poly (lactic acid) blends in biomedical applications. *Advanced Drug Delivery Reviews*, *107*, 47–59.
Saini, P., Ganugula, R., Arora, M., & Kumar, M. N. (2016). The next generation non-competitive active polyester nanosystems for transferrin receptor-mediated peroral transport utilizing gambogic acid as a ligand. *Scientific Reports*, *6*, 29501.
Salatin, S., & Yari Khosroushahi, A. (2017). Overviews on the cellular uptake mechanism of polysaccharide colloidal nanoparticles. *Journal of Cellular and Molecular Medicine*, *21*(9), 1668–1686.
Sastry, S. V., Nyshadham, J. R., & Fix, J. A. (2000). Recent technological advances in oral drug delivery - a review. *Pharmaceutical Science & Technology Today*, *3*(4), 138–145.
Sharma, G., Sharma, A. R., Nam, J. S., Doss, G. P., Lee, S. S., & Chakraborty, C. (2015). Nanoparticle based insulin delivery system: The next generation efficient therapy for type 1 diabetes. *Journal of Nanobiotechnology*, *13*, 74.
Sheng, J., Han, L., Qin, J., Ru, G., Li, R., Wu, L., Cui, D., Yang, P., He, Y., & Wang, J. (2015). N-trimethyl chitosan chloride-coated PLGA nanoparticles overcoming multiple barriers to oral insulin absorption. *ACS Applied materials & interfaces*, 7(28), 15430–15441.
Shi, Y., Sun, X., Zhang, L., Sun, K., Li, K., Li, Y., & Zhang, Q. (2018). Fc-modified exenatide-loaded nanoparticles for oral delivery to improve hypoglycemic effects in mice. *Scientific Reports*, *8*(1), 726.

Shrestha, N., Bouttefeux, O., Vanvarenberg, K., Lundquist, P., Cunarro, J., Tovar, S., Khodus, G., Andersson, E., Keita, A. V., Gonzalez Dieguez, C., Artursson, P., Preat, V., & Beloqui, A. (2018). The stimulation of GLP-1 secretion and delivery of GLP-1 agonists via nanostructured lipid carriers. *Nanoscale, 10*(2), 603–613.

Singer, F. R. (2020). The evaluation and treatment of Paget's disease of bone. *Best Practice & Research. Clinical Rheumatology, 34*, 101506.

Singh, A. K., & Narsipur, S. S. (2011). Cyclosporine: A commentary on brand vs generic formulation exchange. *Journal of Transplantation, 2011*, 480642.

Song, Y., Shi, Y., Zhang, L., Hu, H., Zhang, C., Yin, M., Zhang, X., & Sun, K. (2019). Oral delivery system for low molecular weight protamine-dextran-poly(lactic-co-glycolic acid) carrying exenatide to overcome the mucus barrier and improve intestinal targeting efficiency. *Nanomedicine, 14*(8), 989–1009.

Soudry-Kochavi, L., Naraykin, N., Di Paola, R., Gugliandolo, E., Peritore, A., Cuzzocrea, S., Ziv, E., Nassar, T., & Benita, S. (2018). Pharmacodynamical effects of orally administered exenatide nanoparticles embedded in gastro-resistant microparticles. *European Journal of Pharmaceutics and Biopharmaceutics: Official Journal of Arbeitsgemeinschaft fur Pharmazeutische Verfahrenstechnik e.V, 133*, 214–223.

Sun, T., Dasgupta, A., Zhao, Z., Nurunnabi, M., & Mitragotri, S. (2020). Physical triggering strategies for drug delivery. *Advanced Drug Delivery Reviews*.

Suzuki, H., Hamao, S., Seto, Y., Sato, H., Wong, J., Prud'homme, R. K., Chan, H. K., & Onoue, S. (2017). New nano-matrix oral formulation of nanoprecipitated cyclosporine A prepared with multi-inlet vortex mixer. *International Journal of Pharmaceutics, 516*(1–2), 116–119.

Suzuki, K., Kim, K. S., & Bae, Y. H. (2019). Long-term oral administration of Exendin-4 to control type 2 diabetes in a rat model. *Journal of Controlled Release: Official Journal of the Controlled Release Society, 294*, 259–267.

Svenningsson, P., Palhagen, S., & Mathe, A. A. (2017). Neuropeptide Y and calcitonin gene-related peptide in cerebrospinal fluid in Parkinson's disease with comorbid depression vs patients with major depressive disorder. *Frontiers in Psychiatry, 8*, 102.

Taber, D. J., Baillie, G. M., Ashcraft, E. E., Rogers, J., Lin, A., Afzal, F., Baliga, P., Rajagopalan, P. R., & Chavin, K. D. (2005). Does bioequivalence between modified cyclosporine formulations translate into equal outcomes? *Transplantation, 80*(11), 1633–1635.

Takebe, T., Imai, R., & Ono, S. (2018). The current status of drug discovery and development as originated in United States academia: The influence of industrial and academic collaboration on drug discovery and development. *Clinical and translational science, 11*(6), 597–606.

Tang, H., Zhang, H., Ye, H., & Zheng, Y. (2018). Receptor-mediated endocytosis of nanoparticles: Roles of shapes, orientations, and rotations of nanoparticles. *The Journal of Physical Chemistry. B, 122*(1), 171–180.

Tedesco, D., & Haragsim, L. (2012). Cyclosporine: A review. *Journal of Transplantation, 2012*, 230386.

Thwala, L. N., Delgado, D. P., Leone, K., Marigo, I., Benetti, F., Chenlo, M., Alvarez, C. V., Tovar, S., Dieguez, C., Csaba, N. S., & Alonso, M. J. (2018). Protamine nanocapsules as carriers for oral peptide delivery. *Journal of Controlled Release: Official Journal of the Controlled Release Society, 291*, 157–168.

Torosantucci, R., Brinks, V., Kijanka, G., Halim, L. A., Sauerborn, M., Schellekens, H., & Jiskoot, W. (2014). Development of a transgenic mouse model to study the immunogenicity of recombinant human insulin. *Journal of Pharmaceutical Sciences, 103*(5), 1367–1374.

TRANS-INT Final Report Summary - TRANS-INT (New Oral Nanomedicines: Transporting Therapeutic Macromolecules across the Intestinal Barrier). https://cordis.europa.eu/project/id/281035/reporting (accessed 09/01/2020).

Tsai, L. C., Chen, C. H., Lin, C. W., Ho, Y. C., & Mi, F. L. (2019). Development of multifunctional nanoparticles self-assembled from trimethyl chitosan and fucoidan for enhanced oral delivery of insulin. *International Journal of Biological Macromolecules, 126*, 141–150.

Tseng, C. H., Lee, K. Y., & Tseng, F. H. (2015). An updated review on cancer risk associated with incretin mimetics and enhancers. *Journal of Environmental Science and Health. Part C, Environmental Carcinogenesis & Ecotoxicology Reviews, 33*(1), 67–124.

Tsuda, K., Yamanaka, K., Kitagawa, H., Akeda, T., Naka, M., Niwa, K., Nakanishi, T., Kakeda, M., Gabazza, E. C., & Mizutani, H. (2012). Calcineurin inhibitors suppress cytokine production from memory T cells and differentiation of naive T cells into cytokine-producing mature T cells. *PLoS One*, 7(2), e31465.

Vacha, R., Martinez-Veracoechea, F. J., & Frenkel, D. (2011). Receptor-mediated endocytosis of nanoparticles of various shapes. *Nano Letters, 11*(12), 5391–5395.

Vance, M. E., Kuiken, T., Vejerano, E. P., McGinnis, S. P., Hochella, M. F., Jr, Rejeski, D., & Hull, M. S. (2015). Nanotechnology in the real world: Redeveloping the nanomaterial consumer products inventory. *Beilstein Journal of Nanotechnology, 6*, 1769–1780.

Vecchio, I., Tornali, C., Bragazzi, N. L., & Martini, M. (2018). The discovery of insulin: An important milestone in the history of medicine. *Front. Endocrinol, 9*, 613.

Verma, A., Sharma, S., Gupta, P. K., Singh, A., Teja, B. V., Dwivedi, P., Gupta, G. K., Trivedi, R., & Mishra, P. R. (2016). Vitamin B12 functionalized layer by layer calcium phosphate nanoparticles: A mucoadhesive and pH responsive carrier for improved oral delivery of insulin. *Acta Biomaterialia, 31*, 288–300.

Verma, A., & Stellacci, F. (2010). Effect of surface properties on nanoparticle-cell interactions. *Small (Weinheim an der Bergstrasse, Germany), 6*(1), 12–21.

Viana, R., & Payne, M. W. (2015). Use of calcitonin in recalcitrant phantom limb pain complicated by heterotopic ossification. *Pain Research & Management: the Journal of the Canadian Pain Society = Journal de la Societe Canadienne Pour le Traitement de la Douleur, 20*(5), 229–233.

Vik, A., & Yatham, L. N. (1998). Calcitonin and bipolar disorder: A hypothesis revisited. *Journal of Psychiatry & Neuroscience: JPN, 23*(2), 109–117.

Walkey, C. D., & Chan, W. C. (2012). Understanding and controlling the interaction of nanomaterials with proteins in a physiological environment. *Chemical Society Reviews, 41*(7), 2780–2799.

Wang, Y., & Kohane, D. S. (2017). External triggering and triggered targeting strategies for drug delivery. *Nature Reviews Materilas, 2*(6), 1–14.

Ways, T. M. M., Lau, W. M., & Khutoryanskiy, V. V. (2018). Chitosan and its derivatives for application in mucoadhesive drug delivery systems. *Polymers, 10*(3), 267.

Watson, C. G., Steed, D. L., Robinson, A. G., & Deftos, L. J. (1981). The role of calcitonin and parathyroid hormone in the pathogenesis of post-thyroidectomy hypocalcemia. *Metabolism: Clinical and Experimental, 30*(6), 588–589.

Werle, M., Takeuchi, H., & Bernkop-Schnurch, A. (2009). Modified chitosans for oral drug delivery. *Journal of Pharmaceutical Sciences, 98*(5), 1643–1656.

Wong, C. Y., Al-Salami, H., & Dass, C. R. (2018). Recent advancements in oral administration of insulin-loaded liposomal drug delivery systems for diabetes mellitus. *International Journal of Pharmaceutics, 549*(1–2), 201–217.

Wong, C. Y., Al-Salami, H., & Dass, C. R. (2020). Cellular assays and applied technologies for characterisation of orally administered protein nanoparticles: A systematic review. *Journal of Drug Targeting, 28*(6), 585–599.

Xin, Y., Yin, M., Zhao, L., Meng, F., & Luo, L. (2017). Recent progress on nanoparticle-based drug delivery systems for cancer therapy. *Cancer Biology & Medicine, 14*(3), 228–241.

Yeh, T.-H., Hsu, L.-W., Tseng, M. T., Lee, P.-L., Sonjae, K., Ho, Y.-C., & Sung, H.-W. (2011). Mechanism and consequence of chitosan-mediated reversible epithelial tight junction opening. *Biomaterials*, *32*(26), 6164–6173.

Yu, J., Zhang, Y., Sun, W., Kahkoska, A. R., Wang, J., Buse, J. B., & Gu, Z. (2017). Insulin-Responsive Glucagon Delivery for Prevention of Hypoglycemia. *Small (Weinheim an der Bergstrasse, Germany)*, *13*(19).

Yun, Y., Cho, Y. W., & Park, K. (2013). Nanoparticles for oral delivery: Targeted nanoparticles with peptidic ligands for oral protein delivery. *Advanced Drug Delivery Reviews*, *65*(6), 822–832.

Zhang, L., Shi, Y., Song, Y., Duan, D., Zhang, X., Sun, K., & Li, Y. (2018). Tf ligand-receptor-mediated exenatide-Zn(2 +) complex oral-delivery system for penetration enhancement of exenatide. *Journal of Drug Targeting*, *26*(10), 931–940.

Zhang, L., Shi, Y., Song, Y., Sun, X., Zhang, X., Sun, K., & Li, Y. (2018). The use of low molecular weight protamine to enhance oral absorption of exenatide. *International Journal of Pharmaceutics*, *547*(1–2), 265–273.

Zhang, Y., Qian, P., Zhou, H., Shen, R., Hu, B., Shen, Y., Zhang, X., Shen, X., Xu, G., & Jin, L. (2018). Pharmacological signatures of the exenatide nanoparticles complex against myocardial ischemia reperfusion injury. *Kidney & Blood Pressure Research*, *43*(4), 1273–1284.

Zheng, S. L., Roddick, A. J., Aghar-Jaffar, R., Shun-Shin, M. J., Francis, D., Oliver, N., & Meeran, K. (2018). Association between use of sodium-glucose cotransporter 2 inhibitors, glucagon-like peptide 1 agonists, and dipeptidyl peptidase 4 inhibitors with all-cause mortality in patients with type 2 diabetes: A systematic review and *meta*-analysis. *JAMA*, *319*(15), 1580–1591.

Zhou, Y., Liu, L., Cao, Y., Yu, S., He, C., & Chen, X. (2020). A Nanocomposite Vehicle Based on Metal-Organic Framework Nanoparticle Incorporated Biodegradable Microspheres for Enhanced Oral Insulin Delivery. *ACS Applied Materials & Interfaces*, *12*(20), 22581–22592.

Zhu, C., Peng, T., Huang, D., Feng, D., Wang, X., Pan, X., Tan, W., & Wu, C. (2019). Formation Mechanism, In vitro and In vivo Evaluation of Dimpled Exenatide Loaded PLGA Microparticles Prepared by Ultra-Fine Particle Processing System. *AAPS PharmSciTech*, *20*(2), 64.

CHAPTER FIVE

Novel formulation approaches for gastrointestinal targeting: characterization and animal model considerations

Swapnil Khadke, Fiona MacDougall and Tariq Islam
BDD Pharma, Glasgow, United Kingdom

5.1 Introduction

According to a recent study by Global Burden of Diseases, Injuries, and Risk Factors Study, the global, regional, and national burden of gastrointestinal (GI) disorders such as colorectal cancer (Guren, 2019), pancreatic cancer (Rawla et al., 2019), inflammatory bowel disease (IBD), and stomach cancer (Alatab et al., 2020; Ng et al., 2016) is rising (Fig. 5.1).

Because of the limitations of studying human gastrointestinal disease in a clinical environment, the use of animal models to examine the mechanisms of gastrointestinal disease has been of great importance. Rodent models remain the most commonly used animal model for the study of human disease because of their relatively low cost and maintenance requirements, rapid reproduction rates, and availability of research tools such as murine antibodies (Chung et al., 1997; Low, 2012; Ziegler et al., 2016). However, rodent animal models often fail to fully mimic anatomy, physiology, clinical signs, and significant pathological signs of human diseases (Boivin et al., 2003). Due to this, there has been recent increased interest within the scientific community in developing large animal models that more closely approximate the clinical and pathologic features of human disease (Cibelli et al., 2013; Flisikowska et al., 2012). Of the large animal species used for translational research, the pig offers a marked advantage because this species has essential anatomic and physiological similarities with human beings, particularly regarding the gastrointestinal tract as shown in Fig. 5.2.

Oral Delivery of Therapeutic Peptides and Proteins.
DOI: https://doi.org/10.1016/B978-0-12-821061-1.00005-8

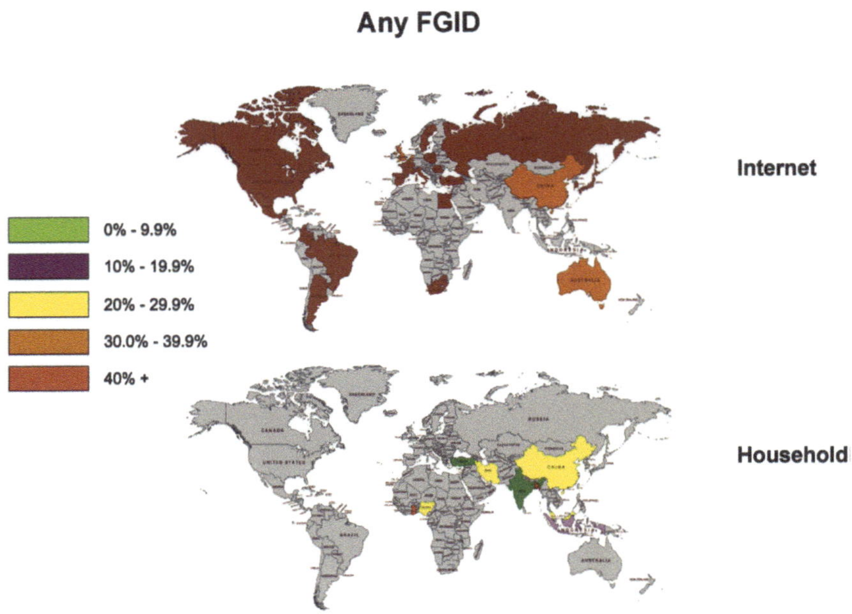

Figure 5.1 Survey of epidemiology of gastrointestinal disorders worldwide (Sperber et al., 2021).

5.2 Oral route and human gastrointestinal anatomy (general physiological considerations for colonic drug delivery)

The oral route is considered the most common route of drug delivery because of patient acceptability, economical, convenience and there is no need for sterile equipment for manufacture. The main disadvantage associated with this route is that dosage forms are subjected to the variable physiological conditions (pH, fluid, transit, and erosive forces) within the GI tract that may affect performance. To clearly understand the in vivo environment to which these dosage forms are subjected, GI tract anatomy and physiology must be considered.

The GI tract is a continuous muscular, digestive organ that consists of the mouth, pharynx, esophagus, stomach, small intestine, and colon, which leads to the terminal opening or anus. The stomach is a "J" shaped organ that lies between the esophagus and the small intestine. The primary function of the stomach is to store food, initiate digestion of proteins, kill bacteria, and move food as chyme to the small intestine.

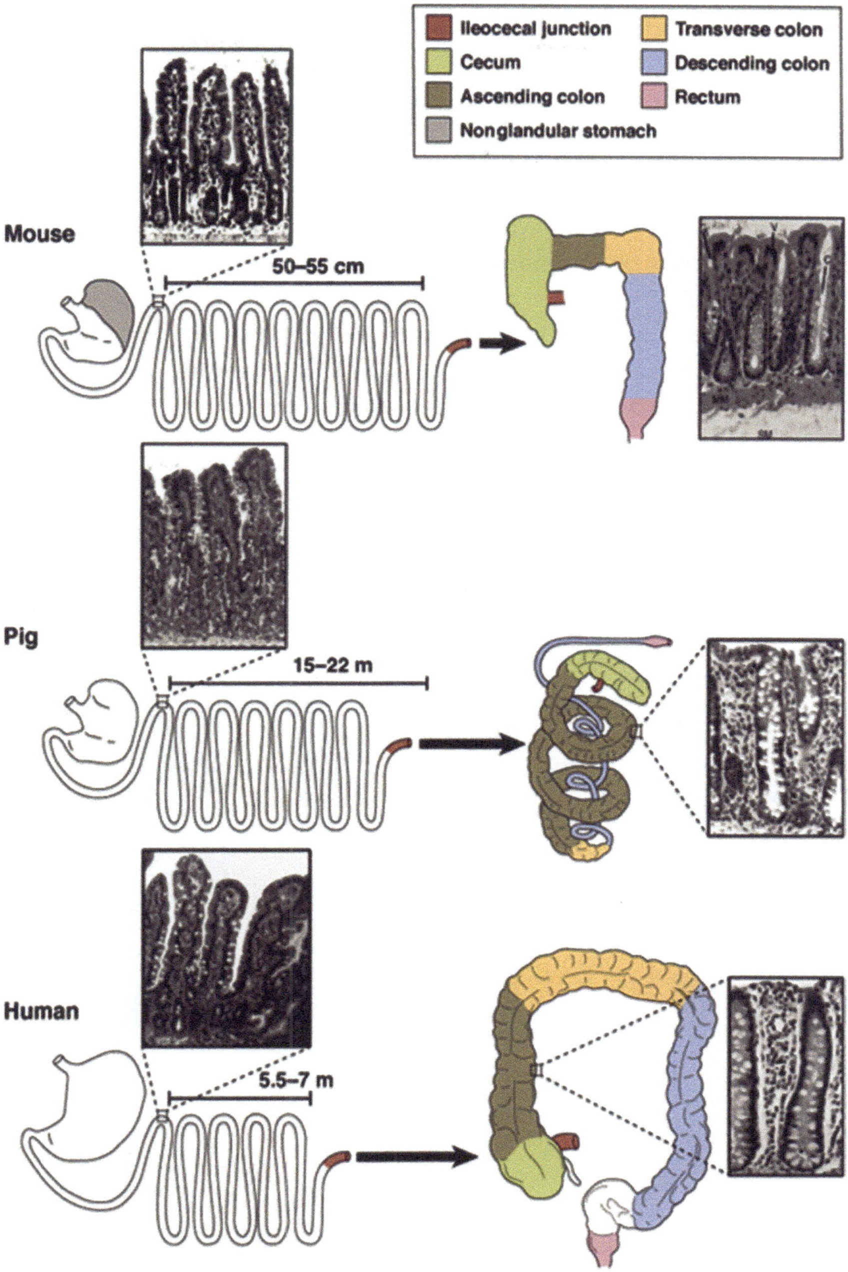

Figure 5.2 Schematic diagram for comparison of mouse, porcine, and human gastrointestinal tract anatomy and histology (Ziegler et al., 2016).

The stomach is divided into the fundus, the body, and the pyloric region, which consists of pyloric sphincter and pylorus. Fundus and body distend to accommodate a large amount of food.

The small intestine (SI) is the longest part of the GI tract and is divided into three sections, namely, the duodenum, jejunum, and ileum. The luminal surface of the small intestine is covered by villi containing fine projections known as microvilli, which increases the active surface area by 600 times. Enterocytes or columnar absorptive cells cover over 90% of the villi surface. The SI is a major site for the absorption of food and drugs.

Unabsorbed food passes to the colon through the ileocecal valve and then follows in sequence through the ascending, transverse, and descending colon. The ascending colon plays a significant role in the absorption of water and minerals.

During targeting drugs to the GI tract for colonic delivery, pH of GI tract, GI motility, gastric emptying time, small intestine, and colonic transit time, GI forces should be considered and studied carefully. The physiological and microbial changes to the GI tract in the IBD patient and healthy are shown in Fig. 5.3.

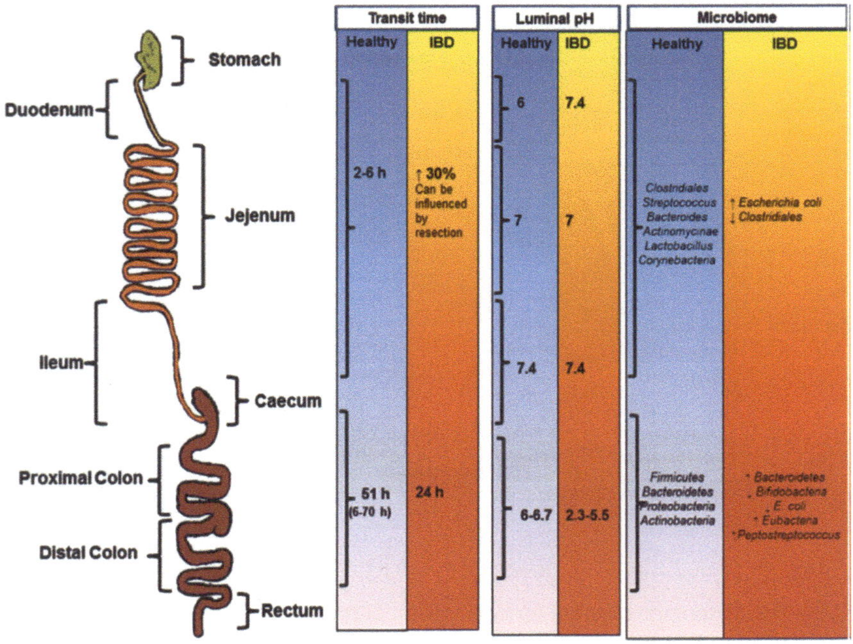

Figure 5.3 Physiological and microbial changes to gastrointestinal tract (Hua et al., 2015).

5.2.1 Intraluminal pH of the human gastrointestinal tract

The intraluminal pH is rapidly changed from highly acid in the stomach to about pH 6 in the duodenum. The pH gradually increases in the small intestine from pH 6 to about pH 7.4 in the terminal ileum. The pH drops in the cecum, but again gradually increases, reaching pH 6.7 in the rectum.

Most of the drugs used as pharmaceuticals are weak acids or bases, and the absorption of these drugs generally occurs in a nonionized form (Pleuvry, 2005). The existence of these weak acids or bases in unionized form largely depends upon the pH of the GI tract (Dressman et al., 2007). Additionally, the performance of some excipients (e.g., alginates, carbomer, carboxymethylcellulose sodium, chitosan, dextrin, guar gum, polymethacrylates, anionic emulsifying wax) used in the formulation of oral drug delivery systems are dependent upon the pH (Rowe et al., 2003).

Different methods have been used to determine the pH with the GI tract. Ovesen et al. (1986) used a microelectrode technique and found that the median fasted pH of the stomach was 2.01 (range 1.62–2.13). Lindahl et al. (1997) used segmental aspiration of GI content using a perfusion tube to determine the pH in the stomach and jejunum. A median fasted gastric pH value of 1.97 (range 1.4–7.5) and a median fasted proximal jejunum pH value of 7.2 (range 5.3–8.1) were reported. However, results obtained from aspiration techniques are often doubtful. It has been previously reported that the traumatic procedure during aspiration alters the secretions of the GI tract (Bresnick et al., 1993). Noninvasive techniques using radiotelemetric capsules have provided the advantage of measuring the pH of the entire GI tract using an orally ingested pH capsule. Using the same method, the mean pH of the fasted stomach was found to be 1.1 (range 0.4–4.0, (Lui et al., 1986)). Even though it has been widely accepted that the pH in the fasted stomach is highly acidic, the results obtained from different sources can vary because of the method used, as well as circadian rhythm observed in the secretion of HCl (Moore & Englert, 1970).

The pH of the stomach also varies between fasted and fed conditions. A higher pH (median 5, maximum of 7) was found in the fed stomach in comparison to the median fasted pH of 1.7 (Dressman et al., 1990). The rapid increase in pH is due to the buffering actions of an ingested meal and dilution. Depending upon the meal properties (pH, volume, composition), gastric pH returns to standard value after 2–240 min.

The variation of pH from the stomach to colon demonstrates that the performance of the dosage forms can be altered if formulation excipients are susceptible to changes in pH. The significant increase in the pH from the stomach to the small intestine has been successfully utilized to target the dosage forms in the small intestine by using the enteric coating. However, the significant fluctuation in pH from the stomach to the colon is still a major challenge to formulate pH-dependent colon drug delivery systems.

5.2.2 Gastrointestinal motility

GI motility is the driving force for the movement of dosage forms within the GI tract (Fig. 5.4), which in turn may determine the duration of exposure of the dosage forms within the GI tract.

Stomach motility acts differently depending upon the presence or absence of food. In the fasted state (interdigestive), its motility is controlled by the migrating motor complex (MMC). Fasted state motility patterns can be divided into four different phases (Washington et al., 2001). During this cycle, if the food is consumed, it moves to fed state motility.

During phase III of MMC, the powerful contractions with maximum frequency and amplitude empty the stomach of digestible food. Phase III wave moves distally from the stomach to the terminal ileum. When one wave reaches the ileum, another is started at the stomach. The mean duration of the MMC was found to be 117 ± 13 min (Luiking et al., 1998). The mean duration of the interdigestive cycle was independent of the day/night pattern. However, the length of phase I was higher at night time while the reverse was the case for phase II, yet the length of phase III remained the same (Keller et al., 2001).

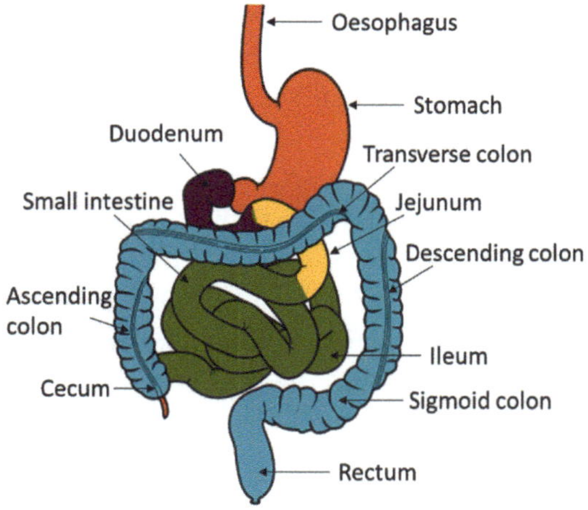

Figure 5.4 Schematic diagram of a GI tract showing the major regions (Remesz, 2022).

In the fed state, mixing, grinding, and emptying all occur at the same time in the antrum of the stomach. All particles are propelled toward the pylorus. Smaller particles flow through the pylorus, while larger particles are retained and refluxed back. This process is repeated when the stomach remains in the fed state. At the end of the fed state, the stomach enters the fasted state, that is, when all food has emptied, and any undigested larger particles are swept away by the phase III of MMC.

In the small intestine, segmentation is the primary mixing pattern and is characteristic of the fed state. Propulsive movement is the mechanism by which luminal chyme is moved down over 5 cm or less. The interdigestive period is seen between meals when the lumen is mostly devoid of contents. During such times, the phase III of MMC contractions propagate from the stomach through the entire small intestine, sweeping it clear of debris.

In the colon, two different types of contractions are encountered, namely nonperistaltic (95%) and high amplitude propagating contractions (5%). A nonperistaltic contraction mixes the contents of the large intestine back and forth but does not move colonic content forward. The second type of contraction which occurs only 6–8 times per day is the high amplitude propagating contraction. This forceful contraction begins in the first part of the large intestine and ends in the rectum.

5.2.3 Gastric emptying time, small intestine, and colonic transit time

When developing delivery systems that are targeted to different regions of the gastrointestinal tract, it is critical to fully understand if they achieve the desired targeted behavior in vivo. Pharmacokinetic profiles alone do not necessarily provide this confidence, particularly where there may be a large population variability in absorption. The predominant imaging techniques used to meet this aim are gamma scintigraphy and magnetic resonance imaging. Gamma scintigraphy can easily be coupled with simultaneous pharmacokinetic blood sampling to correlate formulation location or quantity with resultant blood plasma levels.

Many excellent detailed reviews of the use of gamma scintigraphy in pharmaceutical development have been published (Digenis et al., 1998a,b). However, in essence, the technique involves adding a small quantity of a gamma-emitting radioisotope in a nonabsorbable format (often chelated to DTPA, which prevents absorption across intestinal epithelial membranes), to a component of the formulation to follow its progress through the GI tract. Behaviors such as gastric emptying, small intestinal transit, colon arrival, and dispersion of dosage forms can easily be observed, and where appropriate, the software can be used to obtain quantitative or kinetic information.

In the fasted state, liquids empty continuously, while small multiparticulate formulations will empty as a series of boluses, along with small quantities of gastric mucous/chyme. Any disintegrating oral solid dosage form administered in the fasted state would be expected to empty in a similar manner, either as disintegrated discrete particles or as a liquid if the API rapidly dissolves in the fluid available in the stomach. As a rule of thumb, larger monoliths such as sustained release tablets will be retained in the stomach until phase III of the MMC occurs and the "housekeeper" waves empty out any undigested materials.

The situation becomes more nuanced when food comes into the equation. Following the ingestion of food, the duodenum is responsible for determining the calorific density of the meal, in response to which the rate of gastric emptying is controlled to a constant rate of approximately 200 Kcal/h. It is widely reported that fat has the greatest effect on the rate of gastric emptying of food, which is likely to be a result of a combination of the calorific density of such food, combined with CCK receptors. Food is gradually emptied from the stomach, but any nondisintegrating dosage form will remain in the stomach until all the food has emptied, the fasted MMC cycle is reinstated, and the phase III housekeeper wave is able to eject it from the stomach. The time at which this occurs will therefore be highly dependent on the calorific value of the meal.

Small intestinal transit time is often reported as being the most consistent in the GI tract and the general population, with values of between 3 and 4 h commonly reported. Behavior may be somewhat formulation dependent, and it has been shown that pellet formulations administered at the same time as tablets will travel ahead of the tablet through the small intestine, after which the formulations tend to congregate again around the ileocecal junction before passing into the colon. More recently, scintigraphic studies have unveiled further complexities in the transit in this region, whereby the effect of food ingestion after dosing, on the rate of SI transit was evaluated, and clear evidence of acceleration was demonstrated (Fadda et al., 2009). The gastrocolic response shows that this effect is not limited to the small intestine, but also entry into the colon.

It has also been shown that some commonly used excipients can have a significant impact on transit in this region, with implications for bioavailability (Basit et al., 2002). In the colon, highly variable transit rates have been reported through a range of imaging studies. Transit through this region can also be discontinuous, occurring instead in infrequent propulsive waves.

5.3 Formulation strategies for colon-targeted drug delivery

5.3.1 Time-delayed systems

Pulsatile drug delivery systems are formulations that release drugs after a predetermined lag time. Pulsatile drug release is achieved either by utilizing the principle of site-controlled release or by time-delayed drug release. The performance of the former system is based upon the arrival of the dosage form in the physiological environment within the GI tract, which triggers drug release. Enteric-coated (Nunthanid et al., 2008), enzyme controlled (Fude et al., 2007), pressure-controlled (Muraoka et al., 1998), or colonic bacteria dependent (Wilson & Basit, 2005) formulations are some of the examples of site-specific drug delivery systems. One of the limitations of site-specific drug delivery system is that these formulations do not provide enough opportunity for a formulation scientist to modulate the lag time according to the biological clock because the lag time is dependent upon physiological factors such as gastric emptying and intestinal transit. To overcome the limitation of site-specific drug delivery in chronopharmaceutical applications, time-delayed formulations (TDFs) were developed. The biological clocks of TDFs are initiated once the formulation is administered, and therefore should ideally be independent of the GI environment. Only drug molecules showing uniform absorption throughout the GI tract are suitable candidates for TDFs. This is because if the molecule displays a narrow absorption window or variable absorption profile within the GI tract, the plasma profile may vary depending upon the site of the location of dosage forms. TDFs have been formulated as multiparticulate systems, capsules, and tablets (Stevens, 1998).

5.3.1.1 Multiparticulate systems

Pellets or multiparticulate systems, when used as multiparticulate TDFs, offer some typical advantages that are not often achieved from single unit pulsed release systems. These systems have the specific advantage of having reproducible gastric residence time, less dose dumping, and flexibility of blending pellets of different compositions to achieve desired release profiles. The lengthy manufacturing process is a disadvantage associated with these systems.

Time-controlled explosion system (TES) consists of four different layers, namely nonpareil seed coated by the drug, followed by a swelling layer and finally by a water-insoluble polymer, ethylcellulose as a separate layer (Ueda, Hata, et al., 1994). As the TES was exposed to the dissolution medium or GI fluids, water entered through the insoluble layer causing enough expansion of disintegrant present in the swelling layer. As the swelling force generated by the disintegrant was higher than the tensile strength of the ethylcellulose barrier layer, this caused the rupture of the outer barrier layer leading to rapid drug release. The lag time before drug release was independent of drug solubility, pH of dissolution medium, and pellet size (Ueda, Hata, et al., 1994). As ethylcellulose membrane thickness increased, there was a corresponding increase in lag time (Ueda, Hata, et al., 1994; Ueda, Ibuki, Kawamura, et al., 1994). At the same time, the addition of talc to ethylcellulose in the outer layer decreased lag time by mechanically weakening the membrane (Ueda, Ibuki, Kimura, et al., 1994).

This concept was further modified to develop a tablet-based timed-release system, where the swelling and rupture layers were spray dried onto the core API containing tablet (Fig. 5.5).

5.3.1.2 Capsules

Capsule-based TDFs have been extensively reported in the literature. In most instances, the water penetration into the active constituents within

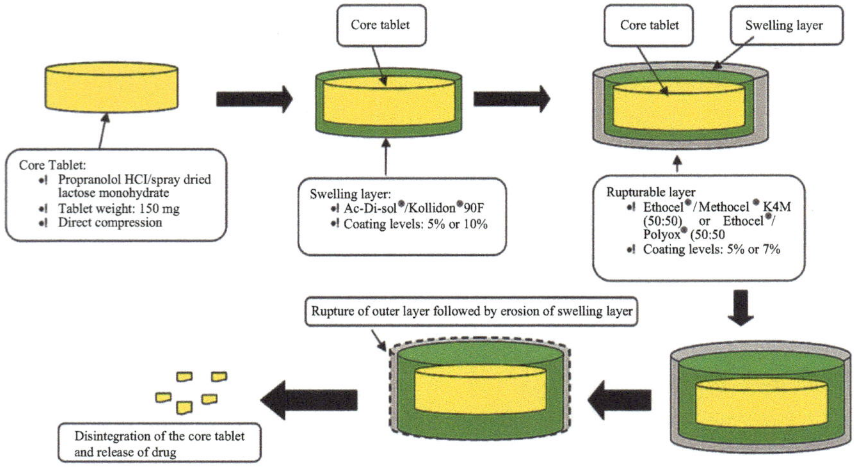

Figure 5.5 Time-controlled explosion system (TES) (Elbatanony, 2019).

the capsule body is either controlled by an insoluble plug, or coating applied to both cap and body of the capsule shell.

A plug-controlled capsule-based drug delivery device, consisting of an insoluble capsule body containing drug with or without other excipients, is closed by a plug. The insoluble capsule body is usually prepared by coating the body of a hard gelatin capsule with water-insoluble polymer (e.g., ethylcellulose). The lag time before drug release from these systems is controlled by the time of entry into the inner capsule, which is in turn controlled by the plug.

The Pulsincap formulation (Stevens et al., 2002; Wilson et al., 1997) and the technology shown in Fig. 5.6, employed a hydrogel plug containing cross-linked polyethylene glycol (PEG). When the formulation came into contact with water, the gelatin cap dissolves and the plug is exposed to fluid, following which the partially swollen plug starts to slide out of the capsule, leading to medium penetration to the inner core.

This swelling, sliding plug was changed to an erodible plug by Ross et al. (2000), which eliminates the requirement of force required to overcome friction and expel the plug from the capsule. The erodible plug consisted of a combination of hydroxypropyl-methylcellulose (HPMC) and lactose prepared by the direct compression method. The lag time of the TDF was modulated by changing the weight ratio of HPMC to lactose. Later, McConville et al. (2004) showed that when the HPMC-lactose plug was manufactured by a wet granulation

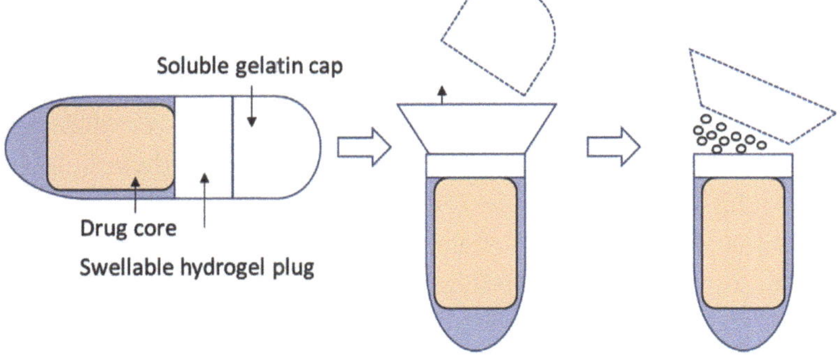

Figure 5.6 Pulsincap technology (Arora, Ali, Baboota, & Qureshi, 2006). *Redrawn from Arora, S., Ali, J., Ahuja, A., Baboota, S., & Qureshi, J. (2006). Pulsatile drug delivery systems: An approach for controlled drug delivery.* Indian Journal of Pharmaceutical Science, 68*(3), 295–300.*

method, the lag time of drug release was increased. This increase was attributed to the increased water mobility during wet granulation that allows more significant swelling and movement of HPMC chains. Recently, Zou et al. (2008) used a hydrophilic plug (HPMC and lactose) to develop a floating time-delayed capsule-based system, where an insoluble capsule body contained drug as well as floating material. The lag time in this system was modulated by changing the viscosity grade of HPMC or the thickness of the plug.

In contrast to the previously discussed hydrophilic plug, Krögel and Bodmeier (1998) used a wax-surfactant erodible plug using Gelucire (fatty acid ester of glycerol with or without polyethylene glycol) and sodium dodecyl sulfate with or without glyceryl mono-oleate. These plugs offered a distinct advantage of tight contact between the capsule body and plug because these substances can be melted and solidified directly into the capsule body. Erosion of these directly solidified wax matrices was dependent on the hydrophilic lipophilic balance (HLB) value of the wax or the concentration of added surfactants, which in turn regulated the lag time before drug release from these capsule devices. The same group later developed a plug comprising the pectin and pectinolytic enzyme (Krögel & Bodmeier, 1999). It was interesting to note that the erosion of the plug was not affected by GI fluids, but by pectin-enzyme ratio, which was used as a lag time modifier.

In all the examples mentioned above of capsule-based TDFs, the lag time before drug release was controlled by a plug acting as a barrier for fluid penetration into the core constituents. In contrast, (Bussemer et al., 2003) developed a capsule-based TDF where the formulation did not consist of a plug; instead, the barrier layer was applied to both capsule body and cap. These types of applications of the barrier layer are more common in tablet-based TDFs that are discussed in the next section. The capsule shell (cap and body) was coated with swelling layer (disintegrant, Ac-di-Sol) followed by coating with a barrier layer (ethylcellulose and HPMC). This modified capsule cell was used to encase an active drug. The lag time was controlled by the expansion of the swelling layer and subsequent rupturing of the polymer coating. Later the same approach was applied to a soft gelatin capsule shell (Bussemer & Bodmeier, 2003); however, the lag time was longer for the hard gelatin capsule at the same coating level. This was attributed to the differences in the directions of pressure generated by the swelling layer in hard gelatin and soft gelatin capsule (Bussemer & Bodmeier, 2003).

5.3.1.3 Tablets

Time-delayed tablet formulations consist of a rapidly dissolving or sustained-release core, coated by one or more barrier layers. A barrier layer is employed to prevent fluid penetration into the core for a predetermined period, providing the desired lag time before rapid/sustained drug release from the inner core tablet (Fig. 5.7).

The components of the barrier layer(s) are either applied to the core as a liquid coating, dry coating (compression coating), or specialized coating like three-dimensional coating (3DP). 3DP is a rapid prototype solid free-form fabrication technique where a computer-controlled printer is used to deposit liquid binder and drug excipients to build 3D layers. This 3DP coating technique offered the capacity of specific spatial deposition of various materials with excellent resolution within the tablet (Katstra et al., 2000; Rowe et al., 2000). In vitro drug release profiles showed consecutive release of drugs from the different layers. The main drawback of this technique is that the manufacturing method requires specialized instrumentation, which is not readily available. Additionally, the cost of large-scale manufacturing using this technique is not known. Therefore compression coating or liquid coating has been extensively reported in the literature to manufacture TDFs.

Compression coating or press-coating techniques have been employed for the physical separation of incompatible ingredients or coating a product where the tablet core cannot tolerate organic solvent or water and yet needs to be coated. This technique allows higher weight gain to the core tablet in comparison to the liquid coating technique. Ozeki and colleagues

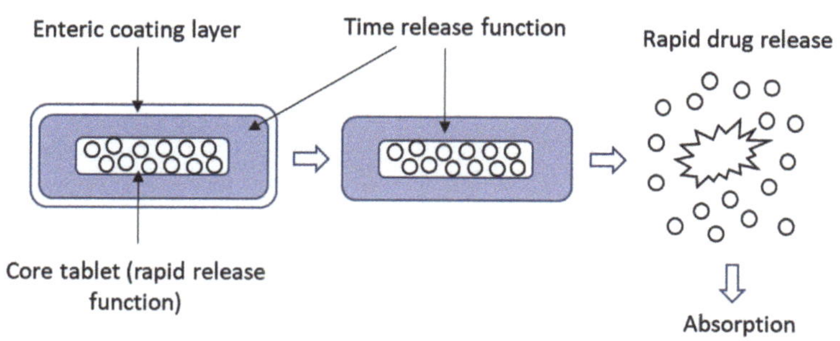

Figure 5.7 Mechanism of drug release from press-coat tablet. *Redrawn from Fukui, E., Miyamura, N., Uemura, K., & Kobayashi, M. (2000). Preparation of enteric coated timed-release press-coated tablets and evaluation of their function by in vitro and in vivo tests for colon targeting.* International Journal of Pharmaceutics, 204*(1), 7–15. https://doi.org/10.1016/S0378-5173(00)00454-3.*

published a series of articles demonstrating that the core tablet can be centrally positioned within the press-coated tablet (PCT) and that the formulation can be manufactured in a single-step process (Ozeki, Ando, et al., 2004; Ozeki, Watanabe, et al., 2004). Furthermore, Ozeki, Ando, et al. (2004) developed a dividable pulsatile release tablet consisting of two cores that are surrounded by an outer barrier layer, as shown in Fig. 5.8.

In tablet-based TDFs manufactured by any technique (3D printing, compression coating or liquid coating), the barrier layer composition plays an essential role in determining the magnitude of the lag time and the mechanism by which it is produced. Generally speaking, the thicker the barrier layer, the longer the lag time to drug release (Fukui, Uemura, et al., 2000; Lin et al., 2004; Ozeki, Ando, et al., 2004; Sungthongjeen et al., 2004). The composition of the barrier layer controls the mechanism of effecting a lag time.

Hydrophilic polymers such as HPMC (Ozeki, Ando, et al., 2004; Ozeki, Watanabe, et al., 2004), polyethyleneoxide (Sawada et al., 2004), or hydroxypropyl cellulose (Fukui, Uemura, et al., 2000; Sangalli et al., 2001), with or without other excipients are commonly used as barrier layer constituents. These hydrophilic polymers swell upon contact with fluids, following which the barrier layer dissolves and/or erodes, allowing fluid access to the inner core tablet. This is discussed in detail in Chapter 8. In general, as the viscosity of these polymers increases, there is an increase in lag time due to the formation of a stronger gel layer that is resistant to both dilutions as well as erosion. This has been illustrated by Sangalli et al. (2003), where the lag time of a PCT was found to increase

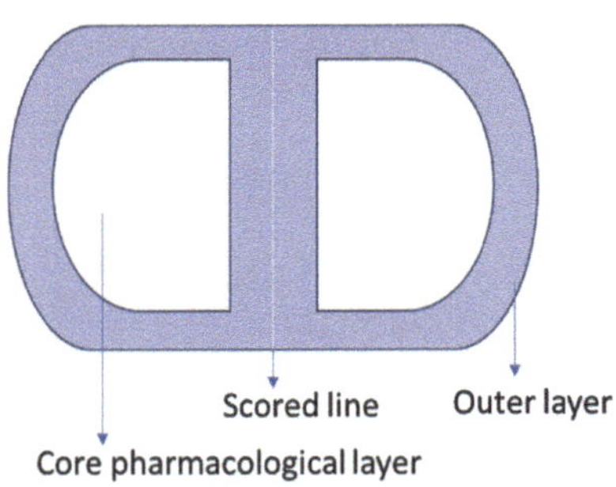

Figure 5.8 Dividable one-step dry-coated tablet (OSDrC). *Redrawn from Ozeki, Y., Watanabe, Y., Okamoto, H., & Danjo, K. (2004). Development of dividable one-step dry-coated tablets (Dividable-OSDRC) and their evaluation as a new platform for controlled drug release.* Pharmaceutical Research, 21*(7), 1177–1183. https://doi.org/10.1023/B:PHAM.0000033004.88953.bc.*

in the order of tablets containing 5 < 50 < 4000 cP HPMC in the barrier layer (Sangalli et al., 2003).

Similarly, in other studies, Fukui, Miyamura, et al. (2000) showed that increasing viscosity grades of HPC from 4.8 to 3740 mPs increased the lag time before drug release. If water-soluble excipients are added to the hydrophilic polymer, the lag time before drug release decreases, probably by making the swelled layer prone to erosion (Sawada et al., 2004). Although these hydrophilic polymers have been successfully used as barrier layer components of TDFs, the major drawback of these systems is the susceptibility of these formulations to hydrodynamics forces (Fukui, Uemura, et al., 2000). This might lead to variable performance when the tablet was in the stomach or small intestine. To overcome the hydrodynamic effect and to target dosage forms to the colon, different researchers have applied an additional enteric coating layer (Fukui, Uemura, et al., 2000; Sangalli et al., 2001; Wilding et al., 1994). This enteric coating layer dissolves at higher pH and avoids exposure of hydrophilic matrix tablets in the stomach.

Sungthongjeen et al. (2004) used water-insoluble polymer (ethylcellulose) as a barrier layer of a TDF. This tablet-based TDF consisted of an immediate-release tablet coated with a swelling layer and finally by water-insoluble polymer where the primary mechanism of producing a lag time of this tablet is similar to TES. The mechanism involves water penetration through a barrier layer followed by expansion of a swelling layer and subsequent breakdown of the barrier layer (Fig. 5.9).

Similarly, Lin et al. (2001a,b) have used ethylcellulose as the directly compressible barrier layer composed of a PCT. The lag time before drug release can be increased either by decreasing the particle size of ethylcellulose in the barrier layer or by increasing the compression pressure during tablet compression because this led to a decreasing porosity, which slows water penetration. It was observed that water penetrates through the lateral surface in these PCTs, and subsequently, the tablet breaks into two halves. The authors suggested that this was because of the loose packing of ethylcellulose particles in the lateral surface in comparison to the upper and lower surfaces. The addition of other excipients (HPMC or lactose) resulted in a complete breakdown of the outer barrier layer. The addition of the water-soluble excipient (spray-dried lactose) to ethylcellulose created larger water-soluble channels, thus decreasing the lag time before release. In contrast, the addition of HPMC decreased the water

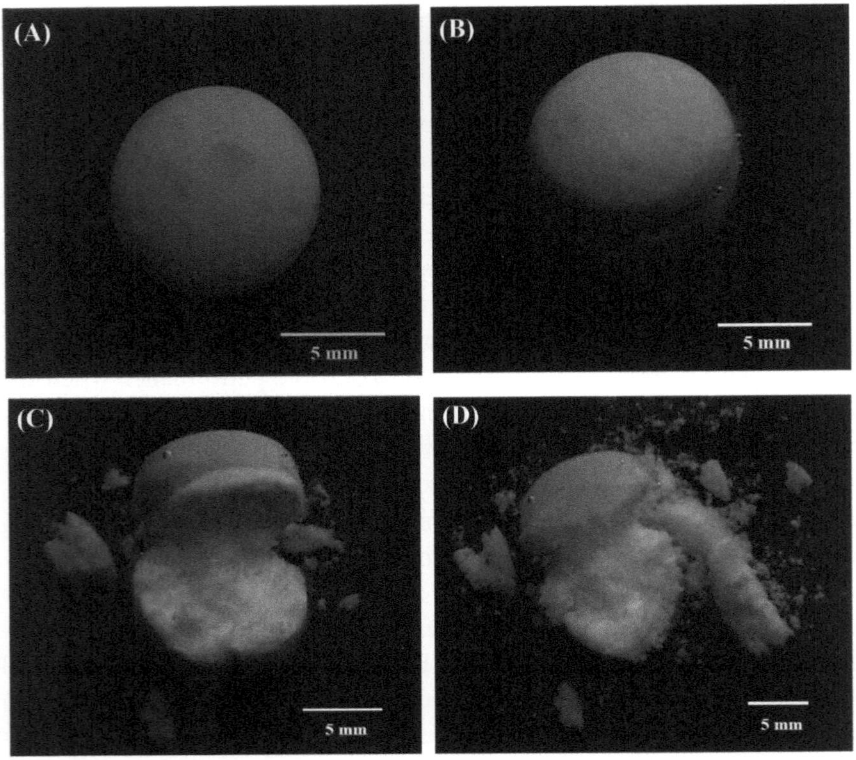

Figure 5.9 Rupture sequence of pulsatile release tablets with a swelling layer and a rupturable ethylcellulose coating (Sungthongjeen et al., 2004).

penetration rate in comparison to a tablet containing lactose due to gel formation. The lag time before drug release was found in the order of PCT containing ethylcellulose and lactose < ethylcellulose and HPMC < ethylcellulose in the barrier layer.

Hydrophobic materials (hydrogenated castor oil, behenic acid, beeswax, carnauba wax) have also been reported as the barrier layer components of the tablet-based TDFs (Ishino & Sunada, 1993; Ishino et al., 1992a,b; Peerapattana et al., 2004). Because of their hydrophobic nature, these materials are used in conjunction with relatively hydrophilic excipients (lactose, HPMC, PEG). The lag time before drug release has been modulated by changing the ratio of these excipients, in a PCT containing water-soluble excipients (lactose, PEG) with hydrogenated castor oil as the barrier layer component, the lag time before drug release from the core tablet as a result of channels created following the dissolution of the water-soluble excipient (Ishino & Sunada, 1993; Ishino et al., 1992a,b). Therefore the higher the percentage of water-soluble excipients; the lower the lag time before drug release. The same approach has been applied to another PCT consisting of behenic acid and lactose as barrier layer components, where the lag time was modulated by changing behenic acid and lactose concentrations in the outer barrier layer (Peerapattana et al., 2004). Pozzi et al. (1994) developed a time-clock system consisting of wax (beeswax and carnauba wax) and surfactant (polyoxyethylene sorbitan monostearate). The removal of the barrier layer in these formulations occurred by dispersion (erosion) of the outer barrier layer when subjected to dissolution media. This system was further explored as colon-targeted drug delivery system using an additional enteric coating layer (Wilding et al., 1994).

The TDFs discussed in the previous examples mostly released drugs from the core tablet in a controlled release fashion, following a lag period in most instances rather than the intended pulsatile release. This is because the mechanism of producing a lag time for these formulations was based upon the hydration of the outer barrier layer, or water penetration through the outer barrier layer, with the outer barrier layer remaining present when the dissolution media reached the core tablet. This prevented rapid drug release.

5.3.2 pH-dependent drug delivery systems

Taking advantage of the gradually increasing pH of the distal gastrointestinal tract, drug-delivery systems using pH-dependent mechanisms have been used to provide a self-controlled mechanism to deliver the drug at specific sites in the gastrointestinal tract.

Eudragit polymers were widely investigated to target colonic release. In most cases Eudragit combined with other Eudragit grades or with other polymers to achieve a specific release profile for rapid pH-based dissolution at a target site such as the upper-to-middle small intestine or colon. For example, Mesalazine was used as a model drug using various ratios of two Eudragit polymers (Eudragit L100 and Eudragit S100) to target pH-dependent colonic delivery (Khan et al., 2000) and the in vitro dissolution studies showed that the drug release can be manipulated by changing the Eudragit L100-Eudgragit S100 ratios in the combinations within the pH range between 6.0 and 7.0. Similar studies were performed to evaluate the release of substances such as 5-aminosalicylic acid (Gupta et al., 2001) and lactose (Khan et al., 1999), using a combination of various ratios of Eudragit polymers to target colonic drug delivery.

5.3.3 Enzyme-sensitive drug delivery systems

The abundance of microflora in the colonic region in comparison with other sections of the gastrointestinal tract has led to researchers investigating the use of the resultant presence of enzymes in that area to trigger drug release in the colon.

5.3.3.1 Polysaccharide-based systems

Polysaccharides such as pectin, guar gum, amylase, inulin, and dextran have been investigated for colon-specific drug delivery, where they can be used to prevent the release of the active ingredient by virtue of the gel structure formed and imparting integrity in the upper GI tract. On arrival in the colonic region, the polysaccharide is degraded by the action of colonic microflora enzymes and subsequently entrapped drug is released (Chourasia & Jain, 2004). For an example, Liu et al. (2006) developed novel complex hydrogel beads prepared from pectin and zein (a protein from corn), which have shown the capability to protect incorporated drugs from premature release into the GI environment (Liu et al., 2006), hence to target the colon-specific release.

5.3.3.2 Phloral technology

Ibekwe et al. (2008) developed a novel technology for precise and consistent delivery to the colon, subsequently marketed as Phloral technology. This technology exploits the changes in gastrointestinal pH described earlier, in combination with the enzymatic activity of the microbiota, as independent but complementary release mechanisms to target site-specific release and enable systemic applications or the treatment of local disorders of the large intestine such as inflammatory bowel disease. The coating solution comprises a mixture of pH-responsive enteric polymer (Eudragit S) and a biodegradable polysaccharide (resistant starch) in a single layer matrix. The dual pH- and bacterially triggered coating was applied to tablets and dosed to healthy volunteers at three different feeding conditions (without food, with breakfast, and 30 min before breakfast), and it showed that 22 out of 23 tablets were disintegrated at the ileocecal junction or in the large intestine, confirming successful targeting with this system.

5.4 Animal models for targeted oral formulations

Animal models have been employed at different stages in pharmaceutical development widely across the industry to evaluate and understand the behavior of novel compounds before moving on to testing in human subjects. It is outside the scope of this review to discuss the ethical considerations of the use of animal models in pharmaceutical research; however, while the use of such models remains necessary, it is important to understand the physiology of different animal models and the implications it may have for evaluating the performance of targeted oral formulations. One review, for example, failed to find a correlation between dogs and men in the reported absorption of 43 drugs (Chiou et al., 2000). The positive impact on animal health can also be considered; the information gained from studies geared toward clinical applications will also be transferrable to animal healthcare products (Song et al., 2016), and many of the physicochemical formulation considerations familiar to the pharmaceutical scientist working in the clinical arena are also of critical importance in animal medicine (Martinez et al., 2002).

5.4.1 General considerations

In light of the known physiological challenges to the oral administration of peptides and proteins, such as local pH, enzymatic degradation, and membrane permeability (Harloff-Helleberg et al., 2017; Tyagi et al., 2018), it follows that the implications of any differences in these across conventional animal models must be carefully considered during the development process, and caution must always be used when extrapolating the data from animal models to man.

Setting aside the challenges in interpreting the bioavailability of the compounds from animal models to man, there are also many challenges that pharmaceutical scientist faces when assessing the behavior of a targeted formulation or delivery vehicle using preclinical models. Initial assessments of molecules may be carried out using smaller animals such as rats, mice, or rabbits, and to gain information on parameters such as bioavailability, toxicity, pharmacodynamics, and dose linearity. This will not necessarily provide a full picture however, particularly as the complexity of modern pharmaceutical formulations increases to address the needs of challenging molecules and delivery targeted delivery to specific regions of the gastrointestinal tract, which is particularly pertinent for peptides and proteins, but also relevant in the increasing trend toward personalized medicines. When dealing with such formulations of complex construction or release behavior, it is essentially necessary to use an animal model that is large enough to be dosed with a human-sized dosage form, otherwise unnecessary scaling down of a tablet or capsule would be required, with release profiles unlikely to mimic exactly those developed for the full-size dose. This effectively narrows the options to larger animals such as dogs and pigs.

5.4.1.1 Use of anesthetic

Smaller animals often require anesthetization to administer oral solid doses or perform the associated sampling procedures while minimizing stress on the animal. Anesthetization can have an effect on many different physiological processes in the animal, and perhaps even the pharmacodynamic outcomes being measured such as blood pressure or blood glucose levels (Harloff-Helleberg et al., 2017), and this can also be expected to have implications for GI transit rates and absorption as physiological processes are generally dampened (Torjman et al., 2005).

5.4.1.2 Circadian considerations

A circadian or biological rhythm is defined as "a self-sustaining oscillation of endogenous origin" (Smolensky, & Lemmer, & Reinberg, 2007; Smolensky & Peppas, 2007) and is prevalent throughout nature in a myriad of biological processes. It is known to result in a greater intensity of symptoms of conditions such as rheumatoid arthritis (Harkness et al., 1982), asthma (Smolensky, & Lemmer, & Reinberg, 2007), cardiac disease (Hermida et al., 2007; Portaluppi & Lemmer, 2007), epilepsy (Gurkas et al., 2016), allergic rhinitis (Smolensky, & Lemmer, & Reinberg, 2007), and pain (Beauchamp & Labrecque, 2007), as shown in Fig. 5.10. Furthermore, the pharmacokinetics and even efficacy of therapeutic compounds have been shown to vary according to time of administration, in both oral and intravenous dosing regimens (Clench et al., 1981; Lévi et al., 2008), and even the side effect profile of a drug can be more or less severe depending on the time of day (Beauchamp & Labrecque, 2007; Lévi et al., 2008).

For example, gastrointestinal transit in humans has been reported as being different earlier in the day when compared to nighttime; a hydrogel matrix tablet administered to volunteers either at 8 am or 8 pm, tended to take longer to empty from the stomach when administered in the evening

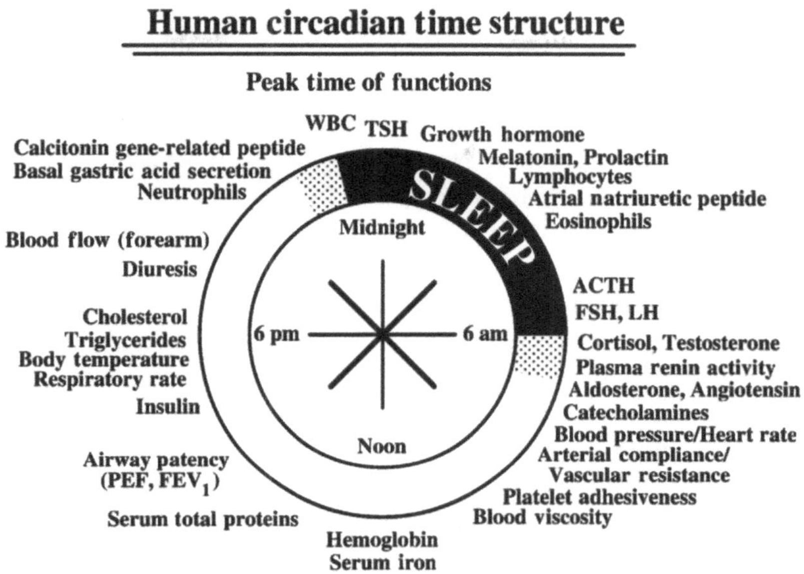

Figure 5.10 Human circadian time structure (Smolensky & Peppas, 2007).

rather than in the morning (Kambayashi et al., 2020), while radiolabeled tablets, radiotelemetry device, and meal were all shown to be delayed when administered at night in a separate study (Coupe et al., 1992).

When performing studies in nocturnal animals, it is likely at most research facilities that dosing would be carried out in the morning to suit the normal working day, which for a nocturnal animal might be considered the same as night-time dosing in a human. Equally, in disease models that show circadian effects, it may be prudent to consider how that might vary in a nocturnal animal. The effect of this on the data obtained should be considered; however, with careful control and understanding; such models can also lead to an enhanced understanding of the chronopharmacology of human disease conditions (Lemmer, 2017).

5.4.2 Gastrointestinal considerations

5.4.2.1 Gastrointestinal transit and forces

Sometimes, the differences can be reflective of the different natural diets of animals, in particular those who are obligate carnivores or herbivores. In the dog stomach, the contracting mixing forces are generally much higher than in humans, making the dog stomach a harsher challenge to the integrity of a sustained release formulation than the human stomach, and as such could result in otherwise promising formulations being rejected unnecessarily. Gastric emptying cycles are similar between dogs and humans, whereas the pylorus of the pig stomach tends to be much narrower than in humans, making the emptying on intact modified release monoliths more challenging. This is particularly problematic when the dosage form is targeted to release lower in the gut, and the onset of a particular function does not begin until it leaves the stomach due to an enteric coat. In one example, it was reported using gamma scintigraphy that an enteric-coated tablet administered to a pig was retained in the stomach for as long as 33 days (Davis et al., 2001; Hossain et al., 1990).

Transit times through the remainder of the gastrointestinal tract can also be quite different in animal models, a factor that becomes further complicated by differences in the physical length of sections of the gut. This becomes critically important in certain types of modified release preparations in terms of site of release, and this in combination with the resultant reduction in mucosal surface area available for absorption could also lead to misleading data on absorption or bioavailability where an API is only absorbed from a specific region. This issue has been overcome with the use of loperamide to slow the rate of gastrointestinal transit in the dog and potentially avoid underestimation of bioavailability (Yamada et al., 1995).

The intestinal physiology of the pig is often regarded as being more similar to that of man (Davis et al., 2001); however, the challenges described earlier remain in administering dosage forms that are expected to remain as a solid monolith as it transits the gut. To overcome this, surgical models have been employed to allow direct administration to the jejunum of the pig for novel insulin delivery devices (Hashim et al., 2019).

The dog model might be considered as means of overcoming the pyloric challenges with gastric emptying of monoliths in the pig model; however, it has been shown that the forces exerted on a dosage form in the canine stomach, while similar in the fed state, are generally higher in the fated state. This correlates well with a study where a sustained release formulation previously found to dose dump in men in the fed state also failed to produce prolonged release in the fasted dog model, which proved a harsh predictive model for a man (McInnes et al., 2008).

5.4.2.2 Gastrointestinal pH and fluid composition

Regional pH differences in animal models also have the potential to significantly impact dosage form performance. As most peptide and proteins tend to be highly soluble in an aqueous environment, the impact of local pH on dissolution rate is less likely to be a significant factor in the performance of these treatments (Harloff-Helleberg et al., 2017). However, a significant proportion of colon-targeted formulation approaches use an aspect of pH dependant behavior to facilitate release in the colon, which may be tailored to human conditions, but not triggered in an animal model. The intestine of the dog model, for example, is generally found to be +1 pH unit higher than for humans for each equivalent section of the gut poststomach. Also, an enteric coat designed to dissolve after gastric emptying from the stomach might be found to dissolve early in the dog model, where the fed and fasted pH values tend to be highly variable (Koziolek et al., 2019). Regional pH differences could also affect protein structure or folding (Brown et al., 2020).

This has led researchers to the use of pentagastrin to modify the dog stomach pH (Akimoto et al., 2000; Zane et al., 2014), which was found to predict the direction of food effects in humans, although not necessarily the magnitude, and it may be considered less ideal to complicate assessment of novel dosages with the administration of additional molecules.

Differences in the fluid volumes and concentration, enzymes, and bile salts have also been reported across animal models (Hatton et al., 2015), and the impact of this should be carefully considered according to the specific dosage. Enzymatic cleavage is a key barrier to peptide and protein administration (Brown et al., 2020), and as such, any species differences in agents such as proteases will influence the performance in that animal model.

5.4.2.3 Mucosal considerations

It is already particularly challenging to achieve absorption of large molecular weight peptides and proteins, so predictions might be particularly sensitive to differences in animal models when compared with humans, and this can be further complicated with the use of absorption/permeation enhancers that might also behave differently in different species. The dog model has shown a poor correlation for absorption of several compounds when compared to humans (Chiou et al., 2000).

A study of absorption modifying excipients in the dog and rat model found that the rat model was more susceptible to the effects of the absorption modifiers, which may have been due to the already more permeable dog intestine being less sensitive to these effects (Dahlgren et al., 2018). The researchers state that without careful consideration of gastrointestinal conditions and animal model variability in the effects of absorption enhancers, the data will be of "limited clinical value," and add nothing to our understanding of the behavior of these compounds (Dahlgren et al., 2019). Indeed, recent work suggests that the rat single-pass intestinal perfusion model may not be suitable for predicting differences in human regional intestinal permeability of low-to-medium permeability drugs (Dahlgren et al., 2020), as the correlation between data from the two was not found.

Depending on the expected route across the mucosal epithelia, any differences in transporters between animal models and humans should also be considered.

5.4.2.4 Bacteria

For colon-targeted oral formulations that employ coatings that are degraded by the microbiota of the colon, the effect of differences between animal models and humans should also be considered. The microflora of the gut can be greatly influenced by diet in animals, particularly whether they are carnivores or herbivores, and species of bacteria exist in the animal gut that has not been found in humans (Gootenberg & Turnbaugh, 2011).

5.5 Conclusion

An array of increasingly complex drug pharmaceutical systems have been developed with differing mechanisms for targeting distal sections of the gastrointestinal tract. Animal models for the evaluation of complex oral targeted formulations can provide useful insight into the likely performance in humans, particularly with novel compounds that have not yet been evaluated clinically for safety; however, there are many physiological differences that must be carefully understood and accounted for, whether in the development of peptide and protein actives, or conventional small molecule APIs. Caution must be exercised when extrapolating such data to the clinical setting, and the traits that are critical to the performance of a particular system carefully considered when interpreting results. Ultimately, clinical administration will provide the only true test of the viability of the formulation as a therapeutic formulation.

References

Akimoto, M., Nagahata, N., Furuya, A., Fukushima, K., Higuchi, S., & Suwa, T. (2000). Gastric pH profiles of beagle dogs and their use as an alternative to human testing. *European Journal of Pharmaceutics and Biopharmaceutics: Official Journal of Arbeitsgemeinschaft fur Pharmazeutische Verfahrenstechnik e.V*, *49*(2), 99–102. Available from https://doi.org/10.1016/S0939641l(99)00070-3.

Alatab, S., et al. (2020). The global, regional, and national burden of inflammatory bowel disease in 195 countries and territories, 1990–2017: A systematic analysis for the global burden of disease study 2017. *Lancet Gastroenterology & Hepatology*, *5*(1), 17–30. Available from https://doi.org/10.1016/S2468-1253(19)30333-4.

Arora, S., Ali, J., Ahuja, A., Baboota, S., & Qureshi, J. (2006). Pulsatile drug delivery systems: An approach for controlled drug delivery. *Indian Journal of Pharmaceutical Science*, *68*(3), 295–300.

Basit, A. W., Podczeck, F., Newton, J. M., Waddington, W. A., Ell, P. J., & Lacey, L. F. (2002). Influence of polyethylene glycol 400 on the gastrointestinal absorption of ranitidine. *Pharmaceutical Research*, *19*(9), 1368–1374. Available from https://doi.org/10.1023/A:1020315228237.

Beauchamp, D., & Labrecque, G. (2007). Chronobiology and chronotoxicology of antibiotics and aminoglycosides. *Advanced Drug Delivery Reviews*, *59*(9), 896–903. Available from https://doi.org/10.1016/j.addr.2006.07.028.

Boivin, G. P., et al. (2003). Pathology of mouse models of intestinal cancer: Consensus report and recommendations. *Gastroenterology*, *124*(3), 762–777. Available from https://doi.org/10.1053/gast.2003.50094, Mar.

Bresnick, W. H., Rask-Madsen, C., Hogan, D. L., Koss, M. A., & Isenberg, J. I. (1993). The effect of acute emotional stress on gastric acid secretion in normal subjects and duodenal ulcer patients. *Journal of Clinical Gastroenterology*, *17*(2). [Online]. Available: https://journals.lww.com/jcge/Fulltext/1993/09000/The_Effect_of_Acute_Emotional_Stress_on_Gastric.6.aspx.

Brown, T. D., Whitehead, K. A., & Mitragotri, S. (2020). Materials for oral delivery of proteins and peptides. *Nature Reviews Materials*, *5*(2), 127–148. Available from https://doi.org/10.1038/s41578-019-0156-6.

Bussemer, T., & Bodmeier, R. (2003). Formulation parameters affecting the performance of coated gelatin capsules with pulsatile release profiles. *International Journal of Pharmaceutics*, *267*(1), 59–68. Available from https://doi.org/10.1016/j.ijpharm.2003.07.008.

Bussemer, T, Dashevsky, A, & Bodmeier, R. (2003). A pulsatile drug delivery system based on rupturable coated hard gelatin capsules. *Journal of Controlled Release*, *93*, 331–339. Available from https://doi.org/10.1016/j.jconrel.2003.08.012.

Chiou, W. L., Jeong, H. Y., Chung, S. M., & Wu, T. C. (2000). Evaluation of using dog as an animal model to study the fraction of oral dose absorbed of 43 drugs in humans. *Pharmaceutical Research*, *17*(2), 135–140. Available from https://doi.org/10.1023/A:1007552927404.

Chourasia, M. K., & Jain, S. K. (2004). Polysaccharides for colon targeted drug delivery. *Drug Delivery*, *11*(2), 129–148. Available from https://doi.org/10.1080/10717540490280778.

Chung, C. S., Lee, S. K., & AY. (1997). Mouse models for human diseases. *Hong Kong Medical Journal = Xianggang yi xue za zhi / Hong Kong Academy of Medicine*, *3*(2), 201–209.

Cibelli, J., et al. (2013). Strategies for improving animal models for regenerative medicine. *Cell Stem Cell*, *12*(3), 271–274. Available from https://doi.org/10.1016/j.stem.2013.01.004, Mar.

Clench, J., Reinberg, A., Dziewanowska, Z., Ghata, J., & Smolensky, M. (1981). Circadian changes in the bioavailability and effects of indomethacin in healthy subjects. *European Journal of Clinical Pharmacology*, *20*(5), 359–369. Available from https://doi.org/10.1007/BF00615406.

Coupe, A. J., Davis, S. S., Evans, D. F., & Wilding, I. R. (1992). Nocturnal scintigraphic imaging to investigate the gastrointestinal transit of dosage forms. *Journal of Controlled Release: Official Journal of the Controlled Release Society*, *20*(2), 155–162. Available from https://doi.org/10.1016/0168-3659(92)90161-J.

Dahlgren, D., et al. (2018). The effects of three absorption-modifying critical excipients on the in vivo intestinal absorption of six model compounds in rats and dogs. *International Journal of Pharmaceutics*, *547*(1), 158–168. Available from https://doi.org/10.1016/j.ijpharm.2018.05.029.

Dahlgren, D., Sjöblom, M., & Lennernäs, H. (2019). Intestinal absorption-modifying excipients: A current update on preclinical in vivo evaluations. *European Journal of Pharmaceutics and Biopharmaceutics: Official Journal of Arbeitsgemeinschaft fur Pharmazeutische Verfahrenstechnik e.V*, *142*, 411–420. Available from https://doi.org/10.1016/j.ejpb.2019.07.013.

Dahlgren, D., Cano-Cebrián, M.-J., Olander, T., Hedeland, M., Sjöblom, M., & Lennernäs, H. (2020). Regional intestinal drug permeability and effects of permeation enhancers in rat. *Pharmaceutics*, *12*(3), 242. Available from https://doi.org/10.3390/pharmaceutics12030242.

Davis, S. S., Illum, L., & Hinchcliffe, M. (2001). Gastrointestinal transit of dosage forms in the pig. *The Journal of Pharmacy and Pharmacology*, *53*(1), 33–39. Available from https://doi.org/10.1211/0022357011775163.

Digenis, G. A., Sandefer, E. P., Page, R. C., & Doll, W. J. (1998a). Gamma scintigraphy: An evolving technology in pharmaceutical formulation development-Part 1. *Pharmaceutical Science & Technology Today*, *1*(3), 100–108. Available from https://doi.org/10.1016/S1461-5347(98)00032-7.

Digenis, G. A., Sandefer, E. P., Page, R. C., & Doll, W. J. (1998b). Gamma scintigraphy: An evolving technology in pharmaceutical formulation development—Part 2. *Pharmaceutical Science & Technology Today*, *1*(4), 160–165, [Online]. Available:. Available from https://doi.org/10.1016/S1461-5347(98)00041-8.

Dressman, J. B., et al. (1990). Upper gastrointestinal (GI) pH in young, healthy men and women. *Pharmaceutical Research*, *7*(7), 756–761. Available from https://doi.org/10.1023/A:1015827908309.

Dressman, J. B., Vertzoni, M., Goumas, K., & Reppas, C. (2007). Estimating drug solubility in the gastrointestinal tract. *Advanced Drug Delivery Reviews*, *59*(7), 591–602. Available from https://doi.org/10.1016/j.addr.2007.05.009.

Elbatanony, R. S. E. (2019). Modified pH independent/time controlled explosion system (TES) for targeted drug delivery in the lower intestinal tract: Formulation and pharmacokinetic evaluation in healthy volunteers. *Journal of Drug Delivery Science Technology*, *50*, 163–173. Available from https://doi.org/10.1016/j.jddst.2019.01.029.

Fadda, H. M., McConnell, E. L., Short, M. D., & Basit, A. W. (2009). Meal-induced acceleration of tablet transit through the human small intestine. *Pharmaceutical Research*, *26*(2), 356–360. Available from https://doi.org/10.1007/s11095-008-9749-2.

Flisikowska, T., et al. (2012). A porcine model of familial adenomatous polyposis. *Gastroenterology*, *143*(5), 1173–1175. Available from https://doi.org/10.1053/j.gastro.2012.07.110, e7.

Fude, C., Lei, Y., Jie, J., Hongze, P., Wenhui, L., & Dongmei, C. (2007). Preparation and in vitro evaluation of pH, time-based and enzyme-degradable pellets for colonic drug delivery. *Drug Development and Industrial Pharmacy*, *33*(9), 999–1007. Available from https://doi.org/10.1080/03639040601150393.

Fukui, E., Uemura, K., & Kobayashi, M. (2000). Studies on applicability of press-coated tablets using hydroxypropylcellulose (HPC) in the outer shell for timed-release preparations. *Journal of Controlled Release: Official Journal of the Controlled Release Society*, *68*(2), 215–223. Available from https://doi.org/10.1016/S01683659(00)00261-3.

Fukui, E., Miyamura, N., Uemura, K., & Kobayashi, M. (2000). Preparation of enteric coated timed-release press-coated tablets and evaluation of their function by in vitro and in vivo tests for colon targeting. *International Journal of Pharmaceutics*, *204*(1), 7–15. Available from https://doi.org/10.1016/S0378-5173(00)00454-3.

Gootenberg, D. B., & Turnbaugh, P. J. (2011). Companion animals symposium: Humanized animal models of the microbiome1. *Journal of Animal Science*, *89*(5), 1531–1537. Available from https://doi.org/10.2527/jas.2010-3371.

Gupta, V. K., Beckert, T. E., & Price, J. C. (2001). A novel pH- and time-based multi-unit potential colonic drug delivery system. I. Development. *International Journal of Pharmaceutics*, *213*(1), 83–91. Available from https://doi.org/10.1016/S0378-5173(00)00649-9.

Guren, M. G. (2019). The global challenge of colorectal cancer. *The Lancet Gastroenterology and Hepatology*, *4*(12), 894–895. Available from https://doi.org/10.1016/S2468-1253(19)30329-2.

Gurkas, E., Serdaroglu, A., Hirfanoglu, T., Kartal, A., Yılmaz, U., & Bilir, E. (2016). Sleep-wake distribution and circadian patterns of epileptic seizures in children. *European Journal of Paediatric Neurology: EJPN: Official Journal of the European Paediatric Neurology Society*, *20*(4), 549–554. Available from https://doi.org/10.1016/j.ejpn.2016.04.004.

Harkness, J. A., et al. (1982). Circadian variation in disease activity in rheumatoid arthritis. *British Medical Journal (Clinical Research Edition).*, *284*(6315), 551–554. Available from https://doi.org/10.1136/bmj.284.6315.551.

Harloff-Helleberg, S., Nielsen, L. H., & Nielsen, H. M. (2017). Animal models for evaluation of oral delivery of biopharmaceuticals. *Journal of Controlled Release: Official Journal of the Controlled Release Society*, *268*, 57–71. Available from https://doi.org/10.1016/j.jconrel.2017.09.025.

Hashim, M., et al. (2019). Jejunal wall delivery of insulin via an ingestible capsule in anesthetized swine-A pharmacokinetic and pharmacodynamic study. *Pharmacology Research Perspectives*, 7(5). Available from https://doi.org/10.1002/prp2.522, e00522–e00522.

Hatton, G. B., Yadav, V., Basit, A. W., & Merchant, H. A. (2015). Animal farm: Considerations in animal gastrointestinal physiology and relevance to drug delivery in humans. *Journal of Pharmaceutical Sciences*, *104*(9), 2747–2776. Available from https://doi.org/10.1002/jps.24365.

Hermida, R. C., Ayala, D. E., & Portaluppi, F. (2007). Circadian variation of blood pressure: The basis for the chronotherapy of hypertension. *Advanced Drug Delivery Reviews*, *59*(9), 904–922. Available from https://doi.org/10.1016/j.addr.2006.08.003.

Hossain, M., Abramowitz, W., Watrous, B. J., Szpunar, G. J., & Ayres, J. W. (1990). Gastrointestinal transit of nondisintegrating, nonerodible oral dosage forms in pigs. *Pharmaceutical Research*, 7(11), 1163–1166. Available from https://doi.org/10.1023/A:1015936426906.

Hua, S., Marks, E., Schneider, J. J., & Keely, S. (2015). Advances in oral nano-delivery systems for colon targeted drug delivery in inflammatory bowel disease: Selective targeting to diseased vs healthy tissue. *Nanomedicine Nanotechnology, Biology Medicine*, *11*(5), 1117–1132. Available from https://doi.org/10.1016/j.nano.2015.02.018.

Ibekwe, V. C., Khela, M. K., Evans, D. F., & Basit, A. W. (2008). A new concept in colonic drug targeting: A combined pH-responsive and bacterially-triggered drug delivery technology. *Alimentary Pharmacology & Therapeutics*, *28*(7), 911–916. Available from https://doi.org/10.1111/j.1365-2036.2008.03810.x.

Ishino, R., & Sunada, H. (1993). Studies on application of wax matrix system for controlled drug release. *Chemical & Pharmaceutical Bulletin (Tokyo).*, *41*(3), 586–589. Available from https://doi.org/10.1248/cpb.41.586.

Ishino, R., Yoshino, H., Hirakawa, Y., & Noda, K. (1992a). Absorption of diltiazem in Beagle dog from pulsatile release tablet. *Chemical & Pharmaceutical Bulletin (Tokyo).*, *40* (11), 3094–3096. Available from https://doi.org/10.1248/cpb.40.3094.

Ishino, R., Yoshino, H., Hirakawa, Y., & Noda, K. (1992b). Design and preparation of pulsatile release tablet as a new oral drug delivery system. *Chemical & Pharmaceutical Bulletin (Tokyo).*, *40*(11), 3036–3041. Available from https://doi.org/10.1248/cpb.40.3036.

Kambayashi, A., Sako, K., & Kondo, H. (2020). Effects of diurnal variation and food on gastrointestinal transit of 111In-labeled hydrogel matrix extended-release tablets and 99mTc-labeled pellets in humans. *Journal of Pharmaceutical Sciences*, *109*(2), 1020–1025. Available from https://doi.org/10.1016/j.xphs.2019.09.025.

Katstra, W. E., Palazzolo, R. D., Rowe, C. W., Giritlioglu, B., Teung, P., & Cima, M. J. (2000). Oral dosage forms fabricated by three dimensional printingTM. *Journal of Controlled Release: Official Journal of the Controlled Release Society*, *66*(1), 1–9. Available from https://doi.org/10.1016/S0168-3659(99)00225-4.

Keller, J., Gröger, G., Cherian, L., Günther, B., & Layer, P. (2001). Circadian coupling between pancreatic secretion and intestinal motility in humans. *American Journal of Physiology Liver Physiology*, *280*(2), G273–G278. Available from https://doi.org/10.1152/ajpgi.2001.280.2.G273.

Khan, M. Z. I., Prebeg, Ž., & Kurjaković, N. (1999). A pH-dependent colon targeted oral drug delivery system using methacrylic acid copolymers: I. Manipulation of drug release using Eudragit® L100-55 and Eudragit® S100 combinations. *Journal of Controlled Release: Official Journal of the Controlled Release Society*, *58*(2), 215–222. Available from https://doi.org/10.1016/S0168-3659(98)00151-5.

Khan, M. Z. I., Štedul, H. P., & Kurjaković, N. (2000). A pH-dependent colon-targeted oral drug delivery system using methacrylic acid copolymers. II. Manipulation of drug release using Eudragit® L100 and Eudragit S100 combinations. *Drug Development and Industrial Pharmacy*, *26*(5), 549–554. Available from https://doi.org/10.1081/DDC-100101266.

Koziolek, M., et al. (2019). Characterization of the GI transit conditions in Beagle dogs with a telemetric motility capsule. *European Journal of Pharmaceutics and Biopharmaceutics: Official Journal of Arbeitsgemeinschaft fur Pharmazeutische Verfahrenstechnik e.V*, *136*, 221–230. Available from https://doi.org/10.1016/j.ejpb.2019.01.026.

Krögel, I., & Bodmeier, R. (1998). Pulsatile drug release from an insoluble capsule body controlled by an erodible plug. *Pharmaceutical Research*, *15*(3), 474–481. Available from https://doi.org/10.1023/A:1011940718534.

Krögel, I., & Bodmeier, R. (1999). Evaluation of an enzyme-containing capsular shaped pulsatile drug delivery system. *Pharmaceutical Research*, *16*(9), 1424–1429. Available from https://doi.org/10.1023/A:1018959327311.

Lemmer, B. (2017). Signal transduction and chronopharmacology of regulation of circadian cardiovascular rhythms in animal models of human hypertension. *Heart Failure Clinics*, *13*(4), 739–757. Available from https://doi.org/10.1016/j.hfc.2017.05.009.

Lin, K., Lin, S., & Li, M. (2001a). Compression forces and amount of outer coating layer affecting the time-controlled disintegration of the compression-coated tablets prepared by direct compression with micronized ethylcellulose. *Journal of Pharmaceutical Sciences*, *90*(12), 2005–2009. Available from https://doi.org/10.1002/jps.1151.

Lin, S.-Y., Lin, K.-H., & Li, M.-J. (2001b). Micronized ethylcellulose used for designing a directly compressed time-controlled disintegration tablet. *Journal of Controlled Release: Official Journal of the Controlled Release Society*, *70*(3), 321–328. Available from https://doi.org/10.1016/S0168-3659(00)00360-6.

Lin, S.-Y., Li, M.-J., & Lin, K.-H. (2004). Hydrophilic excipients modulate the time lag of time-controlled disintegrating press-coated tablets. *AAPS PharmSciTech*, *5*(4). Available from https://doi.org/10.1208/pt050454, e54–e54.

Lindahl, P., et al. (1997). Alveogenesis failure in PDGF-A-deficient mice is coupled to lack of distal spreading of alveolar smooth muscle cell progenitors during lung development. *Development (Cambridge, England)*, *124*(20), 3943–3953. Available from https://doi.org/10.1242/dev.124.20.3943.

Liu, L., Fishman, M. L., Hicks, K. B., Kende, M., & Ruthel, G. (2006). Pectin/zein beads for potential colon-specific drug delivery: Synthesis and in vitro evaluation. *Drug Delivery*, *13*(6), 417–423. Available from https://doi.org/10.1080/10717540500394935.

Low, M. J. (2012). Mouse models in gastroenterology research. *Gastroenterology*, *143*(6), 1410–1412. Available from https://doi.org/10.1053/j.gastro.2012.10.011.

Lui, C. Y., Amidon, G. L., Berardi, R. R., Fleisher, B. D., Youngberg, C., & Dressman, J. B. (1986). Comparison of gastrointestinal pH in dogs and humans: Implications on the use of the beagle dog as a model for oral absorption in humans. *Journal of Pharmaceutical Sciences*, *75*(3), 271–274. Available from https://doi.org/10.1002/jps.2600750313.

Luiking, Y. C., van der Reijden, A. C., van Berge Henegouwen, G. P., & Akkermans, L. M. A. (1998). Migrating motor complex cycle duration is determined by gastric or duodenal origin of phase III. *American Journal of Physiology Liver Physiology*, *275*(6), G1246–G1251. Available from https://doi.org/10.1152/ajpgi.1998.275.6.G1246.

Lévi, F., Altinok, A., Clairambault, J., & Goldbeter, A. (2008). Implications of circadian clocks for the rhythmic delivery of cancer therapeutics. *Philosophical Transactions of the Royal Society a-Mathematical Physical and Engineering Sciences*, *366*(1880), 3575–3598. Available from https://doi.org/10.1098/rsta.2008.0114.

Martinez, M., Augsburger, L., Johnston, T., & Jones, W. W. (2002). Applying the biopharmaceutics classification system to veterinary pharmaceutical products: Part I: Biopharmaceutics and formulation considerations. *Advanced Drug Delivery Reviews*, *54*(6), 805–824. Available from https://doi.org/10.1016/S0169-409X(02)00070-4.

McConville, J. T., Ross, A. C., Chambers, A. R., Smith, G., Florence, A. J., & Stevens, H. N. E. (2004). The effect of wet granulation on the erosion behaviour of an HPMC–lactose tablet, used as a rate-controlling component in a pulsatile drug delivery capsule formulation. *European Journal of Pharmaceutics and Biopharmaceutics: Official Journal of Arbeitsgemeinschaft fur Pharmazeutische Verfahrenstechnik e.V*, *57*(3), 541–549. Available from https://doi.org/10.1016/j.ejpb.2004.01.004.

McInnes, F., Clear, N., Humphrey, M., & Stevens, H. N. E. (2008). In vivo performance of an oral MR matrix tablet formulation in the beagle dog in the fed and fasted state: Assessment of mechanical weakness. *Pharmaceutical Research*, *25*(5), 1075–1084. Available from https://doi.org/10.1007/s11095-007-9462-6.

Moore, J. G., & Englert, E. (1970). Circadian rhythm of gastric acid secretion in man. *Nature*, *226*(5252), 1261–1262. Available from https://doi.org/10.1038/2261261a0.

Muraoka, M., et al. (1998). Evaluation of intestinal pressure-controlled colon delivery capsule containing caffeine as a model drug in human volunteers. *Journal of Controlled Release: Official Journal of the Controlled Release Society*, *52*(1), 119–129. Available from https://doi.org/10.1016/S0168-3659(97)00201-0.

Ng, W. K., Wong, S. H., & Ng, S. C. (2016). Changing epidemiological trends of inflammatory bowel disease in Asia. *Intestinal Research*, *14*(2), 111–119. Available from https://doi.org/10.5217/ir.2016.14.2.111.

Nunthanid, J., Huanbutta, K., Luangtana-anan, M., Sriamornsak, P., Limmatvapirat, S., & Puttipipatkhachorn, S. (2008). Development of time-, pH-, and enzyme-controlled colonic drug delivery using spray-dried chitosan acetate and hydroxypropyl methylcellulose. *European Journal of Pharmaceutics and Biopharmaceutics: Official Journal of Arbeitsgemeinschaft fur Pharmazeutische Verfahrenstechnik e.V, 68*(2), 253–259. Available from https://doi.org/10.1016/j.ejpb.2007.05.017.

Ovesen, L., Bendtsen, F., Tage-Jensen, U., Pedersen, N. T., Gram, B. R., & Rune, S. J. (1986). Intraluminal pH in the stomach, duodenum, and proximal jejunum in normal subjects and patients with exocrine pancreatic insufficiency. *Gastroenterology, 90*(4), 958–962. Available from https://doi.org/10.1016/0016-5085(86)90873-5.

Ozeki, Y., Watanabe, Y., Okamoto, H., & Danjo, K. (2004). Development of dividable one-step dry-coated tablets (Dividable-OSDRC) and their evaluation as a new platform for controlled drug release. *Pharmaceutical Research, 21*(7), 1177–1183. Available from https://doi.org/10.1023/B:PHAM.0000033004.88953.bc.

Ozeki, Y., Ando, M., Watanabe, Y., & Danjo, K. (2004). Evaluation of novel one-step dry-coated tablets as a platform for delayed-release tablets. *Journal of Controlled Release: Official Journal of the Controlled Release Society, 95*(1), 51–60. Available from https://doi.org/10.1016/j.jconrel.2003.10.028.

Peerapattana, J., Otsuka, K., & Otsuka, M. (2004). Time-controlled pulse-drug release from dry-coated wax matrix tablets for colon drug delivery. *Bio-medical Materials and Engineering, 14*, 293–301.

Pleuvry, B. J. (2005). Factors affecting drug absorption and distribution. *Anaesthesia Intensive Care Medicine, 6*(4), 135–138. Available from https://doi.org/10.1383/anes.6.4.135.63632.

Portaluppi, F., & Lemmer, B. (2007). Chronobiology and chronotherapy of ischemic heart disease. *Advanced Drug Delivery Reviews, 59*(9), 952–965. Available from https://doi.org/10.1016/j.addr.2006.07.029.

Pozzi, F., Furlani, P., Gazzaniga, A., Davis, S. S., & Wilding, I. R. (1994). The time clock system: A new oral dosage form for fast and complete release of drug after a predetermined lag time. *Journal of Controlled Release: Official Journal of the Controlled Release Society, 31*(1), 99–108. Available from https://doi.org/10.1016/0168-3659(94)90255-0.

Rawla, P., Sunkara, T., & Gaduputi, V. (2019). Epidemiology of pancreatic cancer: Global trends, etiology and risk factors. *World Journal of Oncology, 10*(1), 10–27. Available from https://doi.org/10.14740/wjon1166.

Remesz, O. Schematic of GI tract (wiki-pl: Orem, commons: Orem), via Wikimedia Commons. <https://commons.wikimedia.org/wiki/File:GISystem.svg> Accessed 13.04.22.

Ross, A. C., Macrae, R. J., Walther, M., & Stevens, H. N. E. (2000). Chrono pharmaceutical drug delivery from a pulsatile capsule device based on programmable erosion. *The Journal of Pharmacy and Pharmacology, 52*(8), 903–909. Available from https://doi.org/10.1211/0022357001774787.

Rowe, C. W., Katstra, W. E., Palazzolo, R. D., Giritlioglu, B., Teung, P., & Cima, M. J. (2000). Multimechanism oral dosage forms fabricated by three dimensional printing. *Journal of Controlled Release: Official Journal of the Controlled Release Society, 66*(1), 11–17. Available from https://doi.org/10.1016/S0168-3659(99)00224-2.

Rowe, Ray, Sheskey, Paul, & Weller, Paul (2003). *Handbook of Pharmaceutical Excipients*. Pharmaceutical Press.

Sangalli, M. E., Maroni, A., Zema, L., Busetti, C., Giordano, F., & Gazzaniga, A. (2001). In vitro and in vivo evaluation of an oral system for time and/or site-specific drug delivery. *Journal of Controlled Release: Official Journal of the Controlled Release Society, 73*(1), 103–110. Available from https://doi.org/10.1016/S0168-3659(01)00291-7.

Sangalli, M. E., Maroni, A., Zema, L., Cerea, M., Conte, U., & Gazzaniga, A. (2003). A study on the release mechanism of drugs from hydrophilic partially coated perforated matrices. *Farm, 58*(9), 971–976. Available from https://doi.org/10.1016/S0014-827X(03)00168-X.

Sawada, T., Kondo, H., Nakashima, H., Sako, K., & Hayashi, M. (2004). Time-release compression-coated core tablet containing nifedipine for chronopharmacotherapy. *International Journal of Pharmaceutics*, *280*(1), 103−111. Available from https://doi.org/10.1016/j.ijpharm.2004.05.004.

Smolensky, M. H., & Peppas, N. A. (2007). Chronobiology, drug delivery, and chronotherapeutics. *Advanced Drug Delivery Reviews*, *59*(9), 828−851. Available from https://doi.org/10.1016/j.addr.2007.07.001.

Smolensky, M. H., Lemmer, B., & Reinberg, A. E. (2007). Chronobiology and chronotherapy of allergic rhinitis and bronchial asthma. *Advanced Drug Delivery Reviews*, *59*(9), 852−882. Available from https://doi.org/10.1016/j.addr.2007.08.016.

Song, Y., Peressin, K., Wong, P. Y., Page, S. W., & Garg, S. (2016). Key considerations in designing oral drug delivery systems for dogs. *Journal of Pharmaceutical Sciences*, *105*(5), 1576−1585. Available from https://doi.org/10.1016/j.xphs.2016.03.007.

Sperber, A. D., et al. (2021). Worldwide prevalence and burden of functional gastrointestinal disorders, results of Rome foundation global study. *Gastroenterology*, *160*(1), 99−114. Available from https://doi.org/10.1053/j.gastro.2020.04.014, e3.

Stevens, H. N. E. (1998). Chronopharmaceutical drug delivery. *The Journal of Pharmacy and Pharmacology*, *50*(Suppl. 9), 5. Available from https://doi.org/10.1111/j.2042-7158.1998.tb02205.x.

Stevens, H. N. E., et al. (2002). Evaluation of pulsincapTM to provide regional delivery of dofetilide to the human GI tract. *International Journal of Pharmaceutics*, *236*(1), 27−34. Available from https://doi.org/10.1016/S0378-5173(02)00012-1.

Sungthongjeen, S., Puttipipatkhachorn, S., Paeratakul, O., Dashevsky, A., & Bodmeier, R. (2004). Development of pulsatile release tablets with swelling and rupturable layers. *Journal of Controlled Release: Official Journal of the Controlled Release Society*, *95*(2), 147−159. Available from https://doi.org/10.1016/j.jconrel.2003.10.023.

Torjman, M. C., Joseph, J. I., Munsick, C., Morishita, M., & Grunwald, Z. (2005). Effects of Isoflurane on gastrointestinal motility after brief exposure in rats. *International Journal of Pharmaceutics*, *294*(1), 65−71. Available from https://doi.org/10.1016/j.ijpharm.2004.12.028.

Tyagi, P., Pechenov, S., & Anand Subramony, J. (2018). Oral peptide delivery: Translational challenges due to physiological effects. *Journal of Controlled Release: Official Journal of the Controlled Release Society*, *287*, 167−176. Available from https://doi.org/10.1016/j.jconrel.2018.08.032.

Ueda, S., Ibuki, R., Kimura, S., Murata, S., Takahashi, T., Tokunaga, Y., & Hata, T. (1994). Development of a novel drug release system, time-controlled explosion system (TES). III. Relation between lag time and membrane. *Chemical & Pharmaceutical Bulletin (Tokyo).*, *42*(2), 364−367. Available from https://doi.org/10.1248/cpb.42.364.

Ueda, S., Ibuki, R., Kawamura, A., Murata, S., Takahashi, T., Kimura, S., & Hata, T. (1994). Development of a novel drug delivery system, time-controlled explosion system (TES). IV. In vivo drug release behavior. *Journal of Drug Targeting*, *2*(2), 133−140. Available from https://doi.org/10.3109/10611869409015901.

Ueda, S., Hata, T., Asakura, S., Yamaguchi, H., Kotani, M., & Ueda, Y. (1994). Development of a novel drug release system, time-controlled explosion system (TES). I. Concept and design. *Journal of Drug Targeting*, *2*(1), 35−44. Available from https://doi.org/10.3109/10611869409015891.

Washington, N., Washington, C., & Wilson, C. G. (2001). *Physiological pharmaceutics: Biological barriers to drug absorption*. Taylor Fr.

Wilding, I. R., Davis, S. S., Pozzi, F., Furlani, P., & Gazzaniga, A. (1994). Enteric coated timed release systems for colonic targeting. *International Journal of Pharmaceutics*, *111*(1), 99−102. Available from https://doi.org/10.1016/0378-5173(94)90406-5.

Wilson, C. G., et al. (1997). Evaluation of a gastro-resistant pulsed release delivery system (Pulsincap) in humans. *Drug Delivery*, *4*(3), 201–206. Available from https://doi.org/10.3109/10717549709051893.

Wilson, P. J., & Basit, A. W. (2005). Exploiting gastrointestinal bacteria to target drugs to the colon: An in vitro study using amylose coated tablets. *International Journal of Pharmaceutics*, *300*(1), 89–94. Available from https://doi.org/10.1016/j.ijpharm.2005.05.010.

Yamada, K., Furuya, A., Akimoto, M., Maki, T., Suwa, T., & Ogata, H. (1995). Evaluation of gastrointestinal transit controlled-beagle dog as a suitable animal model for bioavailability testing of sustained-release acetaminophen dosage form. *International Journal of Pharmaceutics*, *119*(1), 1–10. Available from https://doi.org/10.1016/0378-5173(94)00350-E.

Zane, P., Guo, Z., MacGerorge, D., Vicat, P., & Ollier, C. (2014). Use of the pentagastrin dog model to explore the food effects on formulations in early drug development. *European Journal of Pharmaceutical Sciences: Official Journal of the European Federation for Pharmaceutical Sciences*, *57*, 207–213. Available from https://doi.org/10.1016/j.ejps.2013.09.018.

Ziegler, A., Gonzalez, L., & Blikslager, A. (2016). Large animal models: The key to translational discovery in digestive disease research. *Cellular Molecular Gastroenterology Hepatology*, *2*(6), 716–724. Available from https://doi.org/10.1016/j.jcmgh.2016.09.003.

Zou, H., Jiang, X., Kong, L., & Gao, S. (2008). Design and evaluation of a dry coated drug delivery system with floating–pulsatile release. *Journal of Pharmaceutical Sciences*, *97*(1), 263–273. Available from https://doi.org/10.1002/jps.21083.

CHAPTER SIX

CMC Considerations for conventional oral peptide formulations. Scale up and validation

Ruchit Trivedi
Loxo Oncology at Lilly, Louisville, CO, United States

6.1 Introduction

Protein and peptide therapeutics have been increasingly used since the approval of Human Insulin almost 40 years ago (Leader et al., 2008). Protein therapeutics can play a significant role in modifying disease modalities and improving treatment options. However, until recently, injectable routes of administration (IV, IM, or SC) were most used for peptide therapeutics (Shaji & Patole, 2008). Several alternate routes of administration such as buccal (Caon et al., 2015), intranasal (Meredith et al., 2015), pulmonary (Newman, 2017), rectal (Hua, 2019), transdermal (Mandal et al., 2018; Pena-Juarez et al., 2021), and ocular (Mandal et al., 2018) have been evaluated with varying degree of success. While each of these routes has its own advantages, the oral route of administration remains the most attractive route, especially for chronic administration of drugs. The oral route has several advantages over other routes of administration such as patient compliance, cost of manufacture, and ease of administration, among others.

Designing oral peptide delivery systems presents unique challenges to pharmaceutical scientists due to their physicochemical properties such as large molecular size, susceptibility to enzymatic degradation, short plasma half-life, ion permeability, immunogenicity, and the tendency to undergo aggregation, adsorption, and denaturation (Saffran et al., 1986). All of this leads to negligible oral bioavailability, typically less than 1% (Renukuntla et al., 2013). As previously described within this book, there have been various strategies

Oral Delivery of Therapeutic Peptides and Proteins.
DOI: https://doi.org/10.1016/B978-0-12-821061-1.00006-X

employed to successfully design peptide formulations for oral use. To expand on this knowledge area, this chapter will focus on scaling up of these formulations as well as designing a robust process to reproducibly manufacture the formulations.

6.2 Chemistry, manufacturing, and controls

Chemistry, manufacturing, and controls (CMC) is an integral part of the drug development process as well as any pharmaceutical product submission to the FDA. CMC is typically applicable to the entire product lifecycle, from drug candidate selection to post approval, and CMC ensures drug products are reliably effective and safe and maintain their high quality for the patients. CMC is also responsible to guarantee that drug product quality is maintained from the clinical trial phase through commercialization and marketing. CMC considerations are typically tailored to the drug substance, drug product, platform, and delivery system, and they apply to both drug product and the manufacturing facility.

Drug product considerations

- Drug formulation
- Manufacturing process
- Quality control release testing
- Specifications and stability of the product

Manufacturing facility

- Design
- Qualification
- Operation
- Maintenance

6.2.1 Drug product considerations

6.2.1.1 Drug formulation considerations

CMC considerations for oral peptide formulations, as with small-molecule peptide formulations, are about assessing whether the drug product is reproducibly safe and effective. A key aspect of peptides is that they have different mechanisms of degradation and impurity generation, all of which can be detrimental to the product. Table 6.1 shows some of the pathways of degradation and generation of impurities for peptides.

Table 6.1 Mechanisms of degradation and impurity generation of peptides (Wu et al., 2017).

Impurity	Mechanism
Denaturation	Modification of the peptide structure that changes its physical, chemical, and biological properties; occurs in the presence of denaturing agents
Breakage of peptide bond	Upon exposure to harsh conditions, extreme pH, high temperature, or enzymes
Aggregation	Association of hydrophobic amino acid residues to form aggregates; precipitation occurs if on a macroscopic scale
Deamination	Hydrolysis of the side-chain amide linkage of an amino acid residue with the formation of a free carboxylic acid
RedOx	Induced by temperature, pH, metal ions, and buffers during synthesis and/or storage
Disulfide interaction	Change in conformation due to reaction between disulfides within a peptide chain
Diastereoisomerization	Alteration of L-amino acids to D, L mixtures, with the formation of peptide bonds sensitive to proteolytic enzymes
β-Elimination	Proceeds through a carbanion intermediate; susceptible residues under alkaline conditions are Cys, Lys, Phe, Ser, and Thr
Amino acid removal	Incomplete coupling due to incomplete removal of the protecting group of the last coupled amino acid or insufficient activation of the incoming amino acid. Missing one or more amino acid residues at either the N- or C-terminal end due to precipitation of resin beads during synthesis leads to truncation of the peptide chain
Amino acid insertion	Insertion of an additional amino acid into the peptide sequence due to excess of 9-fluorenylmethoxycarbonyl (Fmoc)-protected amino acids and improper washing
Dimerization	Self-association of hydrophobic amino acids residues in peptides to form aggregates
By-products generated by incomplete deprotection	Results in protecting groups covalently attached to peptide sequences

Source: Adapted from Wu, L. C., Chen, F., Lee, S. L., Raw, A., & Yu, L. X. (2017). Building parity between brand and generic peptide products: Regulatory and scientific considerations for quality of synthetic peptides. *International Journal of Pharmaceutics, 518* (1–2), 320–334. https://doi.org/10.1016/j.ijpharm.2016.12.051.

A number of these degradation pathways can negatively impact the efficacy of the peptide drug. Interestingly, a few of these pathways and mechanisms can be mitigated using appropriate excipients that are commonly used in conventional oral formulations (Bak & Dai, 2015; Bak et al., 2015). For example, an acid-resistant enteric coating can be utilized to prevent acid hydrolysis of peptides (Mehta, 2004), and peptidase inhibitors and pH modifiers can be included in oral peptide formulations to protect against proteolysis, aggregation, and structural changes in the gastric environment (Lang et al., 1997). Another common difficulty in systemic delivery of peptides through the oral route is that the peptide must remain intact when it enters the bloodstream from the GI tract. Several formulation approaches have been evaluated to improve the absorption of intact peptides from the GI tract with varying degrees of success. These approaches include the use of surfactants (BarbaraHaeberlin & ArminMeinzer, 1996), polymeric nano or microparticles (Allemann et al., 1998), solid lipid nanoparticles (Christophersen et al., 2013), microemulsions (Cilek et al., 2006), liposomes (Kirby, 2000), and micelles (Banerjee & Onyuksel, 2012). Additionally, conventional oral formulations can be used for peptide delivery using permeation enhancers or surfactants in combination with endogenous transporters such as transferrin, biotin, or glucose transporters that enable the intact peptide to cross the intestinal barrier (Bak & Dai, 2015). Moreover, the extent of absorption can be improved by increasing the peptide residence time by including mucoadhesive excipients like chitosan or carbomers in the formulation (Harris & Robinson, 1990; Lehr et al., 1992; Manickam, 2012).

CMC dictates that every formulation has specified critical quality attributes (CQAs) that stay consistent throughout the clinical process to ensure the reproducibility of the formulation as well as the manufacturing process. These include appearance (such as lack of visible crystals), size and shape, content uniformity, assay, permeation enhancer content, impurities and related substances, in vitro drug release profile, residual solvents (if applicable), residual monomers, microbial limits, and package integrity. The ratio of active pharmaceutical ingredient (API) to permeation enhancer is especially important for the formulation to be successful and showing reproducibility in maintaining the ratio is of utmost importance when considering an oral peptide formulation for CMC consideration. Besides the ability to ensure high absorption of the peptide, each permeation enhancer also has its own safety profile owing to its permeation enhancement mechanism. This needs to be well tested and considered while the formulation is in development. Apart from just the content, the physical attributes of the API and permeation enhancer like particle size and surface area may also make a significant difference in the absorption of the API. As the safety and efficacy profile of any systemically acting peptide API will depend on the extent of absorption, it is critical for the physical and chemical attributes of an API as well as a permeation enhancer to stay consistent from development through commercial dosage form. Over 250 permeation enhancers have been tested so far to evaluate their effect on intestinal absorption of peptide drugs (Maher et al., 2016) and each of them has its own unique physicochemical properties. One of the most important CMC considerations is to ensure reproducibility of interactions between the API and the permeation enhancer, as well as ensure that the safety and efficacy of the permeation enhancer are the same from development through commercialization.

Overall, there are several components of a successful oral peptide formulation. A key takeaway for CMC consideration is to ensure consistent quality and physicochemical attributes of API, as well as excipients, to ensure dosage form performance reproducibility. There are many dosage forms that have been clinically used for oral delivery of peptides such as enteric-coated tablets, multiparticulates, and enteric powder dosage forms to ensure successful oral delivery of peptides. Each of these approaches will be discussed from a manufacturing and scale-up perspective in the next section.

6.2.2 Manufacturing process considerations

Pharmaceutical manufacturing has evolved over the years and is performed with more emphasis on science and engineering principles than ever. The efficiency of the manufacturing process can be dramatically improved with innovative engineering and pharmaceutical science principles. FDA insists on process analytical technology (PAT) to define an integrated system approach that regulates drug product quality. This approach uses scientific principles and engineering knowhow to assess and ultimately alleviate risks during manufacturing that may result in an unacceptable product or process quality.

Efficient pharmaceutical manufacturing, utilizing process analytical technology, generally follows the below principles:

- Manufacturing processes are designed to be efficient and ensure reproducible product quality and performance

- A scientific understanding of the formulation and how it impacts the manufacturing and product performance which drive the product and process specifications.
- Real-time monitoring is essential to ensure continuous quality.
- Appropriate quality assurance employs process control strategies to prevent or mitigate the risk of poor product quality. A scientific understanding of the formulation and process drives the design of the process control strategies. (Karnieli, 2017).

More information on process analytical technology initiatives by FDA can be found in the FDA's final report of "Pharmaceutical cGMPs for the 21st Century—A Risk Based Approach."

To employ PAT in manufacturing processes, it is important to understand the process flow for pharmaceutical manufacturing as well as scale-up considerations for different drug product manufacturing processes. The flow chart in Fig. 6.1 shows an overview of the pharmaceutical manufacturing process for a typical peptide product (Fig. 6.2).

As shown in the chart earlier, there are multiple ways to manufacture a peptide drug product for oral use. The most common form is an oral solid dosage form comprising a tablet or capsule of a suitable formulation (powder blend, granules, or multiparticulates), which is enteric coated to provide acid protection to the peptide API. Besides solid orals, there are some technologies like spray drying that can be utilized to enteric coat the peptide and deliver it as a powder in sachet dosage form for oral delivery as well (Trivedi et al., 2017 AAPS poster) (Fig. 6.3).

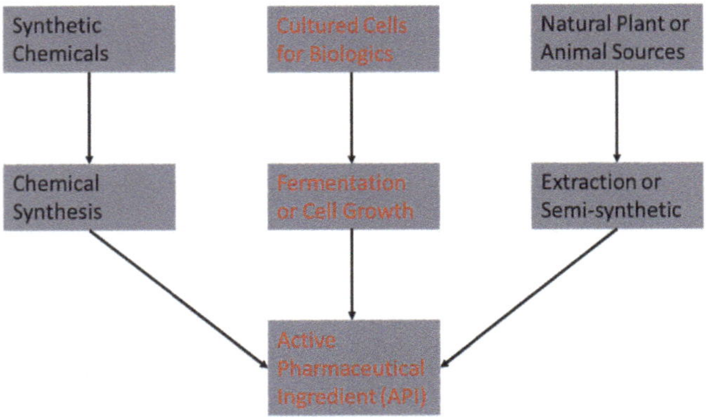

Figure 6.1 Typical process flow for drug substance manufacturing. Red-colored steps indicate the preferred route for peptide APIs. *Adapted from Chapter 79 of Encyclopedia of Occupational Health and Safety, International Labor Organization.*

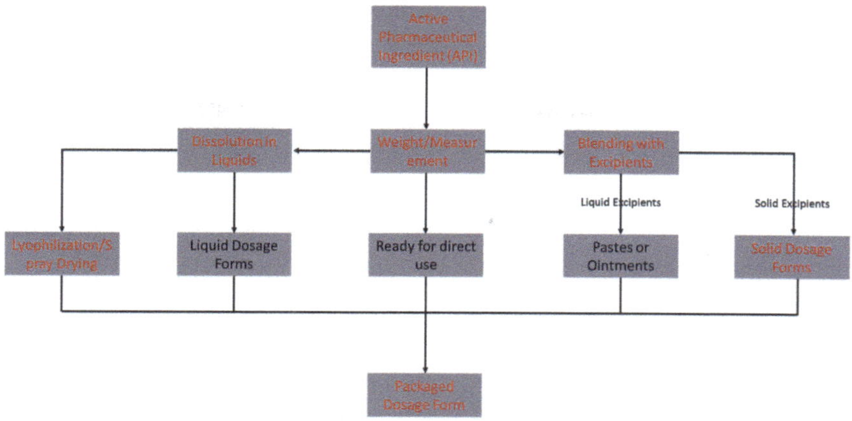

Figure 6.2 Typical process flow for drug product manufacturing. Red-colored steps indicate the preferred route for peptide APIs. *Adapted from Chapter 79 of Encyclopedia of Occupational Health and Safety.*

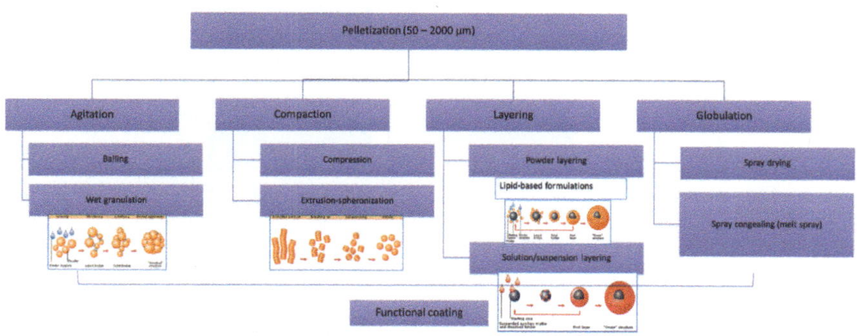

Figure 6.3 Overview of various techniques of multiparticulate manufacture for peptides (Trivedi et al., AAPS 2017 poster).

6.2.2.1 Multiparticulates

Multiparticulates present an attractive option for the oral delivery of peptides. It involves formulating a starting solution or mixture followed by pelletizing the formulation to form granules by wet granulation (proteins and peptides typically lose their secondary structure upon compaction so dry granulation is generally excluded in pelletization techniques), pellets by extrusion–spheronization, layering on a solid support like Microcrystalline Cellulose (MCC) or sugar beads, or globalization by spray drying or spray congealing.

Each of the multiparticulate techniques presents its own challenges during the scale-up of the process, and utilizing PAT is important to ensure the development of a reproducible scaled-up process.

For scaling up the wet granulation process, an attribute-based approach is typically taken when PAT is utilized. Another option is to employ a parameter-based approach that identifies different parts of the process and uses the design of experiments (DOE) to determine the optimum parameters of different parts of the manufacturing process. The chart in Fig. 6.4 shows the scale-up strategies for the wet granulation process.

In addition to wet granulation, Wurster coating is often utilized for drug layering and coating of peptides. Key advantages of Wurster coated formulations for peptides include dose flexibility, ability to coat with enteric, sustained release, and mucoadhesive coatings all in one formulation, and mild conditions to preserve the intact peptide. Wurster coating is a complicated process and involves several stages including solution

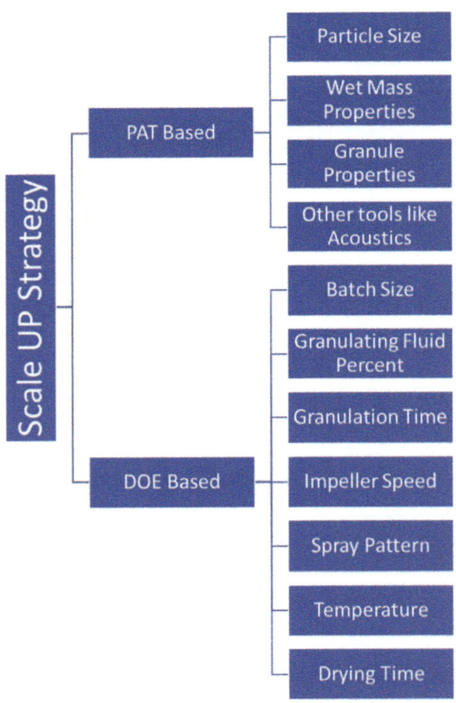

Figure 6.4 Scale-up strategies for wet granulation (Badawi, 2016). *Adapted from Badawi, S. P. P. (2016). Design, development, and scale-up of high-shear wet granulation process. In Y. C. Yihong Qiu, G. Zhang, L. Yu, & V. M. Rao (Eds.),* Developing solid oral dosage forms *(2 ed.). Elsevier.*

preparation, preheating, spraying the solution, and drying of the beads after spraying (Sonar, 2015). Each of these stages has multiple parameters that need to be optimized for a successful scale-up of the process. The following chart shows key considerations for scale-up of Wurster coating process (Figs. 6.5 and 6.6).

An additional technique to wet granulation and Wurster coating is spray drying. Spray drying is typically not employed to create powders containing peptides that can directly be filled into capsules and administered to patients. However, in some instances, peptide APIs can be spray-dried with an enteric polymer like neutralized Hydroxy Propyl Methyl Cellulose Acetate Succinate (HPMCAS) and the resultant Spray Dried Dispersion (SDD) can be granulated with an organic solution of HPMCAS to make peptide granules that provide enteric protection. The end result is that the spray-dried polymer becomes a multivalent salt of the carboxylate-bearing polymer where in vivo, the salt phase separates to form a gel around the peptides that protects them from the gastric environment. Once the formulation is finalized, scale-up of the spray drying process is performed as described in the chart shown in Fig. 6.7.

6.2.2.2 Capsules—granule or powder fill

Capsule filling is an important process in the manufacturing of oral peptides. All the techniques discussed earlier require a final step of filling the peptide containing intermediate into capsules. Besides formulation parameters and the individual intermediate manufacturing scale-up optimization that needs to be performed, capsule filling itself is an intricate process that requires scale-up optimization. The chart in Fig. 6.8 shows scale-up optimization considerations for the capsule filling process.

6.2.2.3 Tablets

Tablets are the most widely used solid dosage form for oral administration. They offer many advantages such as ease of administration, patient compliance, cost of production, possibility of higher doses, elegance, and ease of shipping, among others (Sachin Shivaji Gaikwad, 2020). Tablet design and manufacture is a complex process that involves many steps. The key steps in tablet manufacture are shown in the following chart, each of which, has its own scale-up and process development strategy.

Tablets present a very attractive option for oral delivery of peptides owing to the convenience of chronic dosing and patient compliance with tablets. As mentioned in Section 2.1.1., oral peptide tablets need to

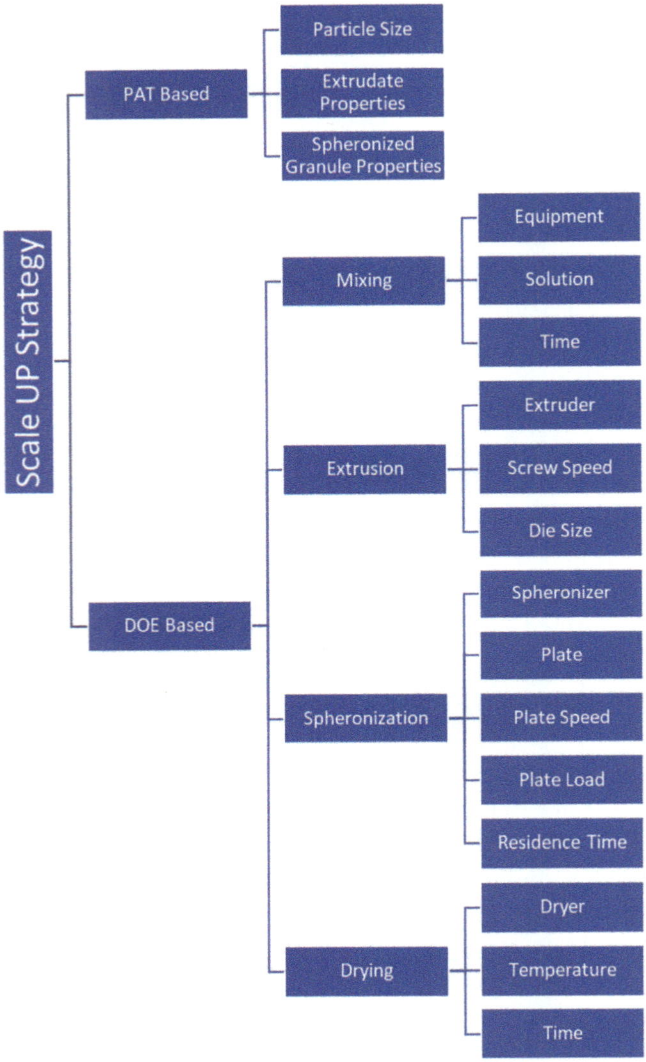

Figure 6.5 Scale-up strategies for extrusion–spheronization (Sinha et al., 2009). *Adapted from Sinha, V. R., Agrawal, M. K., Agarwal, A., Singh, G., & Ghai, D. (2009). Extrusion-spheronization: Process variables and characterization.* Critical Reviews in Therapeutic Drug Carrier Systems, 26*(3), 275-331. https://doi.org/10.1615/critrevtherdrugcarriersyst.v26.i3.20.*

include an enteric coat to protect the peptide from the acidic environment of the stomach, a permeation enhancer to ensure delivery of intact peptide through the intestinal barrier, and potentially a protease inhibitor

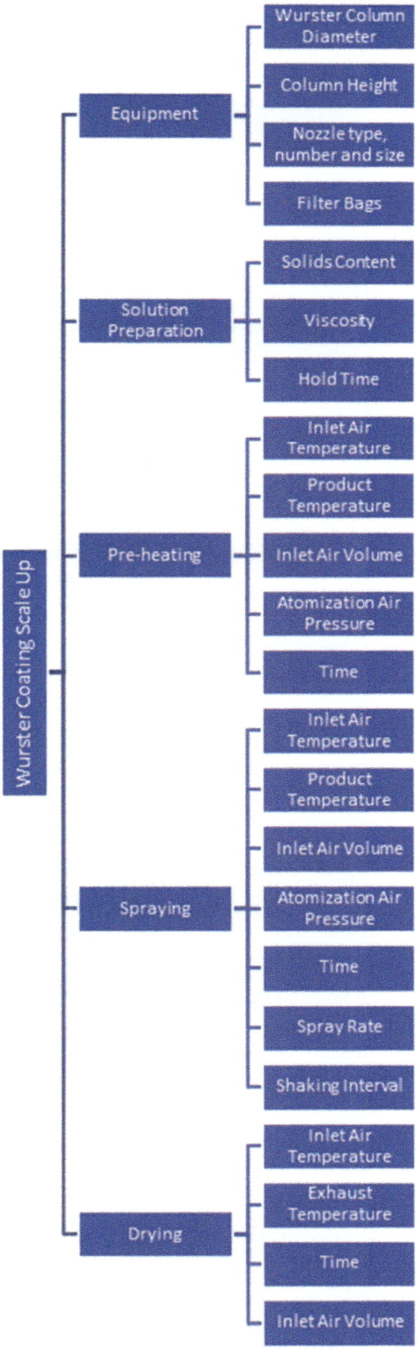

Figure 6.6 Scale-up strategies for Wurster coating process. *Adapted from Sonar, G. (2015). Wurster technology: Process variables involved and scale up science.*

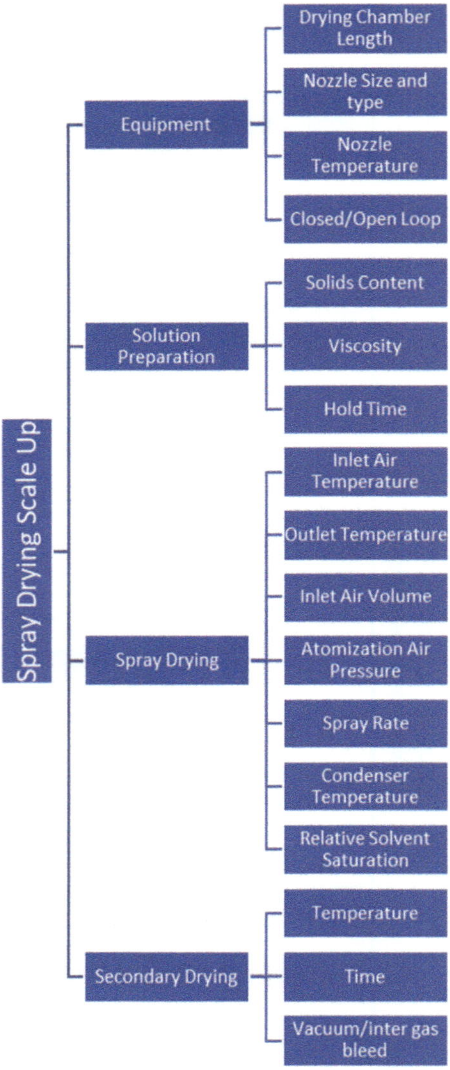

Figure 6.7 Scale-up of spray drying process.

to protect the peptides from proteases. Each of these excipients presents a particular challenge within the tableting process outlined in Fig. 6.9.

Blending is the first step in the manufacture of tablets. Appropriately scaled-up blending process ensures that the initial as well as the final blend that goes into a tablet is uniform throughout the tablet scale-up process. Nonuniform blend typically results in content uniformity issues during release testing, and it results in nonuniform dosage units that result in

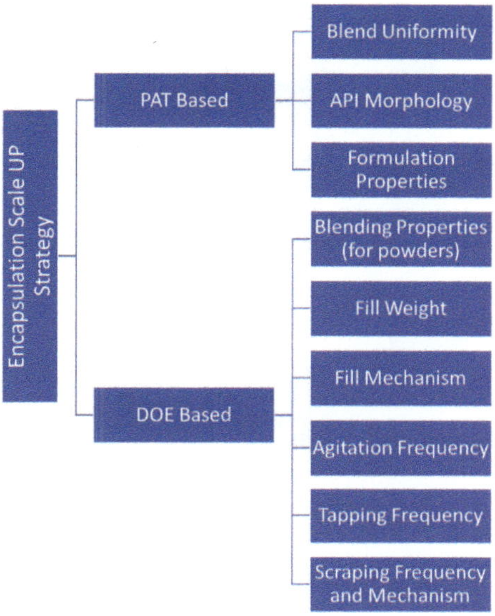

Figure 6.8 Scale-up of capsule filling process.

product failure. The chart in Fig. 6.10 shows the considerations for appropriate blending scale-up. These considerations are applicable in initial as well as final blend manufacture if a granulation step is involved in between.

Granulation follows blending in the tablet manufacturing process. Granulation allows the manufacture of controlled aggregates or granules that enable the blend to flow more freely into the tablet hopper and leads to appropriate dosing through the hopper into the tablet dies that results in uniform dosage units. Solvent-assisted granulation is known as wet granulation that has been described in the "multiparticulates" section. Another, less common option is dry granulation or roller compaction where the blend is gently compacted into uniform ribbons followed by milling of ribbons that results in a free-flowing blend of a controlled particle size. This option is less commonly employed for peptides due to the sensitivity of peptide and protein dosage forms to compaction. Some peptides lose their secondary structure and their activity upon exertion of a force that can lead to loss of activity. In these cases, roller compaction, and indeed tableting, should be avoided. Several considerations are required for successful scale-up of roller compaction, which are listed in Fig. 6.11.

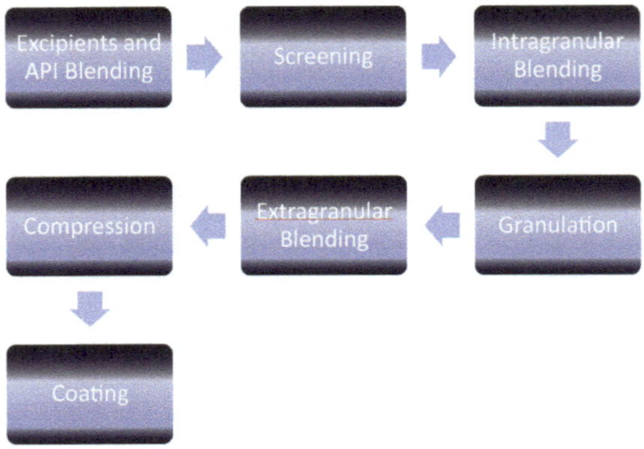

Figure 6.9 Tablet manufacture process.

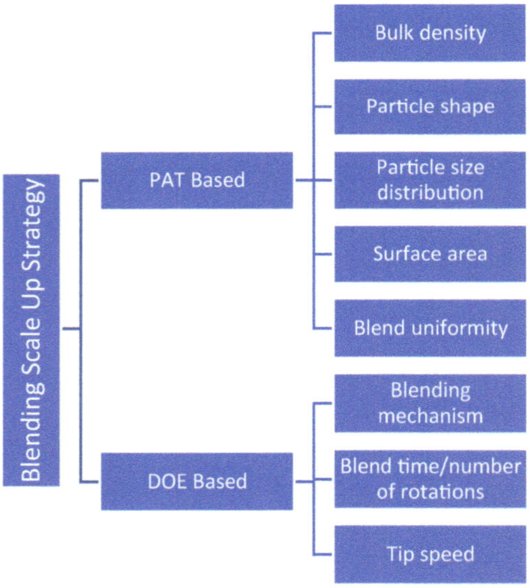

Figure 6.10 Blending scale-up strategy.

Tablet compression requires physical manipulation of the powder or granules, and it compresses them into tablets of the desired shape, size, and weight that requires careful consideration of many CQAs of tablets as well as critical process parameters, which are described in Fig. 6.12.

Tablet scale-up operations are characterized by defining target product profile, ranking process parameters by assessing risks, developing a robust process by a combination of process analytical technology and DOE, conducting process risk assessments and criticality analysis of processing steps

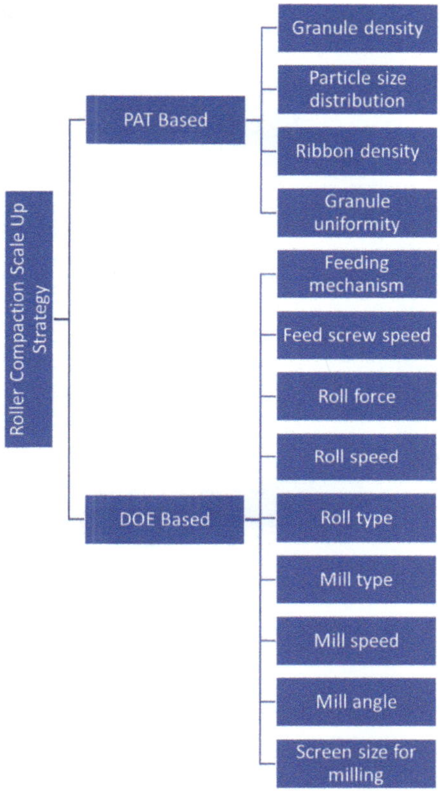

Figure 6.11 Roller compaction scale-up considerations.

that enables defining appropriate process controls, and eventual validation of the process with boundaries and appropriate controls. Once a manufacturing process goes through all these steps, it is ready for regulatory filing. The chart in Fig. 6.13 summarizes the various steps in scale-up and validation of a manufacturing process.

6.2.2.4 Process validation

According to FDA industry guidance on process validation, "process validation is defined as the collection and evaluation of data, from the process design stage through commercial production, which establishes scientific evidence that a process is capable of consistently delivering quality product." (FDA guidance for Industry on Process Validation: General Principles and Practices)

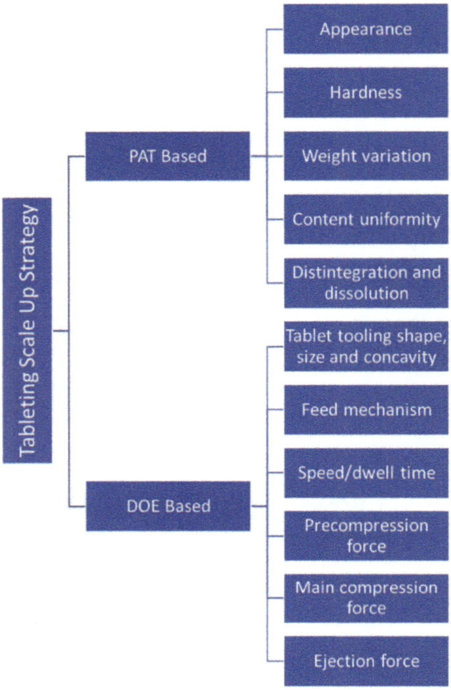

Figure 6.12 Tablet compression scale-up considerations.

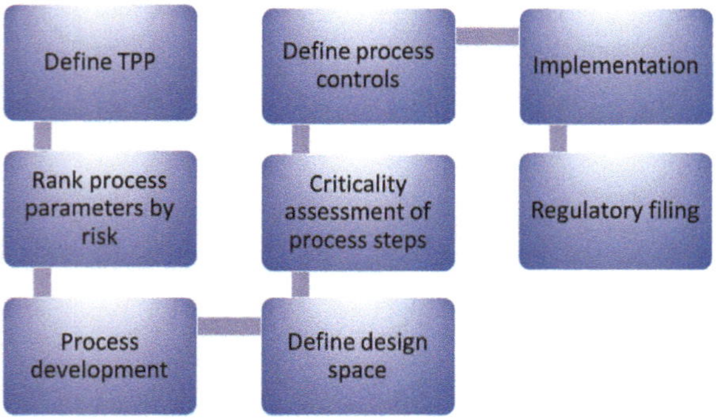

Figure 6.13 Summary of manufacturing design process.

Based on the guidance, process validation for any process relating to pharmaceutical manufacturing can be described in three stages:

- Process design
- Process qualification
- Continued process verification

Process design refers to the definition of the commercial manufacturing process based on the learnings from development and scale-up activities. This has been discussed at length in the previous section involving scale-up.

In the process qualification stage, the process design is verified for its ability to reproduce the commercial manufacturing process. This stage evaluates the process for its ability to produce consistent and reliable quality of products within set specifications. It involves collecting and evaluating data on all the steps of the manufacturing process including building and facilities, raw materials sources, acquisition, and storage, all the processing steps of the manufacturing process, as well as product storage and distribution. It is important to have alternate plans for the process when there are failures observed during the validation process. These contingency plans can be formed early on during the scale-up of the manufacturing process by knowing the boundaries of each parameter that is critical to the manufacturing process.

Continued process verification occurs during routine manufacturing runs where data are continuously collected to ensure that the process design and qualification have been successful, and the process remains sufficiently controlled to ensure product quality. Continuous process verification aims to diagnose and hopefully resolve drift in the process parameters before it reaches a point of failure. The stage has extensive Quality Control involvement in product sampling, analysis, and verification at various points in the manufacturing process to ensure the process is making a product that is compliant with specifications. Extensive record keeping and data analysis are required at this stage to detect anomalies and potential issues with product quality before they result in serious failures.

A detailed guideline for process validation can be found in the "Guidance for industry: Process Validation: General Principles and Practices" issued by FDA in January 2011. Overall, the FDA recommends employing a lifecycle approach to validation that is an ongoing process that maintains the process in a state of control throughout the life of the product.

6.2.3 Quality control release testing, stability, and specifications

Like all pharmaceutical products, oral peptide products also require specifications that define their quality attributes, procedures, and acceptance criteria to ensure the consistent safety and efficacy of both the peptide drug substance and the peptide drug product. The key challenge with quality control of oral peptide products is that the compendial monographs defining the products, as well as the methods governing the testing, are not harmonized between different pharmacopeias that can result in different acceptance criteria throughout the world (Vergote et al., 2009). The key considerations that can be used to determine a harmonized monograph system for peptide drugs are described in this chapter.

6.2.3.1 Peptide drug substance testing

Peptide drug substance (API) testing is a typical part of quality control of oral peptide products, much like any small or large molecule drug substance used in a pharmaceutical product. Developmental characterization is a key part of market authorization application and for peptide drugs, includes, structural identity proof, testing physicochemical properties, biological activity, immunoreactivity, purity, and related substances, optical rotation, counterions, inorganic impurities, amino acid analysis, and enables establishment of relevant specifications for release testing of production lots with a fully characterized reference standard.

6.2.3.2 Drug product testing

Much like the drug substance, oral peptide drug product testing also includes identification, counterions, optical rotation, immunogenicity, inorganic impurities, immunogenicity, and amino acid analysis. Besides these peptide-specific tests, the oral peptide product is also tested for a battery of oral product-related properties like appearance, identity, assay, related substances, dissolution, water content, presence of solvents, and microbiological attributes. Table 6.2 describes the typical tests and methods that can be used for oral peptide drug substance and product release testing along with their acceptance criteria.

Table 6.2 only provides a guideline for the tests typically required for oral peptides. Peptide drugs can have a wide array of specific properties which may prompt additional testing and specifications to ensure quality.

6.2.4 Manufacturing facility considerations

FDA requirements for Good Manufacturing Practices (GMP) manufacturing facilities are described in 21 CFR part 211.42 on buildings and facilities. For commercial products in Europe, the guidelines for European GMP and related validation and qualification of facilities—Regulation No. 1252/2014 and Directive 2003/94/EC—will apply. 21 CFR 211.42

Table 6.2 Typical tests and methods for testing and release of oral peptide drug product

Test	Typical analytical method	Typical specification
Appearance	Visual	Specific color and shape
Solubility	Visual/HPLC-UV	Solubility in water, ethanol, dilute acetic acid
Identification of active moiety	HPLC/LC-MS	Retention time match
Counterion	Colorimetry/ HPLC-UV	Positive
Peptide content/Assay	HPLC-UV	94%–102%
Related substances	HPLC-UV	Depending on use:Individual identified ≤ 1.0% Individual unidentified ≤ 0.5% Total ≤ 5%
Residual reagent	HPLC-UV (IEC)	TFA ≤ 1.0%
Residual solvent	Headspace GC	ICH limits for solvents
Residual catalysts	AAS/ICP-AES	Palladium ≤ 10 ppm
Immunogenicity	Immunochemical tests	Non immunogenic
Peptide-specific purity	Size exclusion/ HPLC-UV	Molecular weight match
Dissolution	Type I or II apparatus/HPLC-UV	≥ 80% release in 60 min for immediate release. Release profile match for controlled release. No release in gastric media
Water content	Karl Fisher	≤ 10%, depends on individual applications
Microbiological attributes (TAMC, TYMC, specific microbes)	Plate counts for TAMC, TYMC, and specific microbes	TAMC ≤ 10^3 cfu/gTYMC ≤ 10^2 cfu/g*E. coli* absent

describes the design and aspects necessary for a facility, including size and construction to qualify it as a pharmaceutical manufacturing facility and to ensure that it complies to GMP requirements. It includes appropriate maintenance, cleaning, and operations as well as size and space requirements to be able to house all the equipment and ensure appropriate material flow, storage, and properly qualified systems to ensure reproducible quality of the products manufactured in the facility. The responsibility for qualification of the facility and its approval for pharmaceutical manufacturing by FDA rests with the end user and is often a critical aspect of CMC submissions as well as commercial production.

6.2.4.1 Facility design

Design of a pharmaceutical manufacturing facility involves understanding the manufacturing process, and each facility design should include the following considerations:

- Flow of manufacturing personnel
- Material flow—raw material, product, waste, equipment (clean and dirty)
- Equipment space and layout requirement
- Operational access requirements
- Maintenance access requirements
- Explosion proofing requirements
- Chemical resistance requirements/surface finishes
- Particulate count requirements
- Cost considerations

Manufacturing facilities typically have an external area where the warehouse, offices, conference rooms, and other administrative requirements are housed. Followed by that, the facility has a transition zone that allows for the input of raw materials, output of products, inflow, and outflow of personnel as well as the outflow of waste. Typically, the transition zone has a clean side and a dirty side, and as the names suggest, inflows happen through the clean side and outflows happen through the dirty side. There is an intermediate zone, which is a clean area that allows personnel to don additional Personal Protective Equipment (PPE) and perform ancillary functions such as weighing and secondary packaging. Finally, the cleanest part of the facility is the critical processing area where the most critical manufacturing steps occur, and it has generally the most stringent particulate requirements.

Pressure differential control is used to prevent cross-contamination between areas, and the facility is designed in a way that the most critical areas have higher pressures and receive more air volume than less critical areas. This provides an air cascade, maintains the desired airflow direction, and prevents contaminants from entering the critical areas. To prevent dispersion of toxic materials that may be present in critical areas, the airflow is typically flipped and the work areas with toxic materials is designed to be negative air pressure.

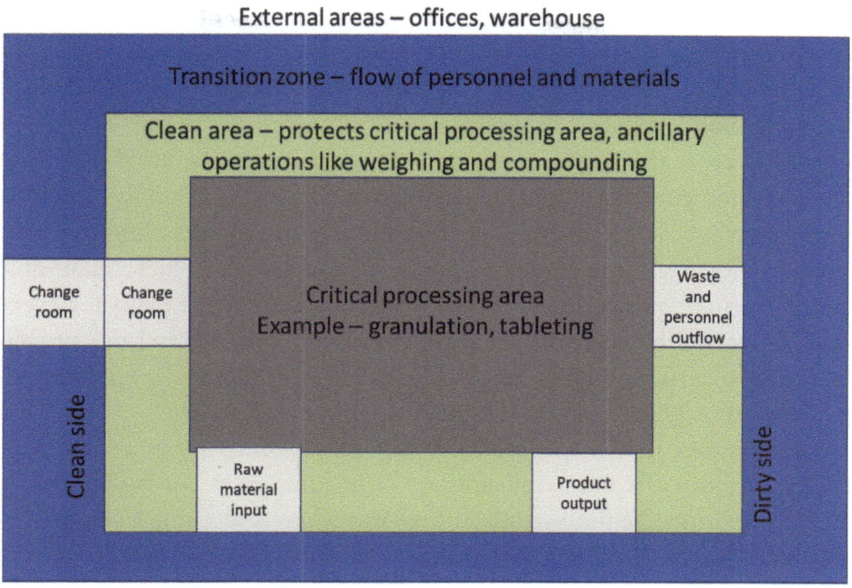

Figure 6.14 Typical oral product manufacturing facility design.

Overall, the entire facility is required to be ISO8 or higher to meet GMP oral product requirements. Fig. 6.14 shows a typical facility suite design.

Facility design can be very challenging to ensure efficiency and regulatory compliance. However, smart facility design can not only drastically improve the efficiency of the process, but also improve product quality, lower product complaints and nonconforming events, as well as ultimately lower the costs of production.

6.2.4.2 Facility qualification

Qualification of a facility is a big part of GMP readiness for pharmaceutical production. Facility qualification includes design qualification (DQ), installation qualification (IQ), operational qualification (OQ), and performance qualification (PQ).

Design qualification involves outlining the roles and responsibilities of everyone involved in the design of a facility from architects and engineers to subcontractors and suppliers. DQ ensures appropriate documentation of the design and makes sure that the appropriate design is implemented. DQ typically involves not only the overall design, but making sure the power, water, Heating, ventilation, and air conditioning (HVAC), and waste removal requirements also meet GMP standards. DQ is not strictly a part of GMP regulations but implementing DQ eventually saves a lot of time, money, and effort in creating a functional pharmaceutical manufacturing facility.

IQ involves making sure equipment is appropriately installed as per manufacturer guidelines. Manufacturer specifications are strictly followed for every equipment that is installed within the facility including air compressors, nitrogen generators, pressure gauges, and all additional manufacturing equipment. IQ ensures that documentation exists for every single aspect of the equipment and its installation process, and this process confirms that the equipment is installed appropriately and meets specifications.

OQ ensures that the equipment and facility function as required. It can be performed at the manufacturer site in some cases but in most cases, OQ is performed at the manufacturing site where the equipment will be commissioned, and operator training is conducted to ensure the correct operation of the equipment using the local utility supply at the location where it is commissioned.

Finally, after DQ, IQ, and OQ are complete, PQ begins with a review of documentation from all previous qualification stages and proceeds to conduct a product manufacture "trial run" of all equipment and facility components by trained operators. This ensures that not only is the facility and equipment adequately installed and runs accurately, but it also makes sure that the operators who are running it are doing it correctly.

6.2.4.3 Facility operation and maintenance

Operation and maintenance of a facility should ideally go hand in hand to ensure consistent quality of products and to prevent downtime as well as delays in manufacturing. A risk-based approach to maintenance is usually beneficial where each equipment and parts of facility are classified based on the direct or indirect impacts they have on the quality of the product. It is also important to have validated maintenance systems and have maintenance plans as a part of equipment introduction, operation, and management. Training programs should be implemented to ensure appropriate staff are trained in facility operation and maintenance as well as to ensure continuous training assessments. Besides equipment, maintenance also includes chemicals, power supplies, fire and security systems, HVAC, water supplies, gas supplies, and other utilities. Finally, all operation and maintenance including cleaning and validations need to be appropriately documented and meticulously maintained so that they can be verified any time it is needed and be presented to a regulatory authority.

6.3 Typical challenges in chemistry, manufacturing, and controls

Appropriately functioning CMC reduces development costs and delays and ensures consistent safety and quality of the drug substance and drug product. Large pharmaceutical companies have the resources to have sufficient CMC oversight to avoid mistakes. Smaller companies, on the other hand, face many CMC challenges. Typically, smaller companies have limited personnel who are responsible for CMC, and it opens the possibility of mistakes. Being aware of the different problems encountered within CMC and being cognizant of the available resources to resolve the challenges is the key to have a smooth functioning CMC group.

Besides typical challenges, oral peptide drug product development also has challenges related to the scientific expertise that is required to guide the development of such a complex product. There are limited resources available for this kind of product, and it is very important to consider all possible challenges that the product might encounter while planning CMC activities. We have described some of the challenges in the previous sections, which, hopefully, will be helpful for CMC planning and considerations.

CMC activities typically are given short timelines in the drug development cycle and therefore can encounter failures due to the speed of execution. Some common failures include missing important process or formulation parameters while defining a TPP, and shortened analytical and process development that can lead to inappropriate limits for a particular process, thereby, opening itself up for downstream failure. Additionally, lack of appropriate stability data due to shortened timelines is another major CMC failure that can jeopardize the drug development program. CMC delays can lead to approval delays that can impact the overall well-being of the company and delay the impact the new drug would have on the patients who need them as well.

Some of these points of failures and risks can be mitigated by employing skilled individuals who can identify and de-risk certain failure modes.

Many small companies utilize the expertise of CMC consultants who are specialized in oral peptide development. Additionally, careful auditing of suppliers and identifying noncompliance issues can go a long way in preventing CMC failures at a later stage in drug product development.

Overall, CMC is an important part of pharmaceutical development which is even more critical when it comes to complex oral peptide drug product development. Having a clear CMC strategy which accounts for the research required of the product, the patient populations, chemical characteristics of the drug itself as well as a well-defined target product profile will indeed be helpful in aligning with the larger clinical and regulatory strategy of the program. By thoroughly assessing, defining, and controlling the CMC aspects of a program, the possibility of making the pharmaceutical product a clinical and commercial success is greatly increased.

References

Allemann, E., Leroux, J., & Gurny, R. (1998). Polymeric nano- and microparticles for the oral delivery of peptides and peptidomimetics. *Advanced Drug Delivery Reviews*, *34* (2−3), 171−189. Available from https://doi.org/10.1016/s0169-409x(98)00039-8.

Bak, A., & Dai, W. (2015). Peptide developability at the discovery-to-development interface—current state and future opportunities. *The AAPS Journal*, *17*(4), 777−779. Available from https://doi.org/10.1208/s12248-015-9755-3.

Bak, A., Leung, D., Barrett, S. E., Forster, S., Minnihan, E. C., Leithead, A. W., Cunningham, J., Toussaint, N., & Crocker, L. S. (2015). Physicochemical and formulation developability assessment for therapeutic peptide delivery—a primer. *The AAPS Journal*, *17*(1), 144−155. Available from https://doi.org/10.1208/s12248-014-9688-2.

Banerjee, A., & Onyuksel, H. (2012). Peptide delivery using phospholipid micelles. *Wiley Interdisciplinary Reviews Nanomedicine Nanobiotechnology*, *4*(5), 562−574. Available from https://doi.org/10.1002/wnan.1185.

BarbaraHaeberlin, T., & ArminMeinzer, G. F. (1996). Cyclodextrins—useful excipients for oral peptide administration? *International Journal of Pharmaceutics*, *137*(1), 103−110.

Caon, T., Jin, L., Simoes, C. M., Norton, R. S., & Nicolazzo, J. A. (2015). Enhancing the buccal mucosal delivery of peptide and protein therapeutics. *Pharmaceutical Research*, *32*(1), 1−21. Available from https://doi.org/10.1007/s11095-014-1485-1.

Christophersen, P. C., Zhang, L., Yang, M., Nielsen, H. M., Mullertz, A., & Mu, H. (2013). Solid lipid particles for oral delivery of peptide and protein drugs I—elucidating the release mechanism of lysozyme during lipolysis. *European Journal of Pharmaceutics and Biopharmaceutics: Official Journal of Arbeitsgemeinschaft fur Pharmazeutische Verfahrenstechnik e.V*, *85*(3 Pt A), 473−480. Available from https://doi.org/10.1016/j.ejpb.2013.07.017.

Cilek, A., Celebi, N., & Tirnaksiz, F. (2006). Lecithin-based microemulsion of a peptide for oral administration: Preparation, characterization, and physical stability of the formulation. *Drug Delivery*, *13*(1), 19−24. Available from https://doi.org/10.1080/10717540500313109.

Harris, D., & Robinson, J. R. (1990). Bioadhesive polymers in peptide drug delivery. *Biomaterials*, *11*(9), 652–658. Available from https://doi.org/10.1016/0142-9612(90)90023-j.

Hua, S. (2019). Advances in nanoparticulate drug delivery approaches for sublingual and buccal administration [Mini Review Article]. *Frontiers in Pharmacology*. Available from https://doi.org/10.3389/fphar.2019.01328.

Karnieli, O. (2017). Bioreactors and downstream processing for stem cell manufacturing. In C. L. D. S. Joaquim, M. S. Cabral, G. Lucas, M. Chase, & Margarida Diogo (Eds.), *Stem cell manufacturing*. Elsevier. Available from https://doi.org/10.1016/C2011-0-05619-3.

Kirby, C. J. (2000). Oil-based formulations for oral delivery of therapeutic peptides. *Journal of Liposome Research*, *10*(4), 391–407.

Lang, V. B., Langguth, P., Ottiger, C., Wunderli-Allenspach, H., Rognan, D., Rothen-Rutishauser, B., Perriard, J. C., Lang, S., Biber, J., & Merkle, H. P. (1997). Structure-permeation relations of met-enkephalin peptide analogues on absorption and secretion mechanisms in Caco-2 monolayers. *Journal of Pharmaceutical Sciences*, *86*(7), 846–853. Available from https://doi.org/10.1021/js960387x.

Leader, B., Baca, Q. J., & Golan, D. E. (2008). Protein therapeutics: A summary and pharmacological classification. *Nature Reviews. Drug Discovery*, *7*(1), 21–39. Available from https://doi.org/10.1038/nrd2399.

Lehr, C. M., Bouwstra, J. A., Kok, W., De Boer, A. G., Tukker, J. J., Verhoef, J. C., Breimer, D. D., & Junginger, H. E. (1992). Effects of the mucoadhesive polymer polycarbophil on the intestinal absorption of a peptide drug in the rat. *The Journal of Pharmacy and Pharmacology*, *44*(5), 402–407. Available from https://doi.org/10.1111/j.2042-7158.1992.tb03633.x.

Maher, S., Mrsny, R. J., & Brayden, D. J. (2016). Intestinal permeation enhancers for oral peptide delivery. *Advanced Drug Delivery Reviews*, *106*(Pt B), 277–319. Available from https://doi.org/10.1016/j.addr.2016.06.005.

Mandal, A., Pal, D., Agrahari, V., Trinh, H. M., Joseph, M., & Mitra, A. K. (2018). Ocular delivery of proteins and peptides: Challenges and novel formulation approaches. *Advanced Drug Delivery Reviews*, *126*, 67–95. Available from https://doi.org/10.1016/j.addr.2018.01.008.

Manickam, B. (2012). Chitosan: A perfect polymer used in fabricating gene delivery and novel drug delivery systems. *International Journal of Pharmacy and Pharmaceutical Sciences*, *4*, 54–56.

Mehta, N. (2004). Oral delivery and recombinant production of peptide hormones Part I: Making oral delivery possible - Enteric-coated capsules or tablets with additional excipients enable intestinal delivery. *Biopharmacy International*, *17*(6), 38–43.

Meredith, M. E., Salameh, T. S., & Banks, W. A. (2015). Intranasal delivery of proteins and peptides in the treatment of neurodegenerative diseases. *The AAPS Journal*, *17*(4), 780–787. Available from https://doi.org/10.1208/s12248-015-9719-7.

Newman, S. P. (2017). Drug delivery to the lungs: Challenges and opportunities. *Therapeutic Delivery*, *8*(8), 647–661. Available from https://doi.org/10.4155/tde-2017-0037.

Pena-Juarez, M. C., Guadarrama-Escobar, O. R., & Escobar-Chavez, J. J. (2021). Transdermal delivery systems for biomolecules. *Journal of Pharmaceutical Innovation*, 1–14. Available from https://doi.org/10.1007/s12247-020-09525-2.

Renukuntla, J., Vadlapudi, A. D., Patel, A., Boddu, S. H., & Mitra, A. K. (2013). Approaches for enhancing oral bioavailability of peptides and proteins. *International Journal of Pharmaceutics*, *447*(1–2), 75–93. Available from https://doi.org/10.1016/j.ijpharm.2013.02.030.

Badawi, S. P. P. (2016). Design, development, and scale-up of high-shear wet granulation process. In Y. C. Yihong Qiu, G. Zhang, L. Yu, & V. M. Rao (Eds.), *Developing solid oral dosage forms* (2 ed.). Elsevier.

Sachin Shivaji Gaikwad, S. J. K. (Ed.), (2020). Review on tablet in tablet techniques. *Beni-Suef University Journal of Basic and Applied Sciences*, *9*(1), 1–7.

Saffran, M., Kumar, G. S., Savariar, C., Burnham, J. C., Williams, F., & Neckers, D. C. (1986). A new approach to the oral administration of insulin and other peptide drugs. *Science (New York, NY)*, *233*(4768), 1081–1084. Available from https://doi.org/10.1126/science.3526553.

Shaji, J., & Patole, V. (2008). Protein and peptide drug delivery: Oral approaches. *Indian Journal of Pharmaceutical Sciences*, *70*(3), 269–277. Available from https://doi.org/10.4103/0250-474X.42967.

Sinha, V. R., Agrawal, M. K., Agarwal, A., Singh, G., & Ghai, D. (2009). Extrusion-spheronization: Process variables and characterization. *Critical Reviews in Therapeutic Drug Carrier Systems*, *26*(3), 275–331. Available from https://doi.org/10.1615/critrevtherdrugcarriersyst.v26.i3.20.

Sonar, G. (2015). Wurster technology: Process variables involved and scale up science. IPPT, Vol 1 (1), 100-109, 2015

Vergote, V., Burvenich, C., Van de Wiele, C., & De Spiegeleer, B. (2009). Quality specifications for peptide drugs: A regulatory-pharmaceutical approach. *Journal of Peptide Science: An Official Publication of the European Peptide Society*, *15*(11), 697–710. Available from https://doi.org/10.1002/psc.1167.

Wu, L. C., Chen, F., Lee, S. L., Raw, A., & Yu, L. X. (2017). Building parity between brand and generic peptide products: Regulatory and scientific considerations for quality of synthetic peptides. *International Journal of Pharmaceutics*, *518*(1–2), 320–334. Available from https://doi.org/10.1016/j.ijpharm.2016.12.051.

CHAPTER SEVEN

Clinical translation of oral peptide delivery technologies

Puneet Tyagi[1] and J. Anand Subramony[2]
[1]Dosage Form Design and Development, BioPharmaceuticals R&D, AstraZeneca, Gaithersburg, MD, United States
[2]Biologics Engineering & Targeted Delivery, Oncology R&D, AstraZeneca, Gaithersburg, MD, United States

7.1 Introduction

Translational studies are key to advancing oral peptides from the bench to the bedside. Successful development of an oral peptide dosage form requires in vitro, preclinical, and nonhuman primate studies, followed by clinical studies in humans to confirm the safety, efficacy, and reproducibility of oral delivery. Although several rounds of promising preclinical studies have been reported in the literature, one of the key impediments to the advancement of oral peptide delivery is the scarcity of human translational studies and nonavailability of sound in vitro–in vivo correlation IVIVC models. Most preclinical studies have reported oral peptide bioavailability in rodents, dogs, and pigs, but those numbers have translated very poorly to human studies and are highly variable. Furthermore, at present, there is no simple rule of thumb that empirically predicts human bioavailability based on readouts from studies in rodents, dogs, pigs, or nonhuman primates. In this chapter, we review some of the studies of oral peptides that have been advanced to the clinic namely salmon calcitonin, octreotide, parathyroid hormone, insulin, and semaglutide and address what can we learn and apply from those studies going forward.

Table 7.1 presents a list of oral peptides that have been advanced to the clinic via human translational studies. Insulin and glucagon-like peptide 1 (GLP-1) are predominant, along with parathyroid hormone, calcitonin, and octreotide. The key reason for the choice of these molecules may be that switching an injectable peptide for a chronic indication to a convenient oral dosage form allows for greater patient centricity by

Oral Delivery of Therapeutic Peptides and Proteins.
DOI: https://doi.org/10.1016/B978-0-12-821061-1.00020-4

Table 7.1 Clinically advanced technologies for oral peptide delivery.

Company	Peptide	Technology	Development stage
Emisphere Technologies	Insulin, GLP-1 analog	Eligen B12; absorption enhancers	Phase 3 FDA approved
Chiasma	Octreotide	Transient permeability enhancer technology; oil-based suspension	Phase 3 FDA approved
Enteris BioPharma	Leuprolide, difelikefalin	Peptelligence; absorption enhancers	Phase 2
Merrion Pharmaceuticals	Insulin	GIPET; absorption enhancers	Phase 1, phase 2
Oramed Pharmaceuticals	Insulin	POD; absorption enhancers and enzyme inhibitors	Phase 2
Oshadi	Insulin	Silica nanoparticles	Phase 1, phase 2
Diasome	Insulin	Hepatic-directed vesicles	Phase 1, phase 2
Biocon	Insulin	HIM-2; absorption enhancers and enzyme inhibitors	Phase 1
Rani Therapeutics	Octreotide	RaniPill capsule	Phase 1
Lyndra			

FDA, US Food and Drug Administration; *GIPET*, gastrointestinal permeation enhancement technology; *GLP-1*, glucagon-like peptide 1; *HIM-2*, hexyl-insulin monoconjugate 2; *POD*, peptide oral delivery.
Source: PharmaCircle. Interested readers are encouraged to refer to the National Library of Medicine's clinical trials database (ClinicalTrials.gov) for the most recent updates.

overcoming needle phobia while also providing freedom from the need for cold-chain management, which is typically required for parenteral formulations. Only limited information is available in the literature describing study designs and published outcomes from these studies, including clinical trial databases and related publications. In this chapter, we describe some new correlations and methodologies to leverage these learnings as the field moves forward to build future oral peptide study designs.

7.2 Clinically translated oral peptides

Until 2010, desmopressin and cyclosporine were the only peptides delivered by the oral route. Since then, understanding of the physiology of the gastrointestinal (GI) tract has grown significantly (Brayden et al., 2020), as has that of the fate of peptides in the GI tract. With this deeper understanding, a number of oral peptide delivery technologies have been advanced to early-stage development, including oral formulations of the GLP-1 receptor agonist semaglutide (Novo Nordisk, Copenhagen, Denmark), which was approved in 2019. A closer look at the developmental history of various oral peptide candidates and their outcomes in clinical trials will improve our scientific understanding and enable further development of oral peptide dosage forms.

7.2.1 Salmon calcitonin

Efforts to orally administer salmon calcitonin (sCT; molecular weight, 3432 Da) to treat postmenopausal osteoporosis and osteoarthritis have contributed greatly to oral peptide research. A series of clinical trials were carried out in the early 2000s using an oral sCT formulation platform known as Peptelligence (Enteris BioPharma). This platform includes an enteric-coated, solid dosage form containing citric acid or another acid that provides buffering capacity to reduce local pH, thereby preventing peptide degradation by proteases.

In a phase 1 trial (NCT00411125), healthy postmenopausal women were studied for the effect of oral sCT on bone resorption biomarkers. This study evaluated serum levels of the C-terminal telopeptide of type I collagen (CTx), a bone resorption marker, from baseline to 3 months and acute 24-h changes in serum CTx and osteocalcin. The corresponding area under the curve (AUC) was assessed at baseline, 1 month, and 3 months (Karsdal et al., 2011). The optimal dose was defined as one with rapid maximum concentration (C_{max}) and elimination times. A dose of 0.8 mg was identified as optimal for rapid absorption, reaching C_{max} in 15–30 min and having an elimination from plasma with a short half-life of 9–15 min. Drug absorption increased when the dose was taken at least 10 min before a meal (and with 50 mL of water) rather than postprandially. This was one of the first examples of the effect of food on oral peptide delivery. This study also reported that the optimal dosing time for osteoporosis therapy was in the evening, when the circadian peak in bone resorption is mitigated. Because bone resorption displays marked diurnal variation, reversible inhibition of bone resorption may result in the best possible efficacy when bone resorption peaks (Karsdal et al., 2008a).

In another phase 1 study (NCT00486369), patients with osteoarthritis were treated with two doses of sCT, 0.6 and 0.8 mg. The two doses were comparably efficient in suppressing serum levels of CTx. As expected, the 0.8-mg formulation resulted in significantly higher C_{max} and subsequently a larger AUC. Dosing in the morning followed by 5 h of fasting resulted in higher C_{max} and a larger AUC than the evening dose, which was followed by supper 1 h l (Karsdal et al., 2010).

A phase 3 study (NCT00525798) evaluated oral sCT with osteoporosis as the target (Henriksen et al., 2016). The formulation, SMC021, consisted of 0.8 mg of sCT combined with Eligen B12 (8-(*N*-2-hydroxy-5-chlorobenzoyl)-amino-caprylic acid [5-CNAC]; Emisphere Technologies), a permeation enhancer, along with vitamin D and calcium. Bone fractures, spinal bone material density, and biomarkers in response to SMC021 were measured over 36 months. Although bone material density increased, the primary endpoint of preventing new fractures was not reached in this study. In addition, the pharmacokinetic data suggested that sCT levels in plasma were at the limit of detection. Although further studies of this formulation for osteoporosis were discontinued, further studies with the same formulation were initiated in two independent phase 3 trials (NCT00486434 and NCT00704847) for knee osteoarthritis (Karsdal et al., 2015). Oral 0.8-mg sCT tablets with 200 mg of 5-CNAC in 50 mL of water were administered twice daily for 24 months to patients with osteoarthritis. The outcomes of these studies were comparable, and the endpoints of joint space narrowing and pain scores were not significantly different from placebo in either trial. Table 7.2 summarizes various clinical studies of oral sCT.

Table 7.2 Clinical trials with oral calcitonin in the Peptelligence formulation.

Study name	Indication	Phase	ClinicalTrials.gov identifier
Pharmacokinetics and Pharmacodynamics Study of Oral Salmon Calcitonin in Healthy Postmenopausal Women	Postmenopausal osteoporosis	1	NCT00411125
SMC021A—Absorption, Efficacy, and Tolerance in Patients with Osteoarthritis. A Placebo-Controlled 14-Days Study	Osteoarthritis	1	NCT00486369
A Study to Evaluate Oral Salmon Calcitonin in the Treatment of Osteoporosis in Postmenopausal Women Taking Calcium and Vitamin D	Osteoporosis	3	NCT00525798
Efficacy and Safety of Oral Salmon Calcitonin in Patients with Knee Osteoarthritis	Osteoarthritis	3	NCT00486434
Efficacy and Safety of Oral Salmon Calcitonin in Patients with Knee Osteoarthritis (OA 2 Study)	Osteoarthritis	3	NCT00704847

7.2.2 Octreotide

Octreotide is a synthetic octapeptide analog of human somatostatin and is used for the treatment of acromegaly and neuroendocrine tumors. Octreotide decreases splanchnic blood flow and inhibits the release of serotonin, gastrin, vasoactive intestinal peptide, secretin, motilin, and pancreatic polypeptide (Yuhico et al., 2012). Clinically, the use of octreotide has been limited to the parenteral route because it exhibits low and variable systemic bioavailability upon oral administration. Transient permeability enhancer technology (Chiasma) that has been used for this formulation combines octreotide with other excipients to form an oily suspension of hydrophilic particles in a lipophilic medium that enhances intestinal permeability. This preparation was shown to facilitate transient paracellular passage of octreotide across the gastrointestinal wall in the small intestines of rats (Cano-Cebrian et al., 2005). In a phase 1 study, the pharmacokinetics of a single oral dose of this formulation was compared with subcutaneous octreotide in healthy volunteers (Table 7.3). The study assessed the safety and tolerability of the new oral octreotide formulation and evaluated whether orally absorbed octreotide is functional and exerts pharmacological suppression of pituitary growth hormone (GH) release (Tuvia et al., 2012). The systemic exposure of orally administered octreotide (20 mg) was similar to that of a 0.1-mg subcutaneous injection in healthy volunteers. Comparable safety profiles were shown for both octreotide formulations. Basal levels of GH and stimulation of GH secretion by GH-releasing hormone were significantly suppressed by a single oral dose of octreotide.

These results led to further testing of octreotide in a phase 3 study (NCT01412424) to determine its efficacy and safety in patients with acromegaly. The objective was to determine the effectiveness of octreotide in maintaining a baseline biochemical response for up to 13 months in patients with acromegaly. Patients receiving injectable somatostatin receptor ligand (SRL) for ≥ 3 months were selected for this study and were switched to daily 40-mg oral octreotide capsules. The dose was increased to 60 mg daily and then to 80 mg daily to control insulin-like growth factor 1 (IGF-1), a major determinant in acromegaly mortality. Subsequent fixed doses were maintained for a 7-month core treatment, followed by a voluntary 6-month extension. Oral octreotide capsules exhibited efficacy in controlling and maintaining IGF-1 and integrated GH levels for ≥ 13 months in subjects with biochemically controlled acromegaly after switching from injectable SRLs. Ninety patients (58%) required an oral octreotide dose of >40 mg to maintain this response (Melmed et al., 2015).

The Chiasma OPTIMAL study (NCT03252353) was designed as a placebo-controlled trial to evaluate the efficacy and safety of oral octreotide in patients with acromegaly who previously demonstrated biochemical control on a stable, long-acting injectable SRL regimen. Owing to its placebo arm, the study also provided an assessment of disease activity after washout of long-term treatment and distinguished between disease-related and treatment-related adverse events (Samson et al., 2020). The formulation was approved by the US Food and Drug Administration (FDA) in 2020 for the treatment of acromegaly (Chiasma announces, 2020) (Table 7.3).

Table 7.3 Clinical trials with oral octreotide in the Chiasma formulation.

Study name	Indication	Phase	ClinicalTrials.gov identifier
Efficacy and Safety of Octreotide (MYCAPSSA [Formerly Octreolin]) for Acromegaly	Acromegaly	3	NCT01412424
Comparison of Oral Octreotide Capsules to Injectable Somatostatin Analogs in Acromegaly (MPOWERED)			NCT02685709
Efficacy and Safety of Octreotide Capsules (MYCAPSSA) in Acromegaly (OPTIMAL)			NCT03252353

7.2.3 Parathyroid hormone

The beneficial effect of parathyroid hormone (PTH) treatment on bone mass and quality, as well as a reduction in osteoporotic fractures, has been well demonstrated in the clinic (Lane et al., 2000). Although injectable PTH forms are approved for the treatment of osteoporosis, their large-scale clinical use remains limited by poor patient acceptance due to the need for long-term daily subcutaneous injections (Verhaar & Lems, 2010). Patients with osteoporosis have found it extremely difficult to self-inject, and hence patient centricity was low for this approach. An oral formulation could bypass these limitations and provide a better-tolerated treatment form that would be expected to enhance patient compliance.

A phase 1 clinical study (NCT02571140) comparing the pharmacokinetics of oral PTH with those of a subcutaneous injection of teriparatide was carried out in healthy male volunteers (Table 7.4). The pharmacokinetic profile of oral PTH demonstrated rapid absorption and

Table 7.4 Clinical trials with oral PTH.

Study name	Indication	Phase	ClinicalTrials.gov identifier
A Study to Assess the Safety, Pharmacokinetics, and Pharmacodynamics of Oral PTH (1–34) Formulations in Healthy Subjects	Healthy subjects	1	NCT02571140
A Three Stage, Study to Assess the Safety, Pharmacokinetics, and Pharmacodynamics of Oral PTH (1–34) in Healthy Subjects	Drug safety	1	NCT02202603
A Pilot Study to Assess the Safety and Efficacy of Oral PTH (1–34) in the Treatment of Hypoparathyroidism	Hypoparathyroidism	2	NCT02152228
A Six-Month Phase 2 Study of Oral hPTH (1–34) (EBP05) in Postmenopausal Women with Low Bone Mass	Osteoporosis	2	NCT04003467
Oral PTH(1–34) PK and PD Study in Patients with Hypoparathyroidism	Hypoparathyroidism	2	NCT03516773
Pharmacodynamics, Pharmacokinetics, Safety, and Tolerability of PTH Analog Tablets in Postmenopausal Women	Postmenopausal osteoporosis	2	NCT01321723

disappearance rates, which led to a short pharmacokinetic exposure to the drug. The total systemic exposure (AUC) after the administration of oral PTH was similar to that of the subcutaneous drug, whereas the C_{max} of the oral PTH formulation was about twofold higher than that of the injection. Oral PTH appeared to be safe and well tolerated, and no drug-related adverse events were reported. Two adverse events (cough and dyspepsia) were reported 1 day after administration of the oral drug (Burshtein et al., An oral).

Oral PTH formulation was further developed in a similar phase 1 study (NCT02202603). That study focused on the control of drug absorption while minimizing its variability. A high C_{max} coefficient of variation of 123% was observed for the original formulation. The pharmacokinetics of two improved oral formulations showed reduced intersubject variability and significantly increased bioavailability. The systemic exposure (AUC) and C_{max} after administration of the modified oral PTH formulations were higher than those of the original formulation; the coefficient of variation was 123% for the basic oral formulation, 91% for modified formulation 1, and 67% for modified formulation 2. Furthermore, serum calcium levels increased from 0.07 mg/dL for the original formulation to 0.12 mg/dL and 0.32 mg/dL, respectively, for the two modified formulations (Burshtein et al., Enhanced).

A phase 2 trial (NCT01321723) aimed to establish the pharmacodynamics, pharmacokinetics, safety, and tolerability of an oral tablet formulation of PTH, matching placebo tablets, and subcutaneous teriparatide as a positive control in postmenopausal women with osteoporosis (Henriksen et al., 2013; Sturmer et al., 2013). The study was conducted to characterize the percent change from baseline in bone mineral density after 24 weeks. No significant differences were observed between weeks 0 and 24 in any pharmacokinetic parameters tested, demonstrating good reproducibility and leading to induction of bone formation and significantly increased levels of bone mineral density.

In a pilot phase 2 study (NCT02152228) to evaluate the safety, tolerability, and efficacy of an oral PTH preparation in adult volunteers with hypoparathyroidism, primary endpoints were monitoring and recording adverse events related to hyper- and hypocalcemia, reduced use of exogenous calcium and/or alpha-D3 supplements, and plasma calcium levels. Another phase 2 study (NCT03516773) with the same formulation was conducted to evaluate the pharmacokinetics and pharmacodynamics of 2.25 mg of oral PTH four times a day versus twice a day. Subcutaneous injection was used as a positive control. The same formulation was tested in a randomized phase 2 (NCT04003467) study to determine the effects of 6 months of treatment with oral PTH or placebo on biochemical markers of bone formation, bone resorption, and bone mineral density. Postmenopausal women with low bone mass received oral PTH doses ranging from 1.5 to 2.5 mg. That study showed promising results for serum calcium levels and a reduction in serum phosphate and urinary calcium levels (Caraco et al., 2019). It is not clear why an oral PTH formulation did not go forward for further development (Table 7.4).

7.2.4 Insulin

Diabetes mellitus is a chronic disease that has assumed global pandemic proportions. The condition is characterized by instability and unpredictability of glycemic readings, resulting in poor glycemic control. Uncontrolled hepatic gluconeogenesis has been suggested to be a central underlying cause of diabetes. Insulin administration is indispensable for the treatment of type 1 diabetes and is required in the later stages of type 2 diabetes. Currently, the most predictable and traditional method for administering insulin is the parenteral route, specifically, subcutaneous injection. The development history of insulin and of experimental research on various dosage forms is illustrated in Fig. 7.1. Subcutaneously administered insulin is sometimes inadequate in inducing normal portal

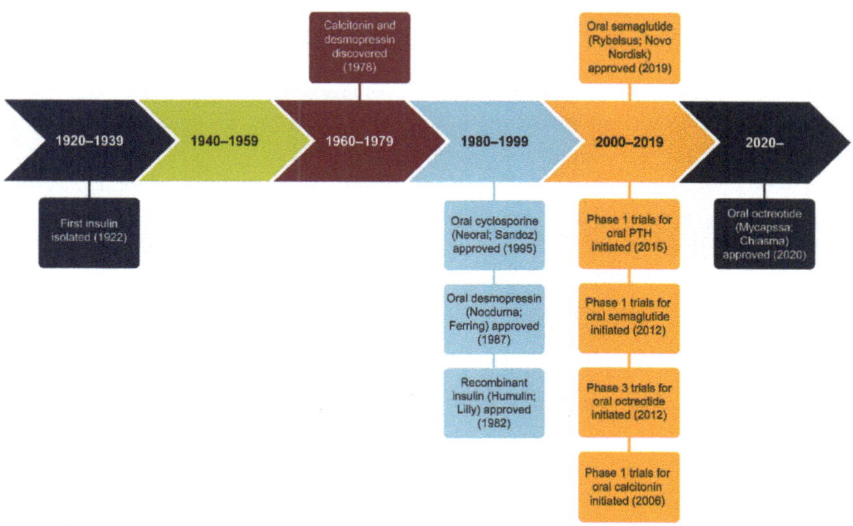

Figure 7.1 Development, discovery, and delivery of oral peptides.

insulin:glucagon ratios (Bischof et al., 2001). Direct intraperitoneal insulin delivery systems have led to improved glycemic control (DeVries et al., 2002), and it has been speculated that oral insulin induces a similar effect.

Oramed Pharmaceuticals has developed a proprietary protein oral delivery technology, named peptide oral delivery (POD), that is composed of protease inhibitors and a chelating agent (Table 7.5). The components are encapsulated in a pH-sensitive capsule to prevent degradation in the stomach and release in the small intestines, where the pH is close to neutral. The protease inhibitors protect insulin from degradation

Table 7.5 Clinical trials with oral insulin.

Study name	Indication	Phase	ClinicalTrials.gov identifier
Effectiveness of Oral Insulin in Unstable Type 1 Diabetes Patients	Brittle type 1 diabetes mellitus	2	NCT00867594
Safety, Pharmacokinetics (PK), and Pharmacodynamics (PD) of Multiple Oral Bedtime Doses of ORMD-0801 in Adult Patients with Type 2 Diabetes Mellitus (T2DM)	Type 2 diabetes mellitus	2	NCT01889667
A Prospective, Randomized, Double-blind, Placebo Controlled Study to Assess the Impact of ORMD-0801 (Insulin Capsules) on the Exogenous Insulin Requirements of Type 1 Diabetics	Type 1 diabetes mellitus	2	NCT02094534
Phase II Safety and Efficacy Study of Oral ORMD-0801 in Patients with Type 2 Diabetes Mellitus	Type 2 diabetes mellitus	2	NCT02496000
A Study of Single and Multiple Doses of Oral Insulin or Placebo in Subjects with Type 2 Diabetes Mellitus	Type 2 diabetes mellitus	2	NCT02954601
A Phase 3 Study to Evaluate the Efficacy and Safety of ORMD-0801 in Subjects with Type 2 Diabetes Mellitus	Type 2 diabetes mellitus	3	NCT04754334
Study to Evaluate the Efficacy and Safety of ORMD-0801 in Subjects with Type 2 Diabetes Mellitus	Type 2 diabetes mellitus	3	NCT04606576
Phase 3 Study to Evaluate the Efficacy and Safety of ORMD-0801 in Subjects with Type 2 Diabetes Mellitus	Type 2 diabetes mellitus	3	NCT04817215

by the ubiquitous proteases in the small intestines. The chelating agent removes calcium, which is an important cofactor for many proteases, thus reducing protease activity and increasing paracellular permeability (Arbit & Kidron, 2017).

In a phase 2 study (NCT00867594), eight patients with uncontrolled type 1 diabetes (hemoglobin A1c [HbA1c], 7.5%–10%) self-administered an oral insulin capsule three times daily, just before meal intake, throughout the 15-day study period, and blood glucose levels were monitored by means of a blind continuous glucose monitoring device (Eldor et al., 2013). Treatment with the ORMD-0801 POD formulation containing 8 mg of insulin was associated with a 24.4% reduction from pretreatment in the frequencies of glucose readings of >200 mg/dL and a significant mean 16.6% decrease in glucose AUC (66,055 ± 5547 mg/dL/24 h pretreatment vs 55,060 ± 3068 mg/dL/24 h with ORMD-0801; $P = .023$).

A phase 2 study (NCT01889667) addressed the safety, pharmacokinetics, and pharmacodynamics of multiple oral bedtime doses of ORMD-0801 in patients with type 2 diabetes. No side effects were observed after dosing up to day 8. Mean daytime glucose levels, as measured by continuous glucose monitoring, were 152.55 and 163.05 mg/dL for the 8- and 16-mg insulin doses, respectively. In comparison, patients who received placebo had mean glucose levels of 175.99 mg/dL.

Phase 3 studies of oral insulin formulations for diabetes were initiated in 2020 and 2021. Two of the three studies have not yet begun recruiting patients as of this writing. In trial NCT04606576, subjects will be randomized to 26 weeks of the double-blind treatment period. Subjects will receive 8-mg ORMD-0801 capsules administered either once daily at night before bedtime or twice daily each morning approximately 45 min before breakfast, or matching placebo. Mean changes from baseline in HbA1C and in fasting glucose will be measured. Trial NCT04817215 will evaluate the efficacy of ORMD-0801 compared with placebo in improving glycemic control (measured by HbA1C) by diet control alone, diet control and metformin monotherapy, or two or three oral glucose-lowering agents (Table 7.5).

7.2.5 Semaglutide

The first orally administered GLP-1 receptor agonist was approved by the FDA in September 2019 (Food and Drug Administration, 2019). The formulation comprised an absorption enhancer, sodium *N*-[8-(2-hydroxy benzoyl)amino]-caprylate (Buckley et al., 2018), to facilitate absorption of the peptide across the GI tract. This was the only successful oral peptide formulation that was approved for large-scale patient use for a chronic disease (type 2 diabetes mellitus).

Initial phase 1 trials of this product were commenced with a focus on investigating optimal dosing conditions for semaglutide (NCT01572753). In a randomized study, healthy subjects received 10 mg of oral semaglutide each morning for 10 days. Subjects were further divided into groups in which the dose was given with 50 or 120 mL of water and the duration of postdose fasting was 15, 30, 60, or 120 min (Donsmark et al., 2017). Clinically relevant semaglutide levels were observed in all subjects, and levels increased with increased postdose fasting times. No significant difference was observed when the volume of water was varied. In another study (NCT02172313), the effect of food intake was evaluated in healthy subjects who were either fed a high-calorie, high-fat diet 30 min predose or fasted overnight for at least 10 h. Limited or no semaglutide exposure was seen in the fed group (Maarbjerg et al., 2017). In other studies, the efficacy of oral semaglutide was evaluated in subjects with renal impairment (NCT02014259) and hepatic impairment (NCT02016911), wherein once-daily oral administration of semaglutide was evaluated with dose escalation, starting at 5 mg for 5 days followed by 10 mg for 5 days. Semaglutide exposure did not vary in a consistent pattern in all groups, and renal or hepatic impairment had no impact on the half-life of semaglutide (Baekdal et al., 2018; Granhall et al., 2018).

In a phase 2 trial (NCT01923181) (Davies et al., 2017), patients with type 2 diabetes were evaluated for the dose–response relationship of five strengths (2.5, 5, 10, 20, and 40 mg) of oral semaglutide compared with placebo as well as once-weekly subcutaneous semaglutide for glycemic control. Other groups included dose escalation to 40 mg for 8 weeks; dose escalation to 40 mg for 2 weeks; oral placebo; or once-weekly subcutaneous semaglutide, 1.0 mg for 26 weeks. The primary endpoint was change in HbA1c from baseline to week 26. Secondary endpoints included change from baseline in body weight and adverse events. Significant reductions in HbA1c were seen with oral semaglutide versus placebo. Reductions in body weight were greater with oral semaglutide (dose-dependent range, −2.1 to −6.9 kg) and subcutaneous semaglutide (−6.4 kg) versus placebo

(−1.2 kg) and were significant for oral semaglutide doses of 10 mg or more versus placebo (Davies et al., 2017).

Another phase 2 study (NCT02453711) (O'Neil et al., 2018) evaluated the efficacy and safety of semaglutide in comparison with liraglutide and a placebo in promoting weight loss. Doses were initiated at 0.05 mg per day and then incrementally escalated to 0.1, 0.2, 0.3, and 0.4 mg per day every 4 weeks. Liraglutide was initiated at 0.6 mg per day and was escalated by 0.6 mg per week up to 3 mg. Estimated mean weight loss was −2.3% for the placebo group versus −6.0% (0.05 mg), −8.6% (0.1 mg), −11.6% (0.2 mg), −11.2% (0.3 mg), and −13.8% (0.4 mg) for the semaglutide groups. Differences in mean body weight reductions for ≥0.2 mg of semaglutide versus liraglutide were all significant (−13.8% to −11.2% vs −7.8%, respectively).

Semaglutide was also evaluated for efficacy and safety in patients with nonalcoholic steatohepatitis (NASH). In a phase 2 trial (NCT02970942) (Newsome et al., 2021) involving patients with biopsy-confirmed NASH and liver fibrosis of stage F1, F2, or F3, patients were assigned to receive once-daily subcutaneous semaglutide at a dose of 0.1, 0.2, or 0.4 mg or corresponding placebo. The primary endpoint was resolution of NASH with no worsening of fibrosis. The percentage of patients meeting this condition was 40% in the 0.1-mg group, 36% in the 0.2-mg group, 59% in the 0.4-mg group, and 17% in the placebo group. The confirmatory secondary endpoint was improvement by at least one fibrosis stage with no worsening of NASH; 43% of the patients in the 0.4-mg group and 33% in the placebo group showed improvement in fibrosis.

At the time of this writing, at least 25 phase 3 studies of oral semaglutide have been completed or are ongoing. This is by far the largest number of phase 3 translational studies ever carried out in which semaglutide is compared with other orally delivered peptide drug products, such as liraglutide, exenetide, dulaglutide, sitagliptin, empagliflozin, and canagliflozin, that are FDA approved for similar indications.

In addition to these comparative studies, phase 3 studies have also been conducted in a Japanese population. EVOKE and EVOKE Plus studies are currently recruiting patients with early Alzheimer disease to evaluate semaglutide. Another study (NCT04596631) is recruiting children and teenagers with type 2 diabetes to evaluate oral semaglutide. In the interest of brevity, not all of these trials are discussed here.

Another study (NCT02863328) (Rodbard et al., 2019) compared the efficacy and safety of oral semaglutide with that of the sodium glucose cotransporter 2 inhibitor empagliflozin in patients with uncontrolled type 2 diabetes who were taking metformin. Oral semaglutide was well tolerated and was superior to empagliflozin in reducing HbA1c at 26 weeks and in reducing HbA1c and body weight at 52 weeks. In a group of phase 3 studies (PIONEER 3, 4, 5, 7, and 8) (Pratley et al., 2021), patients received a once-daily dose of 3, 7, or 14 mg of oral semaglutide or a comparator (placebo, sitagliptin 100 mg, or liraglutide 1.8 mg). Patients were evaluated for semaglutide efficacy (changes in HbA1c and body weight from baseline to planned end of treatment) and safety outcomes. Patients were grouped according to background medication (metformin, a sulfonylurea, thiazolidinedione, a sodium glucose cotransporter 2 inhibitor, insulin, or combinations thereof). Greater reductions in HbA1c and body weight were seen across groups receiving background medication along with oral semaglutide (changes from baseline, −1.0% to −1.5% and −2.2 kg to −5.0 kg, respectively) than with placebo, sitagliptin 100 mg, or liraglutide 1.8 mg (except for similar HbA1c reductions vs liraglutide).

Because type 2 diabetes has a unique phenotype in Japanese patients, novel therapies such as oral semaglutide require evaluation in this population. The PIONEER 10 phase 3 trial (NCT03015220) evaluated once-daily oral semaglutide at 7 mg or 14 mg versus the injectable dulaglutide at 0.75 mg. Health-related quality of life was assessed with the Japanese-specific Diabetes Therapy-Related Quality of Life criteria. The health-related quality of life total score was improved with orally administered semaglutide at 7 and 14 mg compared with dulaglutide at 0.75 mg, as were the domains "anxiety and dissatisfaction with treatment" and "satisfaction with treatment" (Ishii et al., 2021). The PIONEER 9 phase 3 study (NCT03018028) evaluated oral semaglutide in Japanese patients with uncontrolled type 2 diabetes managed with diet or exercise or with oral glucose-lowering drug monotherapy. This study compared once-daily oral semaglutide (3, 7, or 14 mg) or placebo with subcutaneous once-daily liraglutide (0.9 mg). The primary endpoint was change in HbA1c from baseline to week 26. The results showed that oral semaglutide provided significant dose-dependent reductions in HbA1c compared with placebo in Japanese patients with type 2 diabetes and had a safety profile consistent with that of subcutaneous liraglutide (Yamada et al., 2020).

Table 7.6 summarizes the various oral semaglutide clinical studies.

Table 7.6 Clinical trials with oral semaglutide.

Study name	Indication	Phase	ClinicalTrials.gov identifier
Investigation of Optimal Dosing Conditions for a Long-Acting GLP-1 Analogue in Healthy Male Subjects	Type 2 diabetes mellitus Healthy subjects	1	NCT01572753
Effect of Food on the Pharmacokinetics of Oral Semaglutide in Healthy Subjects	Diabetes Healthy subjects	1	NCT02172313
Pharmacokinetics, Safety, and Tolerability of Oral Semaglutide in Subjects with Various Degrees of Impaired Renal Function	Type 2 diabetes mellitus Healthy subjects	1	NCT02014259
Pharmacokinetics, Safety, and Tolerability of Oral Semaglutide in Subjects with Hepatic Impairment	Type 2 diabetes mellitus Healthy subjects	1	NCT02016911
Multiple Dose Trial Examining Dose Range, Escalation, and Efficacy of Oral Semaglutide in Subjects with Type 2 Diabetes	Type 2 diabetes mellitus	2	NCT01923181
Investigation of Safety and Efficacy of Once-Daily Semaglutide in Obese Subjects without Diabetes Mellitus	Obesity Metabolism and nutrition disorder	2	NCT02453711
Investigation of Efficacy and Safety of Three Dose Levels of Subcutaneous Semaglutide Once Daily versus Placebo in Subjects with Nonalcoholic Steatohepatitis	Hepatobiliary disorders Nonalcoholic steatohepatitis	2	NCT02970942
Efficacy and Safety of Oral Semaglutide versus Empagliflozin in Subjects with Type 2 Diabetes Mellitus (PIONEER 2)	Type 2 diabetes mellitus	3	NCT02863328
Efficacy and Long-Term Safety of Oral Semaglutide versus Sitagliptin in Subjects with Type 2 Diabetes (PIONEER 3)	Type 2 diabetes mellitus	3	NCT02607865
Efficacy and Safety of Oral Semaglutide versus Liraglutide and versus Placebo in Subjects with Type 2 Diabetes Mellitus (PIONEER 4)	Type 2 diabetes mellitus	3	NCT02863419

(*Continued*)

Table 7.6 (Continued)

Study name	Indication	Phase	ClinicalTrials.gov identifier
Efficacy and Safety of Oral Semaglutide versus Placebo in Subjects with Type 2 Diabetes and Moderate Renal Impairment (PIONEER 5)	Type 2 diabetes mellitus	3	NCT02827708
Efficacy and Safety of Oral Semaglutide Using a Flexible Dose Adjustment Based on Clinical Evaluation versus Sitagliptin in Subjects with Type 2 Diabetes Mellitus (PIONEER 7)	Type 2 diabetes mellitus	3	NCT02849080
Efficacy and Safety of Oral Semaglutide versus Placebo in Subjects with Type 2 Diabetes Mellitus Treated with Insulin (PIONEER 8)	Type 2 diabetes mellitus	3	NCT03021187
Safety and Efficacy of Oral Semaglutide versus Dulaglutide Both in Combination with One OAD (Oral Antidiabetic Drug) in Japanese Subjects with Type 2 Diabetes (PIONEER 10)	Type 2 diabetes mellitus	3	NCT03015220
Dose–Response, Safety and Efficacy of Oral Semaglutide versus Placebo and versus Liraglutide, All as Monotherapy in Japanese Subjects with Type 2 Diabetes (PIONEER 9)	Type 2 diabetes mellitus	3	NCT03018028
A Trial Comparing the Safety and Efficacy of Semaglutide Once Weekly versus Sitagliptin Once Daily in Japanese Subjects with Type 2 Diabetes (SUSTAIN)	Type 2 diabetes mellitus	3	NCT02254291

7.3 Potential reasons for variable bioavailability in translational studies

We have previously outlined some of the major translational challenges for oral peptide delivery, attributing them to physiological effects

(Tyagi et al., 2018). These effects could be broadly classified into the following five categories:

1. Variability in GI parameters between human subjects
2. Variability in intestinal and gastric pH
3. Effects of gastric emptying
4. Effects of food (meal and water intake)
5. Effect of peptidases

7.3.1 Variability in GI parameters between human subjects

The major contributors to variable intersubject GI parameters are the surface area of the stomach and intestine, the motility and permeability of digestive tract membranes, and variation in gastric volume, gastric mucus, and thickness of the mucous layer.

7.3.1.1 Surface area

The small intestine, with its folds, villi, and microvilli, has a much higher surface area than the stomach. The human stomach mucosal surface area is reported to be in the range of 520–1536 cm^2 (Cox, 1945). The variability in the length of the small intestine falls within the same range and is reported to depend on whether subjects have abdominal disorders, from 160–430 cm in people with no abdominal disorders (measured by radiography) (Fanucci et al., 1988), to 385–538 cm in patients who have undergone laparotomy (Hosseinpour & Behdad, 2008). As the first uptake of peptides occurs within the small intestine, these variations in length and surface area can greatly affect the absorption and hence the bioavailability of peptides.

7.3.1.2 Motility and resulting exposure

The motility of the small intestine plays a key role by altering the site of absorption and/or duration of exposure of peptides. It has been reported that the speed of phase III contractions in the duodenum-jejunum area in healthy volunteers varies from 2.4 to 10 cm/min (Annese et al., 1997). Variation in the duration of phase III contractions, which ranges from 3.6 to 5.7 min in a healthy individual, might be less of a contributing factor. However, this could rapidly change in patients with conditions such as Crohn's disease, in which the speed of phase III contractions can vary from 2.1 to 22.5 cm/min. Although rapid contractions can cause poor peptide absorption, it has also been observed that about 30% of patients with GI disease have no phase III contractions at all, which could lead to a large gradient of peptide localization within the small intestine that results in variability in absorption.

7.3.1.3 Intestinal permeability

In one of the studies of calcitonin discussed in this chapter, it was observed that region-dependent absorption plays a key role in vivo oral absorption of sCT (Sinko et al., 1999). GI diseases such as irritable bowel syndrome (IBS) are associated with modification of intestinal permeability, which is marked by diarrhea and weight loss. Lactulose:mannitol excretion ratio is an indicator of intestinal permeability, and a ratio of ≥ 0.07 is indicative of increased intestinal permeability. In a clinical study comparing healthy volunteers and patients with IBS (Zhou et al., 2010), it was observed that the lactulose:mannitol excretion ratio varied from 0.01–0.06 in healthy volunteers to 0.01–0.24 in patients with IBS.

7.3.1.4 Absorption site transition time

Variability in transit times can result in the dosage form being in an undesirable site, which can lead to poor bioavailability and high variability among patients. The mean $\pm$ standard deviation in intestinal transit time in humans has been reported to be 170 $\pm$ 30 min. After gastric emptying, an enteric dosage form disintegrates in the small intestine within one of the intestinal segments (duodenum, jejunum, or ileum). This is thought to occur within the first 5 min in the duodenum, 5–60 min in the jejunum, and 60–75 min in the ileum (Davis et al., 1988). For various tablet formulations containing calcitonin, it has been observed that disintegration occurred either in the jejunum, with a transit time of ≤ 50 min, or in another site, with a transit time of 20–90 min (Lee et al., 1999), indicating absorption from the ileum. Much of the variability can also be due to the transit time in the human large intestine, which can vary from 8 to 72 h (DeSesso & Jacobson, 2001).

7.3.1.5 Gastric volume

Because solubility is a function of concentration, the volume of liquid secreted in the stomach can affect the solubility of a peptide. The solubility of an active molecule directly affects its absorption and hence its bioavailability. After overnight fasting, healthy volunteers have been observed to have a mean gastric volume of 59.6 $\pm$ 4.2 mL (males) and 46.3 $\pm$ 4.2 mL (females), with a range of 0–225 mL (Brock-Utne et al., 1977).

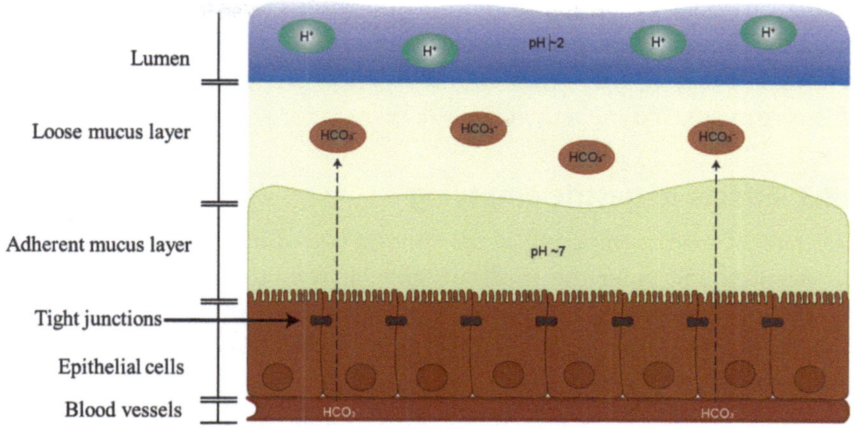

Figure 7.2 Gastric mucosal barrier.

In the case of an ulcerated stomach, however, gastric volume after overnight fasting can be as high as 80 mL or more.

7.3.1.6 Gastric mucus

The mucous layer in the gastric region can affect the absorption of peptides across tight junctions. Variation in the rate of mucus secretion can alter the clearance of an active peptide before it is exposed to the epithelial cell layer. Gastric mucus is composed of an adherent layer, which adheres to the gastric mucosa, and a soluble layer, which is mostly freshly secreted (Fig. 7.2). The mean $\pm$ standard deviation soluble mucin concentration is 93.6 ± 37.8 μg of hexose per mL in healthy volunteers (Iijima et al., 2009). In a retrospective study of normal wall anatomy, Huh et al. reported that the thickness of the mucous layer in the stomachs of 10 patients varied from 1.03 to 1.64 mm (Huh et al., 2003). Mucosal blood flow also plays an important role in the production of mucus and associated bicarbonates, which can affect gastric pH.

7.3.2 Variability in intestinal and gastric pH

Because the optimum solubility of peptides and drug substances are a function of pH, variability in intestinal and gastric pH can have an impact on absorption and bioavailability. In a study of 24 healthy young men and women in a fasted state, the median gastric pH was 1.7 and the median duodenal pH was 6.1. With the administration of a meal, gastric pH increased to a median of 6.7 before returning to pH 1.7 within 2 h

(Dressman et al., 1990). People with cystic fibrosis tend to produce a large amount of mucus, which can block secretions from the pancreas and lead to acid imbalance. One study found that this led to a pH of <5.5 for 25% of patients (Youngberg et al., 1987).

7.3.3 Effects of gastric emptying

The rate of gastric emptying is the time it takes for food to empty from the stomach and enter the small intestine. It is a rate-limiting step in the absorption of peptides, as peptide absorption from the stomach is slow, whereas that from the small intestine is rapid. Many factors, including posture, hormonal activity, and the presence of other drugs, affect gastric emptying. It has been observed that gastric peristalsis is slower in patients who are in an upside-down position than in those in a seated position (Steingoetter et al., 2006). The intestinal absorption of peptides in an enteric-coated formulation largely depends on gastric emptying because peptides are emptied from the stomach in the third phase of the cyclic motility pattern. In terms of disease state, it has been shown that elderly patients with achlorhydria have delayed gastric emptying. In one study, it took approximately 40 min for 50% of 300 mL of orange juice to be emptied from the stomach in elderly patients with achlorhydria, whereas only 18 min was required in an elderly control group (Dressman et al., 1993). It has also been reported that gastric emptying is delayed in patients with gastric ulcers and carcinoma and is accelerated in those with celiac disease and duodenal ulcers (Nimmo, 1976).

7.3.4 Effect of food (meal and water intake)

In studies of 56 healthy women who took 0.8 mg of oral sCT and 200 mg of the 5-CNAC permeation enhancer with water (50 or 200 mL), C_{max} and AUC were significantly dependent upon the volume of water and the amount of time (10, 30, or 60 min) that the formulation was taken before a standard meal (Karsdal et al., 2008b; Tanko et al., 2004). Administration of the dosage form with 50 mL of water showed two- to three-fold higher absorption compared with 200 mL of water.

In another study of calcitonin with 0.8 and 200 mg of the 5-CNAC permeation enhancer, a gradual decrease in bioavailability, to 26% and 59%, was seen with a predose meal at 4 and 2 h before dosing and at 10 min after dosing, indicating the effect of food on oral sCT bioavailability (Karsdal et al., 2009). Other studies on the effect of food intake on oral bioavailability of peptides include an evaluation of LY303366, a lipopeptide antifungal in dogs (Li et al., 2001), and ARD-07 (also known as EP01572), a synthetic ghrelin agonist, in which decreased bioavailability was observed with a high-calorie meal (MacLean et al., 2009). In another study, Tuvia et al. observed lower octreotide levels in plasma when oral formulations were administered with a high-fat, high-calorie meal (Tuvia et al., 2012).

Studies on the effect of food intake on the oral bioavailability of peptides are not extensive, and it is hoped that additional research will become available in the near future as more molecules are advanced through clinical studies.

7.3.5 Effect of peptidases

Degradation of peptides by proteases and peptidases in the GI tract is well known and is a major hurdle to the oral delivery of peptides and biologics. In patients with GI disease, it has been observed that the specific activity of dipeptidases (attacking hydrophobic amino acids) in the small intestinal mucosa of adult humans ranges from 23.7 to 125.2 U/mg nitrogen (each unit is defined as the peptidase activity hydrolyzing 1 μmol of dipeptide per minute at 40°C) (Lindberg, 1966). Although only limited data are currently available in this area, peptide stability in human gastric fluid in the presence of pepsin and pancreatin is another indicator of oral bioavailability. Wang et al. (2015) studied the stability of 17 peptides and observed that only three (cyclosporin, desmopressin, and octreotide) had any stability in small-intestine fluid containing pancreatin.

7.4 Future directions

As of this writing, octreotide and semaglutide are the only two successful oral peptide dosage forms that are approved by the FDA, and their reported bioavailability is in the low single digits. It is possible that solubility at the site of absorption during oral delivery is very low for the various peptides that have been tested in the clinic. Comprehensive translational testing, similar to what has been done with semaglutide, is required for more peptides to perform well in the clinic. It is sometimes important to go back to optimizing the peptide design for improved protease stability and solubility and half-life as well as understand the site of absorption to be able to optimize the delivery. This methodology was systematically followed in two recent publications that showed improved bioavailability in preclinical models (Tyagi et al., 2021). To further improve the uptake and widen the absorption window of peptides, it might be necessary to combine knowledge from small-molecule development, such as gastroretentive delivery systems and other methodologies, to make significant improvements in bioavailability. Use of ingestible devices with either a microneedle or a microjet-based liquid injection mechanism to deliver peptides across tight junctions for systemic delivery could provide larger amounts of the drug. Ingestible devices have been discussed in a separate chapter in this book. It remains to be seen, however, how these ingestible devices can overcome the physiology of the GI tract, as the physiological conditions encountered by the dosage form vary greatly from the in vitro conditions used for the development of these systems. Testing in preclinical whereby ingestible devices are placed either surgically (e.g., endoscopically) within the GI tract or via oral delivery, is being widely followed. Nevertheless, the anatomy and pathophysiology of preclinical species can differ from that of humans. Use of in silico models rather than preclinical models is a growing trend due to the finer control and reproducibility that are possible with these systems. We recommend that studies be conducted in a combination of dogs, pigs, and nonhuman primates, together with sound in silico and in vitro studies, and correlate the results with those from the clinic for pharmacokinetic and bioavailability data to develop sound in vitro/preclinical/clinical correlation to enable the development of future technologies. Finally, digitized clinical trials using continuous monitoring of biomarkers and vital signs for digital endpoints and side effect monitoring, coupled with advanced oral delivery technologies, will enable the widespread development of oral peptide dosage forms in the coming years.

References

Annese, V., et al. (1997). Gastrointestinal motility disorders in patients with inactive Crohn's disease. *Scandinavian Journal of Gastroenterology*, *32*(11), 1107–1117.

Arbit, E., & Kidron, M. (2017). Oral insulin delivery in a physiologic context: Review. *Journal of Diabetes Science and Technology (Online)*, *11*(4), 825–832.

Baekdal, T. A., et al. (2018). Pharmacokinetics, safety, and tolerability of oral semaglutide in subjects with hepatic impairment. *Journal of Clinical Pharmacology*, *58*(10), 1314–1323.

Bischof, M. G., et al. (2001). Effects of short-term improvement of insulin treatment and glycemia on hepatic glycogen metabolism in type 1 diabetes. *Diabetes*, *50*(2), 392–398.

Brayden, D. J., et al. (2020). Systemic delivery of peptides by the oral route: Formulation and medicinal chemistry approaches. *Advanced Drug Delivery Reviews*, *157*, 2–36.

Brock-Utne, J. G., et al. (1977). Fasting volume and acidity of stomach contents associated with gastrointestinal symptoms. *Anaesthesia*, *32*(8), 749–752.

Buckley, S. T., et al. (2018). Transcellular stomach absorption of a derivatized glucagon-like peptide-1 receptor agonist. *Science Translational Medicine*, *10*(467).

Burshtein, G., et al. An oral PTH(1-34) formulation with a pharmacokinetic profile optimized for the treatment of osteoporosis. In *American society for bone and mineral research annual meeting*, Montreal, Canada. Available from: https://www.enterabio.com/images/pdfs/72x42_EnterabioASBMR2018-V3–PDF.pdf.

Burshtein, G., et al. Enhanced bioavailability and reduced pharmacokinetic variability of oral PTH (1-34) in men. In *American society for bone and mineral research annual meeting*, Montreal, Canada. Available from: https://www.enterabio.com/images/pdfs/2017-ASBMR-poster.pdf.

Cano-Cebrian, M. J., et al. (2005). Intestinal absorption enhancement via the paracellular route by fatty acids, chitosans and others: A target for drug delivery. *Current Drug Delivery*, *2*(1), 9–22.

Caraco, Y., et al. (2019). An evaluation of the pharmacodynamic effects of oral hPTH (1–34) formulation in patients with hypoparathyroidism. Abstract no. LB-SUN-972. *Journal of Bone and Mineral Research: The Official Journal of the American Society for Bone and Mineral Research*, *34*(Suppl. 1), 278.

Chiasma announces FDA approval of MYCAPSSA (Octreotide) capsules, the first and only oral somatostatin analog. GlobalNewswire (2020). Available from: https://www.globenewswire.com/news-release/2020/06/26/2054210/0/en/Chiasma-Announces-FDA-Approval-of-MYCAPSSA-Octreotide-Capsules-the-First-and-Only-Oral-Somatostatin-Analog.html.

Cox, A. J. (1945). Variations in size of human stomach. *California and Western Medicine*, *63*, 267.

Davies, M., et al. (2017). Effect of oral semaglutide compared with placebo and subcutaneous semaglutide on glycemic control in patients with type 2 diabetes: A randomized clinical trial. *JAMA: The Journal of the American Medical Association*, *318*(15), 1460–1470.

Davis, S. S., et al. (1988). Gastric emptying of large single unit dosage forms. *The Journal of Pharmacy and Pharmacology*, *40*(3), 205–207.

DeSesso, J. M., & Jacobson, C. F. (2001). Anatomical and physiological parameters affecting gastrointestinal absorption in humans and rats. *Food and Chemical Toxicology: An International Journal Published for the British Industrial Biological Research Association*, *39*(3), 209–228.

DeVries, J. H., et al. (2002). Continuous intraperitoneal insulin infusion in patients with 'brittle' diabetes: Favourable effects on glycaemic control and hospital stay. *Diabetic Medicine: A Journal of the British Diabetic Association*, *19*(6), 496–501.

Donsmark, M., et al. (2017). Evaluation of the effects of water volume with dosing and post-dose fasting period on pharmacokinetics of oral semaglutide. Abstract no. 792. In *Annual meeting of the European Association for the study of diabetes*, Lisbon, Portugal.

Dressman, J. B., et al. (1993). Gastrointestinal parameters that influence oral medications. *Journal of Pharmaceutical Sciences*, *82*(9), 857–872.

Dressman, J. B., et al. (1990). Upper gastrointestinal (GI) pH in young, healthy men and women. *Pharmaceutical Research*, *7*(7), 756–761.

Eldor, R., et al. (2013). Glucose-reducing effect of the ORMD-0801 oral insulin preparation in patients with uncontrolled type 1 diabetes: A pilot study. *PLoS One*, *8*(4), e59524.

Fanucci, A., Cerro, P., & Fanucci, E. (1988). Normal small-bowel measurements by enteroclysis. *Scandinavian Journal of Gastroenterology*, *23*(5), 574–576.

Food and Drug Administration. (2019). FDA approves first oral GLP-1 treatment for type 2 diabetes, October 19, 2021. Available from: https://www.fda.gov/news-events/press-announcements/fda-approves-first-oral-glp-1-treatment-type-2-diabetes.

Granhall, C., et al. (2018). Pharmacokinetics, safety and tolerability of oral semaglutide in subjects with renal impairment. *Clinical Pharmacokinetics*, *57*(12), 1571–1580.

Henriksen, K., et al. (2016). A randomized, double-blind, multicenter, placebo-controlled study to evaluate the efficacy and safety of oral salmon calcitonin in the treatment of osteoporosis in postmenopausal women taking calcium and vitamin D. *Bone*, *91*, 122–129.

Henriksen, K., et al. (2013). Evaluation of the efficacy, safety and pharmacokinetic profile of oral recombinant human parathyroid hormone [rhPTH(1–31)NH(2)] in postmenopausal women with osteoporosis. *Bone*, *53*(1), 160–166.

Hosseinpour, M., & Behdad, A. (2008). Evaluation of small bowel measurement in alive patients. *Surgical and Radiologic Anatomy: SRA*, *30*(8), 653–655.

Huh, C. H., et al. (2003). Individual variations in mucosa and total wall thickness in the stomach and rectum assessed via endoscopic ultrasound. *Physiological Measurement*, *24*(4), N15–N22.

Iijima, K., et al. (2009). Rebamipide, a cytoprotective drug, increases gastric mucus secretion in human: Evaluations with endoscopic gastrin test. *Digestive Diseases and Sciences*, *54*(7), 1500–1507.

Ishii, H., et al. (2021). Effect of orally administered semaglutide vs dulaglutide on diabetes-related quality of life in Japanese patients with Type 2 diabetes: The PIONEER 10 randomized, active-controlled trial. *Diabetes Therapy*, *12*(2), 613–623.

Karsdal, M. A., et al. (2009). Influence of food intake on the bioavailability and efficacy of oral calcitonin. *British Journal of Clinical Pharmacology*, *67*(4), 413–420.

Karsdal, M. A., et al. (2008a). Investigation of the diurnal variation in bone resorption for optimal drug delivery and efficacy in osteoporosis with oral calcitonin. *BMC Clinical Pharmacology*, *8*, 12.

Karsdal, M. A., et al. (2008b). Optimizing bioavailability of oral administration of small peptides through pharmacokinetic and pharmacodynamic parameters: The effect of water and timing of meal intake on oral delivery of Salmon Calcitonin. *BMC Clinical Pharmacology*, *8*, 5.

Karsdal, M. A., et al. (2011). Lessons learned from the development of oral calcitonin: The first tablet formulation of a protein in phase III clinical trials. *Journal of Clinical Pharmacology*, *51*(4), 460–471.

Karsdal, M. A., et al. (2010). The effect of oral salmon calcitonin delivered with 5-CNAC on bone and cartilage degradation in osteoarthritic patients: A 14-day randomized study. *Osteoarthritis and Cartilage/OARS, Osteoarthritis Research Society*, *18*(2), 150–159.

Karsdal, M. A., et al. (2015). Treatment of symptomatic knee osteoarthritis with oral salmon calcitonin: Results from two phase 3 trials. *Osteoarthritis and Cartilage/OARS, Osteoarthritis Research Society*, *23*(4), 532–543.

Lane, N. E., et al. (2000). Bone mass continues to increase at the hip after parathyroid hormone treatment is discontinued in glucocorticoid-induced osteoporosis: Results of a randomized controlled clinical trial. *Journal of Bone and Mineral Research: The Official Journal of the American Society for Bone and Mineral Research*, *15*(5), 944–951.

Lee, Y. H., et al. (1999). Impact of regional intestinal pH modulation on absorption of peptide drugs: Oral absorption studies of salmon calcitonin in beagle dogs. *Pharmaceutical Research*, *16*(8), 1233–1239.

Li, C., et al. (2001). Regional-dependent intestinal absorption and meal composition effects on systemic availability of LY303366, a lipopeptide antifungal agent, in dogs. *Journal of Pharmaceutical Sciences*, *90*(1), 47–57.

Lindberg, T. (1966). Intestinal dipeptidases. Dipeptidase activity in the mucosa of the gastrointestinal tract of the adult human. *Acta Physiologica Scandinavica*, *66*(4), 437–443.

Maarbjerg, S. J., et al. (2017). Evaluation of the effect of food on the pharmacokinetics of oral semaglutide. Abstract no. 148. In *Annual meeting of the European Association for the study of diabetes*, Lisbon, Portugal.

MacLean, C. M., et al. (2009). Effect of food on the pharmacokinetics and pharmacodynamics of an oral ghrelin agonist (ARD-07) in healthy subjects. *Journal of Clinical Pharmacology, 49*(5), 553–559.

Melmed, S., et al. (2015). Safety and efficacy of oral octreotide in acromegaly: Results of a multicenter phase III trial. *The Journal of Clinical Endocrinology and Metabolism, 100*(4), 1699–1708.

Newsome, P. N., et al. (2021). A placebo-controlled trial of subcutaneous semaglutide in nonalcoholic steatohepatitis. *The New England Journal of Medicine, 384*(12), 1113–1124.

Nimmo, W. S. (1976). Drugs, diseases and altered gastric emptying. *Clinical Pharmacokinetics, 1*(3), 189–203.

O'Neil, P. M., et al. (2018). Efficacy and safety of semaglutide compared with liraglutide and placebo for weight loss in patients with obesity: A randomised, double-blind, placebo and active controlled, dose-ranging, phase 2 trial. *Lancet, 392*(10148), 637–649.

Pratley, R. E., et al. (2021). Oral semaglutide reduces HbA1c and body weight in patients with type 2 diabetes regardless of background glucose-lowering medication: PIONEER subgroup analyses. *Diabetes Therapy, 12*(4), 1099–1116.

Rodbard, H. W., et al. (2019). Oral semaglutide vs empagliflozin in patients with type 2 diabetes uncontrolled on metformin: The PIONEER 2 trial. *Diabetes Care, 42*(12), 2272–2281.

Samson, S. L., et al. (2020). Maintenance of acromegaly control in patients switching from injectable somatostatin receptor ligands to oral octreotide. *The Journal of Clinical Endocrinology and Metabolism, 105*(10).

Sinko, P. J., et al. (1999). Biopharmaceutical approaches for developing and assessing oral peptide delivery strategies and systems: In vitro permeability and in vivo oral absorption of salmon calcitonin (sCT). *Pharmaceutical Research, 16*(4), 527–533.

Steingoetter, A., et al. (2006). Effects of posture on the physiology of gastric emptying: A magnetic resonance imaging study. *Scandinavian Journal of Gastroenterology, 41*(10), 1155–1164.

Sturmer, A., et al. (2013). Pharmacokinetics of oral recombinant human parathyroid hormone [rhPTH(1–31)NH(2)] in postmenopausal women with osteoporosis. *Clinical Pharmacokinetics, 52*(11), 995–1004.

Tanko, L. B., et al. (2004). Safety and efficacy of a novel salmon calcitonin (sCT) technology-based oral formulation in healthy postmenopausal women: Acute and 3-month effects on biomarkers of bone turnover. *Journal of Bone and Mineral Research: The Official Journal of the American Society for Bone and Mineral Research, 19*(9), 1531–1538.

Tuvia, S., et al. (2012). Oral octreotide absorption in human subjects: Comparable pharmacokinetics to parenteral octreotide and effective growth hormone suppression. *The Journal of Clinical Endocrinology and Metabolism, 97*(7), 2362–2369.

Tyagi, P., et al. (2021). Targeted oral peptide delivery using multi-unit particulates: Drug and permeation enhancer layering approach. *Journal of Controlled Release: Official Journal of the Controlled Release Society, 338*, 784–791.

Tyagi, P., Pechenov, S., & Anand Subramony, J. (2018). Oral peptide delivery: Translational challenges due to physiological effects. *Journal of Controlled Release: Official Journal of the Controlled Release Society, 287*, 167–176.

Verhaar, H. J., & Lems, W. F. (2010). PTH analogues and osteoporotic fractures. *Expert Opinion on Biological Therapy, 10*(9), 1387–1394.

Wang, J., et al. (2015). Toward oral delivery of biopharmaceuticals: An assessment of the gastrointestinal stability of 17 peptide drugs. *Molecular Pharmaceutics, 12*(3), 966–973.

Yamada, Y., et al. (2020). Dose-response, efficacy, and safety of oral semaglutide monotherapy in Japanese patients with type 2 diabetes (PIONEER 9): A 52-week, phase 2/3a, randomised, controlled trial. *Lancet Diabetes Endocrinology*, *8*(5), 377–391.
Youngberg, C. A., et al. (1987). Comparison of gastrointestinal pH in cystic fibrosis and healthy subjects. *Digestive Diseases and Sciences*, *32*(5), 472–480.
Yuhico, L. S., Gundu, V., & Lenox, R. (2012). Octreotide-induced asystolic events in an intensive care unit patient with gastrointestinal bleeding. *Heart & Lung: The Journal of Critical Care*, *41*(6), e18–e20.
Zhou, Q., et al. (2010). MicroRNA-29a regulates intestinal membrane permeability in patients with irritable bowel syndrome. *Gut*, *59*(6), 775–784.

CHAPTER EIGHT

Regulatory aspects of oral peptide delivery

Eric Gruff
E4 Consulting, Poway, CA, United States

8.1 Basis of regulation and regulatory framework

The regulation of oral peptide (OP) drugs is fundamentally no different from any other medicinal products. Sponsors must demonstrate an acceptable risk-based safety and efficacy profile for an oral peptide drug and must demonstrate suitable control of drug substance and drug product quality by validating production and control processes according to current good manufacturing practices. However, achieving these outcomes for novel therapeutics like oral peptides is often significantly more challenging than for either small molecule orally administered drugs or biologics intended for parenteral administration.

Most proteins, including synthetic peptides consisting of 40 or fewer amino acids, are considered well-characterized biologics, which in the United States fall under the purview of the FDA's Center for Drug Evaluation and Research (CDER). The regulatory strategy for an oral peptide must include consideration of the requirements for both the drug substance and the drug product. The drug substance consists of the pure peptide, which provides the intended therapeutic activity (active ingredient). The drug product (also referred to as the medicinal product) is the form presented to the patient, and usually includes inert materials (excipients) and for capsules, a shell comprised of either gelatin or another pharmaceutically acceptable material.

While it is the drug product that is ultimately approved for marketing, the manufacturing and control of the drug substance are also critically important to the success of the regulatory strategy for the product. The route of administration will not typically have a significant impact on the clinical pathway or design of clinical trials but can impact the details of clinical trial supply and conduct.

Oral Delivery of Therapeutic Peptides and Proteins.
DOI: https://doi.org/10.1016/B978-0-12-821061-1.00010-1

8.2 Investigational studies under an IND

The requirement for initiating human clinical trials in most countries includes the submission and regulatory approval of an investigational new drug (IND) application or equivalent. The content of the IND at a minimum includes extensive information on the chemistry, manufacturing, and controls (CMC) of the drug substance and drug product, nonclinical data [pharmacology, pharmacokinetics (PK), and toxicology], and the protocol for the intended first-in-human (FIH) study.

For an oral peptide, key aspects that are unique to the product will include a demonstration of the quality, stability, and bioavailability of the drug product throughout the life of a batch. Nonclinical data will also be required to demonstrate the relationship between dose and pharmacokinetics, including the relative bioavailability of the oral peptide compared to that obtained from intravenous administration. Clinical studies will require a reliable supply of the OP drug product (DP), and pivotal clinical trials (those that will demonstrate the safety and efficacy of the DP in support of approval) will require that the OP DP is representative of the intended commercial DP.

8.3 Pre-IND meeting

Before submitting the IND package for a novel DP like an oral peptide, a Pre-IND consultation is an excellent way to gain consensus from the regulatory agency regarding the CMC, preclinical and clinical development strategies. A common mistake is to hold the Pre-IND meeting too early in the preclinical program and then realize that there is a need for additional feedback before the IND is submitted. To optimize the utility of the Pre-IND meeting, a common strategy is to hold it just before the design of the pivotal GLP safety studies is finalized. These studies are required in two species (one rodent and one nonrodent) and are critical to establishing a safe pathway to human clinical trials. By obtaining regulatory feedback just before finalizing the protocols, a sponsor can proceed confidently with the GLP toxicology studies.

Sponsors should plan on presenting proposed strategies to the regulatory agency and obtaining feedback rather than asking open-ended questions. Regulators are not shy about offering advice when they sense a potential disconnect with sponsor companies and particularly so with novel drug products. The value of the Pre-IND meeting for products like oral peptide drugs cannot be overstated. In addition to obtaining feedback, the Pre-IND is an opportunity to introduce the company and product to the regulatory agency, so a well-prepared approach is essential to establishing credibility. Asking gratuitous or poorly framed questions can start the relationship with regulators off on a poor footing, and should be avoided. Likewise, a Pre-IND meeting is not intended to guide a sponsor beyond the FIH study. Questions about the pivotal clinical trial design and approvability of the DP when certain approaches are taken will usually result in a regulator stating that it is premature to discuss the approvability and/or that they will make the decision when trial data are available.

CMC questions are frequently asked at Pre-IND meetings, but many are unnecessary. For example, presenting drug substance (DS) or DP specifications that are in accordance with current regulations and guidances and asking if the regulatory agency agrees with them is unnecessary and counterproductive. Sponsor companies should reserve questions for unusual or nuanced situations where the standard CMC practices may not apply or are not clear. In the case of an oral peptide DP, valid areas for discussion may include the use of novel excipients or technologies to produce a suitably stable and bioavailable drug product or the use of novel analytical methods to characterize the DS and/or DP.

Preclinical questions should typically include a request for regulatory agency review of pivotal (GLP) study protocols, as well as feedback on the completeness of the nonclinical data package. The sponsor's target should be to verify that they will be able to submit a complete nonclinical package in the IND that will enable the start of human trials. Before the GLP studies are conducted, dose range finding studies in the target species should be conducted to ensure that the correct doses have been established. This is a critical aspect of the preclinical program as it will ensure that the starting clinical dose(s) are selected and that an acceptable safety margin can be defined. Other preclinical questions may include discussing the need for specialty studies to evaluate potential drug–drug interactions (DDIs), chronic studies to support extended dosing of the OP, and studies intended to understand potential immunogenic responses to OPs.

The clinical portion of a Pre-IND meeting will necessarily focus on the proposed design of the FIH study. Key criteria that should be presented include starting dose and dose escalation scheme, inclusion/exclusion criteria, primary and secondary endpoints, and data treatment. For therapeutic areas or situations where healthy volunteer studies are not possible, the patient population should be defined. It is not necessary to bring a full clinical protocol to the Pre-IND meeting, but a well-thought-out synopsis should be included in the briefing document.

8.4 Investigational new drug application

The IND (or equivalent) allows the study of an unapproved drug in humans. Global regulations for the content and format of an IND have been harmonized to a large degree, but some differences remain. This section will focus on requirements in the United States for new oral peptide drugs.

8.5 Chemistry, manufacturing, and controls regulatory requirements

8.5.1 Drug substance

The DS consists of the peptide that will convey the pharmacological activity in vivo. The regulatory expectation for any drug substance includes a detailed description of how it is made and purified, the controls (test methods and specifications) employed, and an assessment of the stability of the drug substance. The following sections, arranged in the format of Module 3 (quality module) of the common technical document (CTD),[3] which is accepted across the ICH regions, cover the expectations in more detail. By fulfilling the expectations of the CTD, a sponsor will ensure acceptance of the submission and eventual approval of the product.

8.5.1.1 Nomenclature

A summary of the nomenclature used to describe the drug substance should be included. The nomenclature includes company name(s) and codes, the nonproprietary (generic) International Nonproprietary Name (INN)/United States Adopted Names (USAN) name, Chemical Abstracts Service registry number, and proprietary (trade) name, if available. It is not often that the proprietary and generic names are available at the stage of the IND, and so these may be omitted from the submission. Sponsors typically request the nonproprietary name after reaching clinical Proof of Concept.

For US Sponsors, requesting a nonproprietary name typically starts with the USAN Council. After the USAN Council has tentatively accepted the proposed name, it is submitted to the World Health Organization for approval as an INN. Nonproprietary names are unique to the active ingredient (which is typically the drug substance) and are globally recognized. They are considered public property and may not be trademarked or otherwise claimed by the initial sponsor. Another important feature of the INN system is that the names of pharmacologically related substances demonstrate their relationship by using a common "stem."

8.5.1.2 Structure

The structural formula including amino acid sequence, glycosylation sites or other posttranslational modifications, and relative molecular mass should be presented. Other unique features, such as chemical modification or linking groups, should also be described thoroughly.

8.5.1.3 General properties

Relevant physicochemical and other properties of the drug substance should be included here. These properties should be intrinsic to the drug substance and will not typically change from batch to batch or as a function of manufacturing process changes, etc.

For an orally administered peptide, key parameters will include the following:

1. Secondary structure
2. Oligomer and/or aggregation states
3. UV–visible absorption spectrum and molar extinction coefficient of absorption peaks
4. Circular dichroism and fluorescence spectroscopy
5. Isoelectronic point
6. Hydrodynamic properties, including sedimentation velocity, field-flow fractionation, and viscosity
7. Thermodynamic properties, including melting point/range, differential scanning calorimetry, thermal gravimetric analysis, and thermal calorimetry
8. Small-angle X-ray scattering

8.5.1.4 Manufacturer(s)

The name, address, and responsibilities of entities that perform manufacturing activities, laboratories performing testing (in-process, release, and stability), and those responsible for packaging/labeling/storage of the drug substance should be included.

8.5.1.5 Description of the manufacturing process and process controls

This section should include a flow diagram that summarizes the unit operations included in drug substance manufacturing. Then, for each key step in the manufacturing process, a narrative description containing ranges of key process parameters (KPPs) and in-process controls (IPCs) should be included. For a peptide that is synthesized directly, information is presented on the starting materials, reagents, solvents, equipment, and any other critical aspects of the process. The manufacturing process for peptides produced by other means (fermentation or cell-based processes) should be described similarly, with the information included on cell banks, media, harvesting and purification of drug substance, and modification reactions.

8.5.1.6 Control of materials

Materials used in the manufacture of drug substances and the controls for these materials should be described in sufficient detail to demonstrate suitability of use in the manufacturing process. At a minimum, the identity, grade and/or purity, supplier, and how the material is used should be described. For biologically sourced materials (cell systems, precursors, etc.), a description of the manufacturing process and controls should also be provided.

8.5.1.7 Process validation

Validation of manufacturing processes is not normally expected for any orally administered drug substance at the IND stage. However, in the event that aseptic processing, sterilization, or viral inactivation or removal are included in the manufacturing of the drug substance, process validation or evaluation studies should be included for these steps.

8.5.1.8 Manufacturing process development

At the IND stage, there has typically been a very limited amount of manufacturing process development. However, a basic description of the process and analytical development undertaken to support the production of drug substance for use in clinical studies should be included. Sponsors should include significant changes made to the production process from research batches to toxicology and clinical batches. It is also important to describe the evolution of test methods used for the release and stability testing of the drug substance.

8.5.1.9 Characterization of the drug substance

The primary, secondary, and higher-order structures of the peptide should be included along with a description of the analytical data used to determine confirmation. Posttranslational modifications (e.g., glycosylation), immunochemical properties, and biological activity are other important aspects of the drug substance to provide to regulatory agencies.

8.5.1.10 Impurities

Impurities produced as part of the drug substance manufacturing process should be described, including likely sources of origination (process impurity derived during synthesis, raw material impurity carried into final drug substance, etc.) and structure, if known. Because impurities pose a potential safety risk, a thorough understanding of how they arise, their identity, and key properties is critical to ensure adequate control over the peptide drug substance.

As the development program advances into later-stage clinical trials, having authentic impurities reference standards is necessary to quantify the relative response factor and relative retention time of impurities appearing in the drug substance.

8.5.1.11 Specifications

Control of the drug substance relies on establishing pharmaceutically acceptable specifications for the key physicochemical parameters. Key attributes requiring specifications are listed in Table 8.1. For biological products, including peptides, the list of specifications can be daunting compared to small molecule drug substances, but the additional measures ensure the safety, purity, and quality of the peptide.

8.5.1.12 Analytical procedures: description and validation

The test methods used for control of the drug substance must be scientifically sound and demonstrated to be accurate, specific, reproducible, and adequate to fully characterize its properties. Regulatory agencies expect

Table 8.1 Summary of oral peptide drug substance specifications.

Specification	Typical acceptance criteria	Comments
General properties		
Appearance	Describes color and consistency. Requires absence of foreign particles or other obvious defects.	General macroscopic quality measure
Identification (fingerprint)	Matches identity of authentic reference standard	Typically uses HPLC/UPLC or FTIR spectroscopy
Assay	95%–102% (wider ranges should be justified)	Reported on an anhydrous, solvent-free basis. Typically uses HPLC/UPLC
Impurities	Meets the requirements set forth in ICH Guidance Q3A: Impurities in New Drug Substances	Impurity specifications above the ICH limits can be justified by their inclusion in drug substance batches used in preclinical toxicology studies
Water content	Specification should be set based on process capabilities and stability of drug substance	Usually determined by Karl Fischer titration or loss on drying
Residual solvents	Meets the requirements set forth in ICH Guidance Q3C Impurities: Residual Solvents	Specification only needs to include solvents used in the manufacturing process
Elemental impurities	Meets the requirements set forth in ICH Guidance Q3D Elemental Impurities	Specification only needs to include potential impurities likely to be introduced via raw materials, starting materials, and solvents
Particle size distribution (PSD)	Report volume distribution of particles, typically as sizes where 10%, 50%, and 90% of particles have volumes less than the reported size	Additional descriptors, such as whether the distribution is unimodal, bimodal, or multimodal, may be useful.
Bioburden	Total aerobic count NMT 1000 CFU/gTotal yeasts and molds NMT 100 CFU/gAbsence of *E. coli* Absent/1 g	While oral products are not required to be sterile, control of bioburden is critical to ensuring the safety of patients, particularly those with compromised immune systems
Peptide-specific properties		
Identification (sequence)	Amino acid sequence matches theoretical	

phase-appropriate validation of noncompendial analytical methods beginning with the material used for FIH studies.

The description of noncompendial analytical methods should include enough detail for a regulatory body to understand the method, including the type of equipment used, specific information on chromatography columns or other substrates, buffers and solutions used, and key method parameters (temperatures, pressures, detection parameters, sample volumes, flow rates, etc.). The inclusion of sample chromatograms or tables of expected analytical results will assist reviewers in understanding the methods.

Validation of methods for early-stage programs should include a demonstration of system suitability, accuracy, linearity across the range of measurements (e.g., 70%–130% of the nominal range for the main analyte and up to 0%–20% for impurities), intermediate precision/repeatability (different instruments and/or analysts within the same laboratory), limit of detection, limit of quantitation, and sample and standard solution stability. In later development, it is important to demonstrate the interlaboratory precision and robustness of the method when experimental parameters such as mobile phase pH or ionic strength are varied or different columns are used.

Sponsor companies will often state in the initial IND that methods have been validated in a manner suitable for the phase of development, but will not include validation summaries or reports. This approach may be acceptable, but it is advisable to have the validation documents available if requested by regulators to avoid a clinical hold. For material used in registration stability/pivotal clinical studies, the amount of detail should be considerable, both for the analytical methods and their validation.

8.5.1.13 *Batch analyses*

This section of the IND should summarize the analytical results for research and development and preclinical and clinical batches of drug substances. A table is preferable, and in cases where specifications have changed, a footnote to that effect is sufficient. Reviewers will often look at the characteristics of the toxicology lot(s) as compared to the proposed FIH batch to ensure that sufficient safety coverage has been established before human dosing begins.

8.5.1.14 *Justification of specifications*

Sponsors often omit or minimize this section of the IND; however, a reasoned approach to setting specifications is critical to ensuring product safety and quality. At a minimum, an explanation should be included for any specifications that are different from those normally used for similar materials. This is particularly important for DS attributes that relate to safety, including impurities (organic and elemental). As development proceeds, data from additional batches should be used to justify specifications and there should be an aim toward tightening specifications toward the eventual marketed product requirements.

8.5.1.15 Reference standards

A short summary of the characterization of the primary DS reference standard used for testing the drug substance should be provided. Tabular form is acceptable. Additional information may include storage conditions and the retest period.

In later development, individual impurities standards should be prepared and characterized (relative retention times and relative response factors) for specified impurities (those being individually listed in the DS impurities specification). Combinations of impurities standards may be used to prepare system suitability reference standards.

8.5.1.16 Container closure system

The sponsor should provide a brief description of the container closure system used to store drug substances, including materials of construction and dimensions. As development proceeds, additional information (manufacturer, part number(s), and specifications) should be included. If desiccant or other protective materials are utilized, these should also be fully described in this section.

8.5.1.17 Stability

The initial IND requires at least one month of data at the intended (long-term) storage conditions. Typically, sponsors will submit data for varying lengths of storage on multiple research and development and preclinical and clinical batches of drug substances in support of the retest period. DS batches may be used for extended periods of time provided they meet the predetermined specifications at the most recent retest period. In cases where DS is known to become unusable, a shorter-term retest period and expiration period (after which the material cannot be used regardless of testing) may both be assigned.

A summary of the DS stability data should be presented, along with data tables for each batch and condition. Protocols for stability studies containing information on storage conditions, time points, and test methods should also be included as part of the IND.

In later development, a statistical analysis of quantitative measures (assay, impurities, etc.) should be performed on similar batches to develop a statistical representation of the stability behavior of the drug substance.

8.5.2 Drug product

The DP consists of the dosage form intended to be administered to the patient (or in some cases, a precursor requiring dilution or other preparation before use). A key part of the finished DP includes the primary container-closure system, which may contain one or many doses of the formulated drug. DP usually contains excipients, which are inert ingredients that provide a supportive function (solubilizing agent, preservative, diluent, etc.) in the final formulation.

Many of the key properties of the DP are initially controlled via the DS but require continued attention during the manufacturing, storage, and use of the dosage form. A good example is the presence of organic impurities: The DS will contain some impurities that are carried along from starting materials and/or that form during the synthesis of the desired material. These impurities are controlled via the DS specifications, which were derived from the impurities levels that were present in preclinical toxicology studies. In the DP, quantities of existing DS impurities may continue to increase or new impurities may form. These degradation products are also referred to as impurities and must remain below the levels that have been qualified. If a new impurity arises during DP storage and exceeds the ICH qualification limits.

In the following sections, the Module 3 (quality module) CTD requirements[3] for DP are described in more detail. As for DS, by ensuring each section of the CTD is met, a sponsor will ensure acceptance of the IND submission.

8.5.2.1 Description and composition of the drug product

This section should include the composition of the DP, including all components and their amounts on a per unit basis (vial, gram, etc.). A brief overview of the container closure system and any other components (e.g., desiccants) should also be provided.

8.5.2.2 Pharmaceutical development

The pharmaceutical development section should contain information on the development studies conducted to establish that the dosage form, the formulation, manufacturing process, container closure system, microbiological attributes, and usage instructions are appropriate for the purpose specified in the application. [3] In early development, a short description of how the proposed clinical formulation was developed and evaluated will suffice. There should be a description of any differences between DP batches used for nonclinical studies (especially safety/toxicology) and how the differences could affect the safety of the clinical DP.

Subsections of the Pharmaceutical Development section that should be addressed in later-stage development include a description of the drug substance compatibility with formulation (excipients) and container closure materials. Any overages included during manufacturing to achieve the label claim should be discussed and justified. There should be a summary on the development of the manufacturing process, suitability of the container closure system, and formulation compatibility with likely diluents (if applicable).

8.5.2.3 Manufacturer(s)

Similarly to drug substances, the DP manufacturer(s) and contact information should be listed along with a description of the roles of each.

8.5.2.4 Batch formula

The batch formula for a typical batch or for the intended clinical scale should be provided in tabular format. Overages should be indicated, where applicable. For the materials used in the batch, the quality standard (compendial or other) should be provided.

8.5.2.5 Description of the manufacturing process and process controls

In this section, the sponsor should provide a flow diagram and narrative of the compounding, filling, and packaging processes. KPPs should be noted and IPCs described, including sample points and test methods. As development continues, additional KPPs and IPCs usually are implemented to control the manufacturing process.

8.5.2.6 Control of critical steps and intermediates

At the IND stage, it is common to indicate that this section is not applicable. As the process becomes more refined and better understood (and certainly by the time that pivotal clinical batches are manufactured), there will be the implementation of controls during critical steps that should be described (test methods, specifications, provision for reprocessing if IPCs are not met, etc.).

Additionally, control of intermediates must be implemented. Isolated intermediates should have specifications established that are tied to the desired quality of the final DP. For solid oral dosage forms, intermediate controls may be established at the blend stage (uniformity, assay, etc.) and for unfinished products (e.g., tablet cores before film coating). The IND should contain the sampling points, sample sizes, test methods, and specifications for each intermediate.

8.5.2.7 *Process validation and evaluation*

Process validation is not typically conducted at the IND phase, and a simple statement to this effect is usually sufficient. For orally administered peptides, there is typically not a terminal sterilization or other manufacturing process requiring validation before submission of the marketing application.

8.5.2.8 *Control of excipients*

At the initial IND submission, it is only necessary to submit limited information on excipients (brand or type, grade, particle size, etc.). During later development, specific manufacturers and part numbers should be included. For noncompendial excipients, specifications and test methods (including validation) are required. If excipients of human and/or animal origin are used, the means for control of adventitious agents (typically BSE and TSE) must be described.

8.5.2.9 *Novel excipients*

Novel excipients not appearing in other approved orally administered drug products have special requirements. Specifically, the safety of a novel excipient must be established via nonclinical animal studies (and occasionally in human studies) that are similar to those used for new drugs. For this reason, it is very uncommon for a sponsor to begin using a novel excipient in a drug product formulation.

8.5.2.10 *Specifications*

The drug product specifications for an oral peptide DP will include those described in Table 8.2. In the United States, there is a single specification assigned at DP release and during the life of the product (shelf life). In many other regions, including the European Union, the release specifications are tighter than the shelf-life specifications to allow for changes during storage (Table 8.2).

Table 8.2 Typical oral peptide drug product specifications.

Specification	Typical acceptance criteria	Comments
General properties		
Appearance	Describe shape, size, and color. Absence of obvious defects. Capsule products may include appearance of the complete dosage unit and capsule contents	General macroscopic quality measure
Identification (fingerprint)	Matches identity of authentic reference standard	Typically uses HPLC/UPLC or FTIR spectroscopy
Assay (% of label claim)	90%–100% (wider ranges should be justified)	Reported against a reference standard. Typically uses HPLC/UPLC
Impurities	Meets the requirements set forth in ICH Guidance Q3B (R1): Impurities in New Drug Products10	Impurity specifications above the ICH limits can be justified by their inclusion in drug substance or drug product batches used in preclinical toxicology studies
Potency	Acceptable range of biological activity against a specific target	Bioassay may measure relative potency against a standardized target
Water content	Specification should be set based on process capabilities and stability of drug product in the presence of water	Usually determined by Karl Fischer titration or loss on drying
Dissolution	Amount released at predetermined time points in defined media	Specifications should be set in accordance with the nature of DP (immediate release, modified release, etc.)
Uniformity of dosage units	Complies with USP <905> and/or Ph. Eur. 2.9.40	Demonstrated by chemical testing or weight variation depending on total dosage unit size and composition
Preservative content	Typically provided as acceptable range expressed as a percent of target formulated quantity	Acceptable range should be established using bacterial challenge studies and stability data
Bioburden	Total aerobic count NMT 1000 CFU/gTotal yeasts and molds NMT 100 CFU/gAbsence of *E. coli* Absent/1 g	While oral products are not required to be sterile, control of bioburden is critical to ensuring the safety of patients, particularly those with compromised immune systems

8.5.2.11 Analytical procedures: description and validation

As for the DS, test methods used for control of the drug product must be sufficiently robust to enable accurate and reproducible characterization of its identity, safety, and quality. Phase-appropriate validation of noncompendial analytical methods is required for any DP batches to be used in human studies.

Description of noncompendial analytical methods should be included. The detail should include the type of equipment used, specific information on chromatography columns or other substrates, buffers and solutions used, and key method parameters (temperatures, pressures, detection parameters, sample volumes, flow rates, etc.). The inclusion of sample chromatograms or tables of expected analytical results will assist reviewers in understanding the methods.

Method validation for early-stage programs should follow the same level of rigor as for drug substances with full validation completed by the time that registration stability studies are tested.

8.5.2.12 Batch analyses

The Batch analysis section includes a summary of the analytical results for research and development and preclinical (if made) and clinical batches of drug products. A table is preferable, and in cases where specifications have changed, a footnote to that effect is sufficient. If multiple strengths or other permutations of the DP are included in the IND, data should be grouped in a manner that allows the comparison of similar batches.

8.5.2.13 Justification of specifications

For drug products, specification setting is a relatively straightforward process, but in cases where unconventional specifications are proposed, a clear justification should be provided. For example, wider ranges than the typical 90%–110% of label claim for assay could be justified by clinical or other data showing that the required therapeutic levels of the drug are still reached in vivo despite a lower content in the dosage form.

For impurities that can increase upon storage, justification for a higher shelf-life specification compared to that at release should reference toxicology information or another basis.

8.5.2.14 Reference standards

A short summary of the characterization of the reference standard used for testing the drug product should be provided. In the event that the same reference standard is used for DS and DP, it is sufficient to simply note that in the DP section.

8.5.2.15 *Container closure system*

As mentioned earlier, the container closure system for the drug product is critical to maintaining the purity and quality of the material during storage. The container closure section of the IND should describe materials of construction, dimensions, and other critical aspects. For most oral peptide DPs, the container closure system will consist of a bottle, cap, and seal (often supplied as part of the cap, then induction-sealed to the bottle during packaging). It is important that stability data obtained on the drug product are collected using a representative container closure system for the clinical drug product batches.

If desiccant or other protective materials are utilized, these should also be fully described in this section including the quantities used per DP container.

8.5.2.16 *Stability*

For the initial IND, a minimum of 1 month of stability data at the intended (long term) storage conditions is required on a drug product batch that is representative of the intended clinical batch. Where data for research and development batches of drug products are available, it should be provided to establish an initial shelf-life (expiration) period for drug product batches until sufficient data can be obtained on clinical batches.

Sponsors may consider presenting a table in the IND (see Table 8.3 for an example) that proposes increasing expiry periods for DP as longer-term stability data are collected. The advantage of providing this proposal in the IND is that separate submissions will not be required each time

Table 8.3 Example proposal for extrapolating drug product shelf-life from available stability data.

Stability data available (months)	Proposed expiration period (months)
<6	6
6	9
9	12
12	18
18	24
24	36
36	36

new stability data become available to enable a shelf-life extension. Regulatory authorities expect that any extrapolated shelf life is eventually confirmed using real-time data.

In later development, a statistical analysis of quantitative measures (assay, impurities, dissolution, etc.) should be performed on similar batches to develop a statistical representation of the stability behavior of the drug substance. There are several ICH guidances that provide best practices for the collection and analysis of stability data.[10–14]

8.6 Nonclinical requirements

The nonclinical requirements for an IND fall into three related areas: pharmacology, pharmacokinetics, and toxicology. For an oral peptide drug, there are specific studies required to support human clinical studies that differ from peptide drugs administered via parenteral routes. These differences are described in the following sections.

8.7 Pharmacology

The pharmacology studies described in the IND submission provide information on the pharmacologic effect(s) and potentially the mechanism (s) of action of the oral peptide in relevant animal models. While most international regulations do not provide detail on the expectations for pharmacology studies, the sponsor should ensure that sufficient information is provided to enable a safe conduct of clinical trials. This will include an understanding of drug levels required to produce the intended pharmacological effect, off-target pharmacology that modifies unintended in vivo targets, and understanding potentially different effects in a range of patient populations (e.g., pediatric, geriatric, pregnant or nursing mothers, and renally or hepatically impaired subjects). The determination of appropriately scaled human doses requires a complete understanding of the concentrations needed for the desired pharmacological effect in various tissue compartments.

Orally delivered peptides will usually have the same pharmacology as the same entity administered via other routes, but the concentration in various tissue compartments may differ greatly after dosing. Sponsors are expected by regulators to have an understanding of these differences to aid in the determination of dose regimens used in clinical trials.

8.8 Pharmacokinetics

Pharmacokinetic studies provide crucial information on the absorption, distribution, metabolism, and excretion/elimination (ADME) of drugs. For oral peptide drugs, there will be a need to fully understand the differences in ADME in comparison to parenterally administered peptides, which do not need to be absorbed by the body. Distribution and metabolism will also differ after oral administration as first-pass hepatic circulation will have an impact on the active moiety before it reaches the intended target organ(s).

The absorption of orally administered drugs is a complex topic, and this is no different for peptide drugs. Oral peptides must be designed to survive passage through the gastrointestinal (GI) tract in the presence of pH extremes and enzymes and then allow for the absorbance of the active moiety in quantities sufficient to yield the desired pharmacological effect. A variety of techniques have been utilized to confer the appropriate properties to peptide drugs. For new oral peptide DPs, the sponsor should present the dose-PK dependence of the formulations and other peptide modifications studied during early development, as well as detailed ADME information on the proposed clinical formulation.

The time-dependent absorption of an oral peptide drug in both fed and fasted animals (typically rodents or dogs) should be described. As development proceeds, a correlation can be established between animal PK models and data obtained from human subjects in clinical trials. The absorption of a peptide drug begins with its dissolution in gastric or intestinal fluid depending on the peptide's properties. As the solution containing peptide transits into the upper intestine, the drug is absorbed into the portal vein after permeating through the intestinal wall. The drug then reaches the liver via portal vein circulation before entering the systemic circulation. This first pass allows the metabolism of the peptide via a wide range of enzymes that can oxidize, hydrolyze, and modify the peptide in varying degrees.

As previously mentioned, the distribution of the peptide drug will differ considerably from that obtained after parenteral administration where distribution initially follows blood circulation. In addition to first-pass hepatic metabolism and recirculation, the active peptide moiety may distribute very differently than after parenteral administration. Nonclinical ADME studies using radiolabeled (typically ^{14}C) drugs are often conducted early in development to accurately quantitate the tissue distribution and biological half-life in target organs. Later in development, human ADME studies will provide similar information in support of the marketing application.

In addition to first-pass hepatic metabolism, drugs will often undergo metabolism that will reduce the quantity of active moiety available, as well as potentially generating metabolites that may convey the intended pharmacological activity and/or undesirable adverse effects on the target organ(s) or in other parts of the body. To characterize the behavior of peptide drugs, animal studies should include metabolite identification and routes of elimination, as well as screening for carcinogenic, teratogenic, and/or mutagenic (often collectively referred to as "DNA reactive") potential. Metabolites possessing structures that trigger in silico alerts should be further studied in in vitro or in vivo studies designed to confirm the actual behavior of the metabolites.

An important aspect of the nonclinical behavior of a new oral peptide drug is interaction with enzymes, transporters, and proteins. Many drugs inhibit metabolic enzymes, which can lead to unexpectedly high levels of other drugs that are metabolized by the same enzymes. Conversely, enzyme induction caused by a drug can lead to a concomitant reduction in the level of other drugs that are subject to metabolism by the enzyme. Drugs can be taken up by transporters (e.g., P-glycoprotein, BCRP, and many others) that are expressed in different tissues, with results that are often undesirable. Inflammatory conditions in mammals often lead to markedly different expressions of transporters, which can significantly alter the ADME of drugs that are transporter substrates. Plasma protein binding, often to albumin, can impact the pharmacokinetics of an oral peptide. Tightly bound drugs cannot exhibit the desired pharmacological effect, and the protein-bound fraction may serve as a drug depot within the body.

Finally, excretion of a drug and/or its metabolites can occur through many routes including urine, bile/feces, sweat, tears, and saliva. Urine (renal) and feces (hepatic via biliary recirculation) are the most commonly observed routes, and peptide drugs by virtue of their relatively high molecular weight and polar nature are most likely to be excreted in the bile/feces.

Taken as a whole, ADME of oral peptides is usually described using several key measures. These are the elimination half-life ($t_{1/2}$), volume of distribution (V_d), total clearance (Cl), free fraction in plasma (f_u), and relative oral bioavailability compared to IV administration (F%).

8.9 Toxicology

To meet the IND requirements, a sponsor must provide a suitable body of safety (toxicology) data from experiments conducted in vitro and in representative animal species. Fortunately, the international requirements for safety data have been largely harmonized by the ICH. The goals of the preclinical safety program include assessing toxic effects on target organs, determining the dose dependence of these effects, determining reversibility of safety signals, and using the information to set an appropriate starting human dose and maximum safe dose in clinical studies.

For an oral peptide, the preclinical (pre-IND) toxicology studies must be of sufficient duration to cover the intended duration of dosing in the initial human trial at the maximum planned dose (usually referred to as the maximum tolerated dose) scaled for the difference in species. For most drugs, a limit of 1000 mg/kg/day is considered appropriate for acute dosing provided it provides at least a 10-fold margin over the maximum intended clinical exposure. Ideally, a 50-fold exposure margin is desirable for both acute and repeat-dose toxicity studies.

8.9.1 Safety pharmacology studies

The intent of safety pharmacology studies is to provide an assessment of the effect of the drug on key organs and tissues, including those in the cardiovascular, central nervous system, and respiratory systems. Specialized tests, such as the hERG assay, provide key data on potential cardiotoxicity and are required as part of the IND safety package.

8.9.2 Toxicokinetic studies

Toxicokinetic (TK) studies provide a correlation between the systemic exposure achieved in animals and the relationship to dose level and time course of any toxic effects. In the event that metabolites of the peptide drug are suspected of causing additional safety concerns, separate TK studies may be warranted using the individual metabolites.

8.9.3 Acute toxicity studies

Toxicology studies usually begin with single-dose toxicology studies in two mammalian species: one rodent (rat or mouse) and one nonrodent. The nonrodent species should be chosen based on the most relevant to the pharmacodynamic action of the peptide drug. A review of target sequence homology between species is a good starting point, followed by in vitro assays to examine the cross-species comparison of target binding affinity, receptor-ligand occupancy, and kinetics. The use of species-specific cell systems can often provide valuable information before beginning more complex studies in animals.

A limitation of toxicology data for many biologic products, potentially including peptide drugs, is the lack of tissue cross-reactivity and/or pharmacological activity in nonhuman species. In the event that some safety aspects of the peptide are not readily assessed in animal species, a risk assessment and mitigation plan should be provided to address these gaps before starting human clinical trials.

8.9.4 Repeat-dose toxicity studies

Depending on the intended duration of human dosing, repeat-dose toxicity studies may be required. The species (rodent and nonrodent) will be the same as for the acute studies and the route of administration and dosing regimen should be reflective of the intended scheme in humans. For truly chronic indications, studies of 6 months duration will be required before extended dosing in clinical trials is permitted. It is common for sponsors to initiate shorter duration clinical studies in parallel with repeat-dose chronic toxicity studies, and once the long-term safety of the drug has been demonstrated, conduct longer-term dosing in humans.

8.9.5 Immunogenicity studies

It is common for peptides intended for human administration to cause an immunogenic response in animal species. For this reason, it is often necessary to measure antibodies in toxicity species to enable an accurate assessment of the safety studies. Often, antidrug antibodies are generated, which reduce or even eliminate circulating peptide drugs. Other antibodies may lead to pharmacological and/or toxicological effects in the animals. These effects should be understood and reported as part of the IND package.

Importantly, the formation of animal antibodies to human peptides should not cause a sponsor to stop the toxicity studies unless a significant or complete reduction of circulating peptides occurs in a large proportion of the animals. Conversely, the absence of antibody formation in animals is not predictive of human safety or lack of immunogenicity. Careful analysis of the potential for human antibody responses is required to ensure the safety of human trial participants.

8.10 Clinical requirements

From a regulatory perspective, the clinical development of an oral peptide drug is not unlike that for other types of therapeutics. Clinical development progresses from Phase 1 (exploratory safety/tolerability in humans) through Phase 2 (dose rangefinding and preliminary efficacy) and eventually Phase 3 (pivotal demonstration of safety and efficacy). At each phase, the clinical program builds upon the learnings of prior studies in humans and in animal (nonclinical) studies. Regulators expect that clinical studies with novel drugs like oral peptides address potentially new safety/toxicity concerns associated with the type of drug being studied.

For oral peptides, there are several considerations not seen for parenteral peptide drugs. The conditions encountered in the gut include strongly acidic (stomach), bile acids/salts (duodenum), and enzymatic (liver). Conversion of the peptide to other forms may lead to new species (metabolites) with unknown toxicities. It may be necessary to conduct additional nonclinical studies to address these issues if previously unseen metabolites are found during clinical trials.

8.11 Drug–drug interactions

DDIs can lead to significant changes in the blood levels of either the oral peptide drug or other medications. DDIs occur when one or more drugs affect the expression or function of a metabolic enzyme or transporter. Nonclinical studies often inform potential DDIs before the clinical program begins, but the final determination of the need for dose adjustments or other accommodations is often a result of formal human DDI studies. FDA has issued guidances to address both nonclinical and therapeutic protein DDI studies, and these documents should be consulted before designing DDI studies.

8.12 Immunogenicity

Immune responses to therapeutic proteins can lead to a physiological response to the peptide itself or to immune-related adverse events. These occurrences can lead to a loss of efficacy or safety concerns, respectively. In some cases, a therapeutic protein can engender immune responses to related proteins. The route of administration plays a role in peptide immunogenicity, and oral delivery may reduce the propensity for some types of immune response.

Immune responses to peptides are usually mediated by humoral mechanisms, where circulating antibodies specific to the peptide can be found in patients. In these cases, detection of the antibodies will unequivocally confirm the immunogenic response. Circulating levels of the peptide drug will be greatly reduced or eliminated when antibodies exist, and this will greatly impact the efficacy of the drug.

From a safety perspective, there are several potential issues to be addressed. These include anaphylaxis, cytokine release syndrome, serum sickness (delayed hypersensitivity), and cross-reactivity to endogenous proteins. Each of these conditions can range from mild to life-threatening and must be addressed via risk mitigation. Dose reduction and/or supportive care can often mitigate immune responses, and a thorough risk-based approach is required for new peptide therapeutics.

8.13 Regulatory affairs

As discussed throughout this chapter, the overall regulatory framework is the same for a therapeutic oral peptide as for other protein drugs.

Within this context, there are novel aspects when delivering a peptide via oral administration, but they can be addressed by considering oral administration might impact the safety and/or efficacy of the product.

As noted above, the period between oral administration and systemic distribution of the peptide offers the potential for new impurities (degradation products) and metabolites of the parent compound. These moieties must be addressed from a toxicological standpoint, and it is also important to determine if any possess pharmacologic effects similar to the parent.

The CMC differences needed to create a suitable oral peptide product may be in the drug substance (peptide modifications) or the drug product (formulation components, novel protective dosage forms such as microbeads or enteric coatings, etc.). Regulatory strategies should include a plan to ensure the consistent quality of these products, and whether additional nonclinical (or clinical) studies will be needed to justify the use of new excipients.

8.14 Regulatory meetings

The US Food and Drug Administration and European regulatory agencies (and country authorities) provide the opportunity for free or relatively low-cost consultation throughout the drug development process. Sponsors should approach these meetings with careful planning as there is usually only one opportunity for consultation at each phase of development.

The US FDA designates the following "Type B" meetings, which are designed to offer advice to sponsors throughout the development continuum:

1. Pre-IND
2. End of Phase 1
3. End of Phase 2
4. Pre-New Drug Application (NDA) (or Pre-BLA)

In addition to Type B meetings, FDA offers a variety of other meeting types to address issues or sticking points during drug development. These meetings can be requested at any time during product development and can occur face to face, as teleconferences or videoconferences, or in the form of written responses only.

Regulatory meetings typically cover clinical, nonclinical, CMC, and administrative subject matter. It is important to carefully plan the topics to be covered at the meeting well ahead of time. Questions for the regulator should not be open-ended or gratuitous; specifically, questions that can be answered by consulting existing regulations and guidances should be avoided. For an oral peptide, CMC questions regarding novel formulations may be appropriate. Nonclinical and clinical topics should focus on the suitability of the respective data packages for the stage of development, including whether additional studies will be required to address specific patient populations (pediatric, geriatric, pregnant women, etc.) or other situations.

An important aspect of regulatory meetings is the opportunity to obtain regulatory agency feedback on proposed clinical protocol designs. While a formal protocol is not required for these meetings, a synopsis should be presented with a description of proposed patient population, study size (number of treatment arms and patients in each arm), inclusion and exclusion criteria, primary and secondary endpoints, clinical and hematological measurements, and other key parameters. Agency feedback will often highlight issues or offer recommendations not considered by sponsors before the actual submission has been made.

8.15 Marketing authorization

Well-defined oral peptides produced via synthetic means will be reviewed in the United States by FDA's CDER. The marketing authorization request takes the form of a NDA, which must be submitted electronically in the form of the CTD. European Marketing Authorization Application (MAA) submissions are also in the CTD format, with the content of the administrative module (Module 1) being the main difference between differing countries.

The marketing authorization generally contains the same information as the IND application, but with considerably more detail. Additionally, the NDA or MAA contains the proposed materials, processes, and controls for the drug substance and drug product after approval with little room for flexibility or variation once approved. Regulators expect that sponsors will have locked down most aspects of the production and controls post-approval, and this will be reflected in the approved product labeling.

8.16 Pre-NDA meeting

As noted under regulatory meetings, the Pre-NDA meeting is the final opportunity for sponsors to gain FDA feedback on the content of the submission, as well as to ensure that there will be no surprises during the review period. In addition to covering the NDA content, the Pre-NDA meeting is a good opportunity to discuss and negotiate postmarketing requirements or commitments that might be imposed as conditions of approval. Having the Pre-NDA meeting at least 60–90 days ahead of the planned NDA submission will give the sponsor adequate time to address issues raised during the meeting.

The Pre-NDA meeting should cover the entire body of CMC, nonclinical and clinical data, and whether there are any deficiencies that would preclude NDA approval. Consideration should be given to whether the existing data package for the oral peptide fully supports the intended commercial label. Regulatory agencies expect that each aspect of the product labeling has a basis in data obtained during the development process. For example, the patient populations included in the label should have been included in pivotal clinical trials. It is not uncommon to have regulatory agencies indicate that they will require postmarketing studies in underserved populations (pediatric or geriatric patients, nursing mothers, etc.) and to request a postmarketing plan to be submitted before initial product approval is granted.

For an oral peptide therapeutics, some of the specific considerations to address in the Pre-NDA meeting include the stability behavior of both the drug substance and drug product in the intended container-closure system and under typical conditions of use, final specifications to ensure the quality, safety, and efficacy of the drug product, potential drug interactions with commonly used medications, and postmarketing surveillance (pharmacovigilance) to track safety in the uncontrolled setting of a marketed product.

8.17 NDA/MAA submission

The final regulatory stage gate for an oral peptide drug is the marketing authorization. In the United States, this is the New Drug Application and, in the EU, it is the MAA. These submissions are the culmination of years of work by experts in many different disciplines, and assembling them requires a team of regulatory experts with extensive input from subject matter experts across the development continuum.

To maximize the likelihood of a successful review process and rapid approval, the sponsor should treat the NDA/MAA authoring process like any other intensive process. A team should be assembled with representation from CMC (DS and DP), Pharm/Tox, Clinical Research, Medical Affairs, Regulatory Strategy and Operations, and Corporate Management. Before authoring begins, there should be a clear timeline for the submission with a well-defined delineation of responsibilities for authoring, review, dispute resolution, and final publishing and submission. Having an effective project manager will greatly streamline the process.

Because the CTD format is essentially the same for the ICH regions (United States, European Union, and Japan), the following sections will follow the CTD layout and common requirements for Modules 2 through 5. Module 1 contains the regional administrative information, which will differ somewhat depending on where the submission is made.

8.17.1 Module 1: administrative information

The administrative module of a CTD intended for the United States will contain a variety of documents in the following areas:

1. Forms
2. Cover letter
3. Corporate name, address, contact information, certifications, patent information
4. References (drug master files, letter of authorization, previously submitted information)
5. Communications (prior meetings and submissions)
6. Fast track designation request, rolling review request
7. Pediatric waiver requests, requests for pediatric exclusivity
8. Environmental impact statements
9. Product labeling (package insert, draft carton, and container labels) and labeling comprehension studies
10. Promotional materials
11. Proprietary name requests

Each of these items is described in detail in a variety of FDA and ICH Guidances. There will generally not be a need for any significant changes or additions to Module 1 for an oral peptide drug, but the sponsor should discuss the overall content of the CTD with the FDA before submission to ensure that all regulatory expectations will be met during the initial review.

8.17.2 Module 2: summaries

Module 2 is a compilation of quality, safety, and efficacy summaries. It is intended to give reviewers a fast overview of the contents of the CTD in each of the three key areas. For an oral peptide drug, there should be a focus on the unique aspects of delivering the therapy via a relative new route of administration and the differences that have been observed from peptides administered via parenteral routes.

8.17.3 Module 3: quality

Module 3 of a CTD is an extensive summary of the development, manufacture, analysis, and properties of the drug substance and drug product. There are extensive descriptions of the content of Module 3 available on the FDA's website (http://www.fda.gov), so the following sections will describe parts of Module 3 where particular attention should be paid for an oral peptide drug or where the conventional approach may require revision to accommodate the unique properties associated with the drug substance and/or drug product.

8.17.4 Drug substance

The DS consists of the peptide that will convey the pharmacological activity in vivo. The regulatory expectation for any drug substance includes a detailed description of how it is made and purified, the controls (test methods and specifications) employed, and an assessment of the stability of the drug substance. The following sections, arranged in the format of Module 3 (quality module) of the CTD,[3] which is accepted across the ICH regions, cover the expectations in more detail. By fulfilling the expectations of the CTD, a sponsor will ensure acceptance of the submission and eventual approval of the product.

8.17.4.1 General information

A summary of the unique properties of the drug substance that allow oral administration should be described. These may take the form of unnatural amino acids or protecting groups that confer resistance to the conditions encountered in the GI tract or to a variety of physiological enzymes.

8.17.4.2 Manufacture

For an oral peptide drug substance, a description of chemical modification steps to confer stability should be carefully described. In-process controls should provide assurance that the intermediates used to produce the final drug substance are sufficient, particularly where they differ from those used in the manufacture of conventional peptides.

In the event that control of physical characteristics (e.g., particle size distribution, polymorphic form, etc.) is required, a full description of the process parameters and controls used to ensure the correct outcome should be included. For the CTD, a significant amount of detail is expected when describing each process step. Unit operations should include descriptions of times, temperatures, mixing speeds, equipment configuration, raw material and solvent grades, and other relevant details.

8.17.4.3 Characterization of the drug substance

The impact of solid-state properties, including polymorphic forms and particle size distribution, on the key properties of the drug substance should be described. As many physicochemical characteristics of the DS can impact oral bioavailability, it is critical to describe the ranges for these parameters that have been demonstrated to provide suitable exposures during human trials, particularly the pivotal clinical studies used to demonstrate safety and efficacy.

8.17.4.4 Control of the drug substance

Pharmaceutically acceptable specifications for physicochemical parameters will include standard measures used for peptide drugs, and as noted earlier, acceptance criteria for other properties of the DS that may impact oral bioavailability are essential. By this stage of development, a significant body of data on multiple batches of DS should exist. These data should be thoroughly analyzed and the outcomes used to establish a set of proposed commercial specifications. Regulators will look at a variety of factors when approving specifications, including industry standards, the effect of each parameter on the safety and efficacy of the final drug product, and manufacturing process capability. Sponsors should present a solid justification of each specification to maximize the probability of regulatory acceptance.

Validation of noncompendial methods should include interlaboratory comparisons and the effect of varying method parameters (e.g., mobile phase pH or ionic strength, temperature, etc.) on the results.

8.17.4.5 Reference standards

By this point, regulators expect that sponsors will have prepared reference standards not only for the active moiety, but also for specified impurities. A combination of impurity standards can be combined to establish system suitability, and other standards may be employed for test methods such as physical form.

8.17.4.6 Container closure system

The proposed commercial container closure system used to store drug substance should be described in detail, including manufacturer, part number(s), materials of construction and dimensions. In the event that desiccants or molecular sieves are used, full details of each should be described.

8.17.4.7 Stability

Registration stability samples should be stored in representative containers that are typically made of the same materials. It is important to include desiccants or molecular sieves in the stability containers if they are to be used for commercial drug substance batches.

A summary of the DS stability data should be presented, along with data tables for each batch and condition. Protocols for stability studies containing information on storage conditions, time points, and test methods should also be included as part of the CTD. A proposed commercial stability program should be provided, which includes the batches to be placed on stability, the storage conditions to be studied, and the test methods and acceptance criteria that will be employed.

Statistical analysis of quantitative stability-indicating measures (assay, impurities, etc.) should be performed on similar batches to develop a statistical representation of the stability behavior of the drug substance. For an oral peptide DS, it may be necessary to include additional assay data to capture the stability behavior of properties such as particle size distribution and solid-state form.

8.17.5 Drug product

The drug product section of the marketing authorization contains the same general information as was originally presented in the IND application, but updated to include the proposed marketed DP form, process and controls, and with considerably more detail. The following sections describe the expectations for an NDA covering an oral peptide drug product.

8.17.5.1 Description and composition of the drug product

The section should clearly describe the final composition of each strength of drug product, with the quantities of capsule shells (if used) and any added desiccants or molecular sieves. This information should be provided on a per-dosage unit basis.

8.17.5.2 Pharmaceutical development

Subsections of the Pharmaceutical development section should carry a narrative along with supporting summary data (tables and figures) of how the various components of the oral peptide drug were developed.

8.17.5.2.1 Components of the drug product

This section should describe how each DP ingredient (drug substance and excipients) was selected. For an oral peptide drug, there should be a focus on the role of excipients in stabilizing the active moiety against chemical and enzymatic degradation. Detail can include how the various grades, particle sizes, and other key properties of the DS and excipients were selected. Where possible, use data summary tables and figures to elaborate.

8.17.5.2.2 Formulation development

In this section, a detailed narrative of the development studies that were done to establish the formulation should be included. For oral peptide DPs, a particular emphasis should be made on work conducted to provide product stability, along with summary data supporting the final formulation that has been selected. If other approaches were tried and abandoned during development, it is important to include these for comparison and to possibly include an explanation of why some approaches were not selected.

8.17.5.2.3 Manufacturing process development

For many peptide drugs, the manufacturing process may specifically confer properties to the final dosage form that will enable the drug to be administered orally. Examples include encapsulation of the formulation in a protective shell or stabilization of the active moiety in a physical matrix. These processes often require careful control of the process parameters, and the established acceptable ranges should be presented along with supporting data.

8.17.5.2.4 Container closure system

There may not be a significant difference in the container closure system for an oral peptide drug, but a summary of any work done to select a suitable primary packaging system is important. Where desiccants and/or molecular sieves are part of the product package, there should be a narrative describing the rationale for the selection of a particular material/grade/particle size and supporting data to demonstrate the protective effect. Studies demonstrating protection from moisture, oxygen, light, and other environmental factors should also be included here.

8.17.5.2.5 Microbiological attributes

Because patients receiving oral peptide drugs may be immune-compromised (e.g., cancer patients), in these cases it is important to demonstrate that microbial controls have either been implemented or are not necessary. The selection of antimicrobial preservatives should be described if done, and data from microbial limits testing should be presented if available.

8.17.5.2.6 Compatibility

A summary of compatibility and stability studies conducted with excipients, packaging components, desiccants or molecular sieves, and other product contact materials should be included here if not described in other sections. Where potentially reactive materials are used, specific data (e.g., extractable studies) will be needed to satisfy regulators that the use of these materials will not have an adverse effect on the final DP.

8.17.5.3 Manufacture

8.17.5.3.1 Batch formula

The batch formula for the intended commercial batch size of the drug product should be provided. A tabular listing of the drug substance, excipients, processing aids and solvents, and capsule shells or coatings is a key part of this section.

8.17.5.3.2 Description of manufacturing process and process controls

This section should contain a flowchart of the overall manufacturing process for the drug substance. Following the flowchart, each manufacturing step should be provided in narrative form. Each step description will include quantities (absolute masses or volumes and molar equivalents), equipment descriptions, times and temperatures, and expected yields and purities where applicable. There should be a description of samples taken and whether the products from each step are isolated or taken directly to the next step.

KPPs and ranges of manufacturing parameters should be described in detail. IPCs should be described, including sample points, quantities required, and test methods. It is important to describe how KPPs and IPCs relate to the finished DP specifications.

8.17.5.3.3 Control of critical steps and intermediates

Where processing steps are critical to the quality or performance of the final DP, a thorough description of how these steps are controlled should be included. For an oral peptide drug, this will often include steps designed to confer stability to the product. It is common to tie this section to the previous one as there are many common elements.

8.17.5.3.4 Process validation and evaluation

Although process validation is not typically conducted before the NDA is submitted in the United States, regulators in other regions often expect it as part of the MAA.

When process validation has not been conducted, it is still important to present the prevalidation activities that have been conducted in support of the drug product manufacturing process. Studies designed to establish acceptable control ranges for KPPs and to establish equipment operating ranges should be described along with a presentation of summary data.

For submissions containing process validation data, this section can be extensive. Regulatory guidances, including FDA's Guidance for Industry: Process Validation: General Principles and Practices, should be consulted for full details on the conduct and reporting of validation studies.

8.17.5.4 Control of excipients

Where compendial excipients are used, the specifications and analytical procedures can be tied to the compendial requirements. For an oral peptide drug using novel (noncompendial) excipients, it is necessary to include a full description of the excipient materials, analytical procedures used to control the excipients (and a summary of the validation of the methods), justification of the excipient specifications and an overview of the method of manufacture.

As noted earlier, novel excipients not appearing in other approved orally administered drug products have additional safety/toxicology requirements. Specifically, the safety of a novel excipient must be established via nonclinical animal studies (and occasionally in human studies) that are similar to those used for new drugs.

8.17.5.5 Control of drug product

8.17.5.5.1 Specifications

At the NDA stage, specifications for the drug product should be supported by a body of data collected on the batches made throughout development. In the case of an oral peptide DP, a particular focus should be applied to parameters that ensure the stability of the drug in the gut and the release of the active moiety according to the intended time course. In particular, the dissolution specification will often be expressed as several measures: the dosage form retains integrity at low pH (typically simulated gastric fluid or 0.1 N HCl) and then releases the active peptide in simulated intestinal fluid.

Typical specifications and common acceptance criteria are summarized in Table 8.4.

8.17.5.5.2 Analytical procedures: description and validation

As for the DS, test methods used for control of the drug product must be sufficiently robust to enable accurate and reproducible characterization of its identity, safety, and quality. Phase-appropriate validation of noncompendial analytical methods is required for any DP batches to be used in human studies.

Table 8.4 Typical specifications for an oral peptide drug product.

Attribute	Test method(s) typically used	Common acceptance criteria
Appearance	Visual	Describe the dosage form's color, shape, and size. For capsules, the contents should also be described
Identification	HPLC, FT-IR, and/or NMR	Typical to include two orthogonal methods of identification. Sample should exhibit comparable characteristics to a reference standard or reference value
Assay (% label claim)	HPLC (measured against reference standard)	At release: 95%–105% Shelf life: 90%–105%
Impurities	HPLC (area% corrected for relative response factor of each compound)	Individual specified impurities specification should be based on levels qualified in toxicology studies or ICH Q3B(R2)[15] and process capability. Total impurities should be based on batch history and process capability
Dissolution	Dissolution bath with HPLC or UV–visible spectroscopy quantitation	

Description of noncompendial analytical methods should be included. The detail should include the type of equipment used, specific information on chromatography columns or other substrates, buffers and solutions used, and key method parameters (temperatures, pressures, detection parameters, sample volumes, flow rates, etc.). The inclusion of sample chromatograms or tables of expected analytical results will assist reviewers in understanding the methods.

Method validation for early-stage programs should follow the same level of rigor as for drug substances with full validation completed by the time that registration stability studies are tested. The MAA should include details of the validation sufficient to fully describe the work conducted, the acceptance criteria, and summary of results. Many companies include validation reports in an appendix to Module 3.

8.17.5.5.3 Batch analyses

The Batch analysis section includes a summary of the analytical results for research and development and nonclinical and clinical/registration batches of drug products. Tables are preferable and in cases where specifications have changed, a footnote to that effect is sufficient. If multiple strengths or other permutations of the DP are included in the IND, data should be grouped in a manner that allows the comparison of similar batches. For an oral peptide DP, there should be an analysis and discussion of batch results in the context of the proposed marketed product specifications, particularly for attributes that impact safety and efficacy (e.g., impurity profile and dissolution).

8.17.5.6 Characterization of impurities

Recurring impurities at or above the ICH identification limits must be designated as specified and included in the drug product specifications. In addition, a full assessment of the source(s) of each impurity and the potential for toxicity should be discussed in this section. Identification should be presented along with analytical data supporting the proposed structures. In the drug product, the focus should be on impurities (also referred to as degradation products or degradants) that form or increase after the drug substance is combined with excipients in the dosage form.

8.17.5.7 Justification of specifications

The proposed commercial specifications should be supported by a combination of the required product safety and efficacy, along with data from batches produced throughout the development process. Statistical analysis and the resultant implications on the proposed specifications should be conducted for attributes such as assay, impurities, and dissolution.

In most regions, but not the United States, separate specifications exist for release and shelf-life. The justification of specifications should describe how the release specifications and stability behavior relate to the proposed shelf life acceptance criteria.

8.17.5.8 Reference standards

Oral peptide drugs will typically require several reference standards by the NDA stage. The primary reference standard, consisting of fully characterized, purified drug substance, will be used for the primary assay, content uniformity, and dissolution testing. It should be fully characterized using a battery of methods, including NMR, HPLC, elemental analysis, gas chromatography, and water content. The proposed retest period for the standard should be included along with a list of tests conducted for new reference standard batches and at each retest interval.

Additional reference standards for specified impurities are also common and typically are the same as used for the drug substance.

8.17.5.9 Container closure system

The container closure system for the proposed marketed drug product should fully describe the manufacturer(s), part numbers, materials of construction, dimensions, and other critical aspects. Dimensional drawings are often helpful. If coilers, desiccants, and/or molecular sieves are part of the primary container closure system, they should be described in similar detail (Table 8.5).

Table 8.5 Typical stability proposal for regulatory submissions.

Stability data available (months)	Proposed expiration period (months)
<6	6
6	9
9	12
12	18
18	24
24	36
36	36

8.17.5.10 *Stability*

The stability section consists of three subsections: a stability summary and conclusions, postapproval stability protocol and commitment, and data tables. At the NDA stage, data from the drug product registration batches are most important to support the proposed shelf life, but if prior (supporting) batch data are consistent, it may also be combined statistically with the registration batch data.

The stability summary should contain a detailed overview of the stability behavior of the drug product in the proposed commercial container closure system at the intended long-term storage condition, as well as at the accelerated and intermediate conditions. For each attribute (assay, impurities, dissolution, etc.), describe the stability behavior and how it impacts the overall quality, safety, and efficacy of the drug product. Statistical analysis is described in ICH Q1E (FDA, 2004). For drugs that will be dispensed to the patient to take home, stability should include open container studies under worst-case conditions (typically 40°C/75% relative humidity (RH)).

The proposed storage conditions and stability labeling should also be included in this section. If the storage conditions differ for long-term (warehouse) and postdispensing (pharmacy) situations, this should also be noted.

The postapproval stability commitment should follow the ICH Q1A (R2) guidance (FDA, 2003). For most new drugs, including oral peptide products, the postapproval commitment will also enable the extension of shelf life beyond what is obtained at NDA approval.

Stability data should be summarized in tables, with the layout consistent from batch to batch. A sample template for stability data is depicted in Table 8.6. Each table's caption should include the batch number, batch type (preclinical, clinical, registrational, etc.), storage conditions, batch size, manufacturer, and other relevant data.

8.17.6 Module 4: nonclinical

Unlike the IND stage, where nonclinical studies are used to support the study of the investigational drug in human clinical trials, the requirements for an NDA are intended to support the labeling requirements for the drug. ICH M3(R2) Nonclinical Safety Studies for the Conduct of Human Clinical Trials and Marketing Authorization for Pharmaceuticals (ICH, 2010) describes the nonclinical requirements for the NDA. For an

Table 8.6 Sample template for stability data.

	Acceptance criteria	Initial time point (T0)	Month 3	Month 6	Month 9	Month 12
Appearance						
Assay (% label claim)						
Impurities (%)Specified						
Impurity 1 Specified						
Impurity 2 Specified						
Impurity 3 Individual						
Unspecified Impurities						
Total Impurities						
Dissolution (% dissolved after xx min)						
Water content (%)						
Microbial limits						

oral peptide drug, specific studies such as immunogenicity are often required (ICH, 2006), and in cases where the drug is coformulated or copackaged with other drugs, combination drug toxicology studies may be required.

Module 4 should contain copies of the nonclinical study reports that have been collected throughout the development process for the drug.

8.17.7 Module 5: clinical

From a regulatory perspective, the clinical development of an oral peptide drug is not unlike that for other types of therapeutics. Clinical development progresses from Phase 1 (exploratory safety/tolerability in humans) through Phase 2 (dose rangefinding and preliminary efficacy) and eventually Phase 3 (pivotal demonstration of safety and efficacy). At each phase, the clinical program builds upon the learnings of prior studies in humans and in animal (nonclinical) studies. Regulators expect that clinical studies with novel drugs like oral peptides address potentially new safety/toxicity concerns associated with the type of drug being studied.

For oral peptides, there are several considerations not seen for parenteral peptide drugs. The conditions encountered in the gut include strongly acidic (stomach), bile acids/salts (duodenum), and enzymatic (liver). Conversion of the peptide to other forms may lead to new species (metabolites) with unknown toxicities. It may be necessary to conduct additional nonclinical studies to address these issues if previously unseen metabolites are found during clinical trials.

8.18 Drug–drug interactions

Drug-drug interactions (DDIs) can lead to significant changes in the blood levels of either the oral peptide drug or other medications. DDIs occur when one or more drugs affect the expression or function of a metabolic enzyme or transporter. Nonclinical studies often inform potential DDIs before the clinical program begins, but the final determination of the need for dose adjustments or other accommodations is often a result of formal human DDI studies. FDA has issued guidances to address both nonclinical and therapeutic protein DDI studies, and these documents should be consulted before designing DDI studies.

8.19 Immunogenicity

Immune responses to therapeutic proteins can lead to a physiological response to the peptide itself or to immune-related adverse events. These occurrences can lead to a loss of efficacy or safety concerns, respectively. In some cases, a therapeutic protein can engender immune responses to related proteins. The route of administration plays a role in peptide immunogenicity, and oral delivery may reduce the propensity for some types of immune response.

Immune responses to peptides are usually mediated by humoral mechanisms, where circulating antibodies specific to the peptide can be found in patients. In these cases, detection of the antibodies will unequivocally confirm the immunogenic response. Circulating levels of the peptide drug will be greatly reduced or eliminated when antibodies exist, and this will greatly impact the efficacy of the drug.

From a safety perspective, there are several potential issues to be addressed. These include anaphylaxis, cytokine release syndrome, serum sickness (delayed hypersensitivity), and cross-reactivity to endogenous proteins. Each of these conditions can range from mild to life-threatening and must be addressed via risk mitigation. Dose reduction and/or supportive care can often mitigate immune responses, and a thorough risk-based approach is required for new peptide therapeutics.

References

FDA. (2003). FDA Guidance for Industry: Q1A(R2) Stability Testing of New Drug Substances and Products.

FDA. (2004). FDA Guidance for Industry—Q1E: Evaluation of Stability Data.

ICH. (2006). International Conference on Harmonisation (ICH) Guidance: S8 Immunotoxicity Studies for Human Pharmaceuticals.

ICH. (2010). International Conference on Harmonization (ICH) Guidance: M3(R2) Nonclinical Safety Studies for the Conduct of Human Clinical Trials and Marketing Authorization for Pharmaceuticals.

Index

Note: Page numbers followed by "*f*" and "*t*" refer to figures and tables, respectively.

FSC
www.fsc.org
MIX
Papier | Fördert gute Waldnutzung
FSC® C083411

Kolibri 360 GmbH
Ferdinand-Jühlke-Straße 7
99095 Erfurt, Deutschland
produktsicherheit@kolibri360.de